Natural Products and their Bioactives in Antidiabetic Drug Discovery

Natural Products and their Bioactives in Antidiabetic Drug Discovery

Edited by

Kanti Bhooshan Pandey
CSIR-Central Salt & Marine Chemicals Research Institute, Bhavnagar, Gujarat, India

Maitree Suttajit
University of Phayao, Phayao Province, Thailand

Pinar Atukeren
Istanbul University-Cerrahpaşa, Istanbul, Turkey

WILEY

This edition first published 2024

Registered Office
John Wiley & Sons, Inc., 111 River Street, Hoboken, NJ 07030, USA

For details of our global editorial offices, customer services, and more information about Wiley products visit us at www.wiley.com.

Wiley also publishes its books in a variety of electronic formats and by print-on-demand. Some content that appears in standard print versions of this book may not be available in other formats.

A catalogue record for this book is available from the Library of Congress

Hardback ISBN: 9781119983316; ePub ISBN: 9781119983330; ePDF ISBN: 9781119983323; oBook ISBN: 9781119983347

Cover Image: © Yuichiro Chino/Getty Images
Cover Design: Wiley

Set in 9.5/12.5pt STIXTwoText by Integra Software Services Pvt. Ltd, Pondicherry, India

Contents

List of Contributors

Muveddet Emel Alphan
Department of Nutrition and Dietetics
Faculty of Health Science
Istanbul Atlas University
Istanbul, Turkey

Bavani Arumugam
Department of Biomedical Science
Faculty of Medicine
University of Malaya
Kuala Lumpur
Malaysia

Pınar Atukeren
Department of Medical Biochemistry
Cerrahpaşa Faculty of Medicine
Istanbul University-Cerrahpaşa
Istanbul, Turkey

Yasemin Aydin
Department of Biology
Faculty of Science
Istanbul University
Istanbul, Turkey

Sahar Bakht
Department of Pharmaceutical Chemistry
Government College University
Faisalabad
Faisalabad, Pakistan

Onur Baykara
Department of Medical Biology
Cerrahpaşa Medical Faculty
Istanbul University-Cerrahpaşa
Turkey

Bilge Nur Col
Department of Nutrition and Dietetics
Faculty of Health Science
Istanbul Atlas University
Istanbul, Turkey

S. Asha Devi
Laboratory of Gerontology
Department of Zoology
Bangalore University
Bangalore
Karnataka, India

Ravindra Dhar
Centre of Materials Science
University of Allahabad
Prayagraj, India

Prachee Dubey
Government G.I. College
Malwan, Fatehpur
Uttar Pradesh, India

Mansi Gandhi
Department of Chemistry
School of Advanced Sciences
Vellore Institute of Technology
Vellore
Tamil Nadu, India

Institute of Chemistry
The Hebrew University of Jerusalem
Jerusalem, Israel

Waseem Hassan
Department of Pharmacy
COMSATS University Islamabad
Lahore campus
Lahore, Pakistan

Faraza Javaid
Quaid-e-Azam College of Pharmacy
Sahiwal, Pakistan

Gülen Ecem Kalkan
Department of Nutrition and Dietetics
Faculty of Health Science
Istanbul Okan University
Istanbul, Turkey

Ahu Korkut
Department of Obstetrics and Gynecology
Faculty of Medicine
Eskisehir Osmangazi University
Turkey

Umah Rani Kuppusamy
Department of Biomedical Science
Faculty of Medicine
University of Malaya
Kuala Lumpur, Malaysia

Subir Kumar Mandal
CSIR-Central Salt & Marine Chemicals Research Institute
Bhavnagar
Gujarat, India

Suparna Mandal
Laboratory of Gerontology
Department of Zoology
Bangalore University
Bangalore
Karnataka, India

Sonam Mishra
Centre of Materials Science
University of Allahabad
Prayagraj, India

Jigeesha Mishra
Jacob Institute of Biotechnology & Bio-Engineering
Sam Higginbottom University of Agriculture Technology & Sciences
Prayagraj, India

David J. Newman
NIH Special Volunteer
Wayne, Pennsylvania
USA

Shashank Kumar Ojha
Medicine Department
School of Pharmacy
Hebrew University of Jerusalem
Israel

Burcu Ateş Özcan
Department of Nutrition and Dietetics
Faculty of Health Science
Marmara University
Istanbul
Turkey

Murali Krishna Paidi
Academy of Scientific and Innovative Research (AcSIR)
Ghaziabad
India

Bishnu Kumar Pandey
Department of Physics
SPM College, University of Allahabad
Prayagraj, India

Kanti Bhooshan Pandey
CSIR-Central Salt & Marine Chemicals Research Institute
Bhavnagar, Gujarat, India

Academy of Scientific and Innovative Research (AcSIR)
Ghaziabad
India

S. Raja Sekhar
Pathology Section, MPMMCC
Banaras Hindu University Campus
Varanasi
U.P., India

Andleeb Shahzadi
Department of Medical Pharmacology
Faculty of Medicine
Istanbul University-Cerrahpaşa
Istanbul, Turkey

Sandeep Singh
Hadassah Biological Psychiatry Laboratory
Hadassah - Hebrew University Medical Center
Jerusalem, Israel

Haktan Sonmez
Department of Medical Pharmacology
Faculty of Medicine
Istanbul University-Cerrahpaşa
Istanbul, Turkey

Shailendra Kumar Srivastava
Jacob Institute of Biotechnology & Bio-Engineering
Sam Higginbottom University of Agriculture
Technology & Sciences
Prayagraj, India

Maitree Suttajit
Thai Vegetarian Association
Changpuak
Chiang Mai, Thailand

Brahm Kumar Tiwari
Department of Paramedical Sciences
SGT University
Gurugram
Haryana, India

Kanchan Siddaprasad Udata
CSIR-Central Salt & Marine Chemicals Research Institute
Bhavnagar
Gujarat, India

Geeta Watal
Department of Chemistry
University of Allahabad
Prayagraj
India

Karolin Yanar
Istanbul University-Cerrahpaşa
Cerrahpaşa Medical Faculty
Department of Medical Biochemistry
Istanbul, Turkey

Zeynep Mine Coşkun Yazıcı
Demiroğlu Bilim University
Department of Molecular Biology
and Genetic
Istanbul, Turkey

Burcu Yeşilkaya
Department of Nutrition and Dietetics
Faculty of Health Science
Okan University
Istanbul, Turkey

Banu Orta-Yilmaz
Department of Biology
Faculty of Science
Istanbul University
Istanbul, Turkey

Preface

Diabetes mellitus (DM), which is commonly known as "sugar disease", has now become "bitter" for health and this metabolic disease is increasing worldwide at an alarming rate. The combination of chronic pathological disorders in pancreatic β-cells leading to persistent hyperglycemia is connected to serious organ damage followed by failure including kidneys, eyes, feet, nerves, brain, gastrointestinal tract, and cardiovascular system in DM. DM may be of different types; a deficiency of insulin production by the genetic failure of the pancreas at birth (type 1 diabetes mellitus: T1DM), or insufficiency of insulin production in the face of insulin resistance (type 2 diabetes mellitus: T2DM), or insulin resistance through the influence of different hormones during pregnancy (gestational diabetes mellitus: GDM). Progress and development of DM can be prevented/controlled by pharmacological treatments such as insulin therapy and medication by synthetic drugs, but these therapies are either costly or significantly associated with the risk of other health complications and all-cause mortality.

The need to salvage the havoc of chronic complications of DM by bioactive compounds from natural products (NPs) is presently gathering a lot of attention due to their being affordable, clinically effective, and having relatively infrequent side effects than current synthetic drugs. Over the last few decades a large number of natural products have been sought and discovered for their functional values to develop anti-hyperglycemic drugs. However, due to enormous scaffold diversity, structural complexity, and other special features in comparison to conventional synthetic molecules, NPs confer both advantages and challenges for the drug discovery process. NPs enriched with bioactive compounds cover a wider area for optimization of the anti-diabetic drug spectrum by stimulating the proliferation of β-pancreatic cells to produce insulin, modulating enzymes involving glucose uptake and glucose transport and metabolism, modifying gene expression, and hormone activities relating to glucose homeostasis, oxidative stress, and other modifications of mechanisms.

Written by a global team of experts, this book aims to provide readers with an overview of recent research in new drug discovery against diabetic complications based on bioactives from NPs. This book has 20 chapters and is divided in four parts to provide a clearer picture for readers. Part 1 introduces the fundamental concepts of DM and the recent NP-based drug discovery progress, the various druggable targets, and challenges. Part 2 provides a review of bioactive compounds and healthy diets and natural non-insulin drugs against T1DM. Part 3 reviews some effective bioactive compounds from plants against T2DM; and Part 4 presents the updated drug discovery on GDM prevention and management by NPs.

Chapter 1 extends an expert opinion on DM and novel directions towards NP-based drug discovery. The marine ecosystem is one of the major natural treasures for various products; it can play a crucial role in biomedical research and drug development against DM. However, scattered and few studies have bypassed this field. Chapter 2 explores more on this topic. Chapter 3 has been incorporated in the book to explore the carbohydrate-based antidiabetic natural agents. Functional foods rich in bioactive compounds are recently gaining much attention in health promotion. Keeping this rationale in mind, a dedicated Chapter 4 is included in the book that discusses the effectiveness of functional foods in the intervention of diabetic complications and the realistic results in clinical trials.

The implementation of nanotechnology in improving the bioavailability and reducing the threshold dose of bioactive compounds and thus overcoming the associated biopharmaceutical obstacles, and refining the antidiabetic effects of plant bioactives, is discussed in Chapter 5. In understanding the genetics of DM, associated risk genes may introduce new avenues for NP-based drug discovery in preventing complications. Chapters 6 and 20 discuss the genetics of T1DM and GDM and, along with Chapter 16, highlight their interaction with NPs and dietary patterns. Recently, plant polyphenols have gained global interest in DM intervention due to their strong antioxidative, redox restoring and β-cells' proliferation inducing effects. The book contains dedicated Chapters 7, 11, 14, 17, and 19 on the various antidiabetic mechanisms of action of different polyphenols. The immunomodulatory role of NPs' bioactives' relevance in T1DM therapy is incorporated in Chapter 8.

Obesity is now a global public health concern that affects populations of all ages and is linked with the onset of many chronic health impairments; however, the connection between obesity and T1DM is not fully understood. Chapter 9 deals with the discussion on obesity risk factor in patients with T1DM and the possible role of nutritional therapy in its management. Insulin therapy is the most frequently used medication against T1DM; however, its side effects and risks limit its safety. Chapter 10 explores the use of natural non-insulin drugs as a novel approach to enhance therapeutic outcomes against T1DM. Impaired wound healing in T2DM often leads to chronic lesions in these patients, which results in vascular dysfunction, neuropathic abnormalities, immune irregularities, and various deviations in biochemical attributes. Chapter 12 focuses on the potential of clove oil in wound healing and dressing in T2DM. Chapters 13 and 15 highlight nutritive features and bioactivities of thymoquinone and cannabis in the management of T2DM. Chapter 18 discusses the exploration of medicinal and aromatic plants in preventing the pathological events of GDM.

We extend our thanks to all our contributors who provided us with splendid chapters of the field and John Wiley & Sons, Inc., for making every effort to publish this book, when a large group of populations worldwide is concerned with the diabetes associated health issues. Lastly, we hope that readers including professional scientists and experts in various areas of basic, pharmaceutical, and medical research will acknowledge this comprehensive and valuable composition.

Kanti Bhooshan Pandey
Bhavnagar, Gujarat, India

Maitree Suttajit
Chiang Mai, Thailand

Pinar Atukeren
Istanbul, Turkey

Part I

Fundamental Concepts of Diabetes Mellitus and Drug Discovery Process

1

Diabetes Mellitus and Natural Product-based Drug Discovery

Novel Directions

David J. Newman

NIH Special Volunteer, Wayne, PA 19087, USA

1.1 Introduction

In terms of saving lives and/or helping people suffering from this disease, I will demonstrate the history of treatments for the diseases known as "diabetes" in which I include both diabetes 1 (autoimmune destruction of insulin-producing β-cells in the pancreas) and diabetes 2 (inadequate insulin production and/or insulin resistance, though exact mechanisms are not understood). The data presented will exemplify the role(s) of natural products in deriving both the earlier (mainly insulin) and current drugs that ameliorate these diseases. I have included both "versions" of the disease as nowadays there are reports demonstrating adult onset of what would be considered type 1 (T1DM) and a very significant number of type 2 (T2DM), frequently included under the catchall term, "metabolic syndrome" [1] and, later on, some of the current factors that may well lead to such a diagnosis are listed. The chapter will first focus on insulin and its current sources, and will then give details of other agents, some very slightly modified natural products from manifold sources, and then the synthesis of what can be best described as natural product spatial mimics that are in use nowadays, particularly in the treatment of T2DM.

Natural Products and their Bioactives in Antidiabetic Drug Discovery, First Edition.
Edited by Kanti Bhooshan Pandey, Maitree Suttajit and Pinar Atukeren.

1.2 A Brief History of Insulin

The story surrounding the discovery and utilization of insulin in Canada at the University of Toronto linking the utility of insulin to treat diabetes, was the first clinically reported linkage of this human hormone to the treatment of diabetes. However, reports of this disease went back many centuries, even millennia, with initial reports around 552 BC in the *Ebers papyrus* (which can be considered the first written description of diseases and potential treatments) by Egyptian scholars. Similar reports from both Chinese and Indian "academics" in a relatively similar time frame also noted the disease, with all reports noting the common link that patients had a constant flow of "sweet urine". In the time frame of roughly 100 years BCE, Demetrius of Apameia was credited with the introduction of the term "diabetes" (from the Greek word for siphon) since he noted the constant flow of urine from patients Then, in the second century CE, the Greek physician Aretaeus of Cappadocia first described the first accurate clinical description of diabetes [2]. Galen, at approximately the same time as Aretaeus, associated diabetes with a disease of the kidneys, and this persisted for the next 1500 years.

Then, following a number of varied experiments using dogs (with or without their pancreas) and other data from diabetic human patients who had damaged islets of Langerhans in their pancreas, led to the suggestion in 1916 by Sharpey-Schafer of a linkage between a secretion from the pancreas and "control" of carbohydrate metabolism by an as yet unidentified hormone that regulated glucose that he named insulin [2]. The complex interactions, both scientific and personal, that led to the purification of insulin and its initial clinical success are well described by Rosenfeld in 2002 [3]. This report also mentions the first production of crystalline insulin in 1926 by Abel at Johns Hopkins and then, in 1951, Sanger sequenced insulin, the first protein hormone to have its formal structure reported [4, 5]. Until the production of human insulin via biotechnological methods, which will be discussed later in this chapter, all supplies of insulin were effectively from utilization of bovine or porcine sources and, at times, chemical modification to aid in stability and/or ease of utilization.

1.3 Biosynthesis of Insulin

The biosynthetic process produces an inactive 110 residue, preproinsulin, which on translation into the endoplasmic reticulum has the signal peptide removed leaving proinsulin, which folds to give three disulfide linkages between the precursors of the A and B chains. This proinsulin then transits through the Golgi apparatus into the secretory granules. Further removal of selected portions by a number of hydrolytic enzymes produces the active insulin composed of

A (21 amino acids) and B (30 amino acids) chains connected via two disulfide bonds. It is stored as an "inactive" hexamer and is now available for release of the monomer following the correct metabolic stimulus. It should also be pointed out that this is a very ancient molecule and is found across many eukaryotic genera including fungi, teleost fish and, in a modified form, in the venom of cone snails.

1.3.1 Humulin®, the Second Human Hormone Produced by Biotechnology

In 1977, somatostatin was the first peptide hormone to be produced by the then new and novel recombinant DNA process [6, 7]. This was also effectively the foundation product of the company Genentech, which had been founded the previous year. Since the human genes corresponding to *in vivo* production of insulin had not been identified/cloned at that stage of rDNA operations, what could be considered "synthetic genes" were made chemically, corresponding to the formal structure of insulin components. The success of this was demonstrated in a 1979 paper by Goeddel et al. [8], which used the "synthetic human genes" published late the previous year by Crea et al. [9]. Proof of the validity of this work was shown by the approval by the US FDA of Humulin. Just to put the timing in perspective, I have used the following quote from Riggs' article [7]. "For example by 2020 the genes for insulin can be made in a few hours by an automated instrument and then cloned and expressed by a single person in about a week."

1.3.2 Further Derivatives of Insulin Using rDNA Techniques

Once the utility of this process was proven, over time since the early 1980s, insulins with improved pharmacology have been approved. In a recent (2019) open access journal, Guney [10] described the number of agents that are currently available and classified them by their pharmacological parameters.

As of the date of that review the following were available (though not perhaps in all countries). I have omitted the premixed and biphasic analogs in Table 1.1.

Table 1.1 Insulin preparations mainly for T1DM treatment.

Type	Generic name	Brand name
Rapid-acting	Glulisine	Apidra
	Lispro	Humalog
	Aspart	NovoRapid
Short-acting	Insulin isophane	Humulin R

(Continued)

Table 1.1 (Continued)

Type	Generic name	Brand name
	Neutral insulin	Actrarapid
	Regular and isophane human insulins	SciLin N
Intermediate	Human insulin isophane	Humulin N
	INN insulin human	Protophane
Long-acting	Detemir	Levemir
	Glargine	Lantus
	Insulin degludec	Tresiba

Thus, there are multiple variations designed to help T1DM patients control their glucose levels before and after meals.

1.4 Current Possible Insulin-based Treatments for T2DM

In a recent paper by Sebastian et al. [11] there is significant discussion of various insulin-based treatments for T2DM with one of the more promising being the Technosphere (TI) inhaled insulin (Afreeza®), which was approved by the FDA in 2014 for both T1DM and T2DM patients, with further details in the 2021 paper by Levin et al. in 2021 [12]. This is in contrast to the current treatments of a glucagon-like peptide (GLP)-1 receptor agonist after metformin, and variations on other mixed oral/injectable treatments.

1.4.1 Metabolic Syndrome (As of June 2022)

In an article in the American Chemical Society's journal *Central Science* [13], if a patient has any three of the following five criteria as described in Table 1.2 for metabolic syndrome then they have a high risk of T2DM and cardiovascular diseases.

However, there is at least one major caveat in the values listed in Table 1.2 that needs to be taken into account. A significant number of the values listed were taken from the Framingham heart study where less than 5% of the participants were of other than European ancestry. When current data from more diverse populations are included, particularly in Black populations, the figures for HDL need

Table 1.2 Symptoms leading to metabolic syndrome diagnosis.

Criterion
Elevated waist circumference (female ≥88 cm; male ≥102 cm)
Elevated triglycerides (≥150 mg.dL^{-1})
Low HDL (female <50 mg.dL^{-1}; male <40 mg.dL^{-1})
Elevated blood pressure (systolic ≥130 mm Hg, or diastolic ≥85 mm Hg or both)
Elevated fasting blood glucose (≥100 mg.dL^{-1})

to be adjusted. In addition, there are data that East and South Asians are more resistant to insulin than other major populations including Hispanic. Thus, absolute reliance on the figures given in Table 1.2 can lead to incorrect over- or under-counting of populations that need treatment.

1.5 Non-insulin-linked Treatments for T1DM and T2DM

Although insulin is the gold standard for T1DM and recent data demonstrated that various insulin modifications may well be the newer treatment for T2DM, over the last 20 or so years, as some, though not all, of the "causes" of carbohydrate metabolic changes that led to T2DM variations in patients are being identified, the pharmaceutical industries in a number of countries began to explore a variety of non-insulin agents as ameliorants of this multi-organ problem, with a significant number being either based on natural products or modifications thereof.

1.5.1 Guanidines: Agents that Began as Herbal Remedies

There is one very well-known series of agents that were too early for the analyses referred to in the next subheading that covered the 1997 onwards reviews of drug sources made by the author and his colleagues. These agents were directly derived from the anecdotal usage of the perennial herb *Galega officinalis* Linn. as an herbal remedy in the Middle Ages in Europe with versions of *Culpeper's Complete Herbal* suggesting that it had antidiabetic properties (effectively the "sweet urine" commented on earlier in this chapter). *Culpeper's Complete Herbal* as of 1850 is available as part of the "Project Gutenberg" ebook series [14]. In addition to this ebook, there are other much more recent articles that give information on the use of both the natural product galegine, which was used over centuries but not as the pure chemical, though there are reports in the French literature as late as the mid-1930s of usage of partially purified extracts as treatments [15].

The fundamental chemistry report on the usage of guanidino compounds, in what we now know as T2DM, was a paper in the *Journal of Biological Chemistry* in 1918 where Watanabe demonstrated that guanidino compounds, including simple derivatives related to galegine (Figure 1.1–1) could reduce blood sugar levels [16]. The initial synthesis of metformin (Figure 1.1–2) was published in 1922 by Werner and Bell [17]. However, even though its glucose lowering potential was published, it was not used as an antidiabetic agent at that time. In the early 1940s, metformin was "rediscovered" in a search for antimalarial agents and

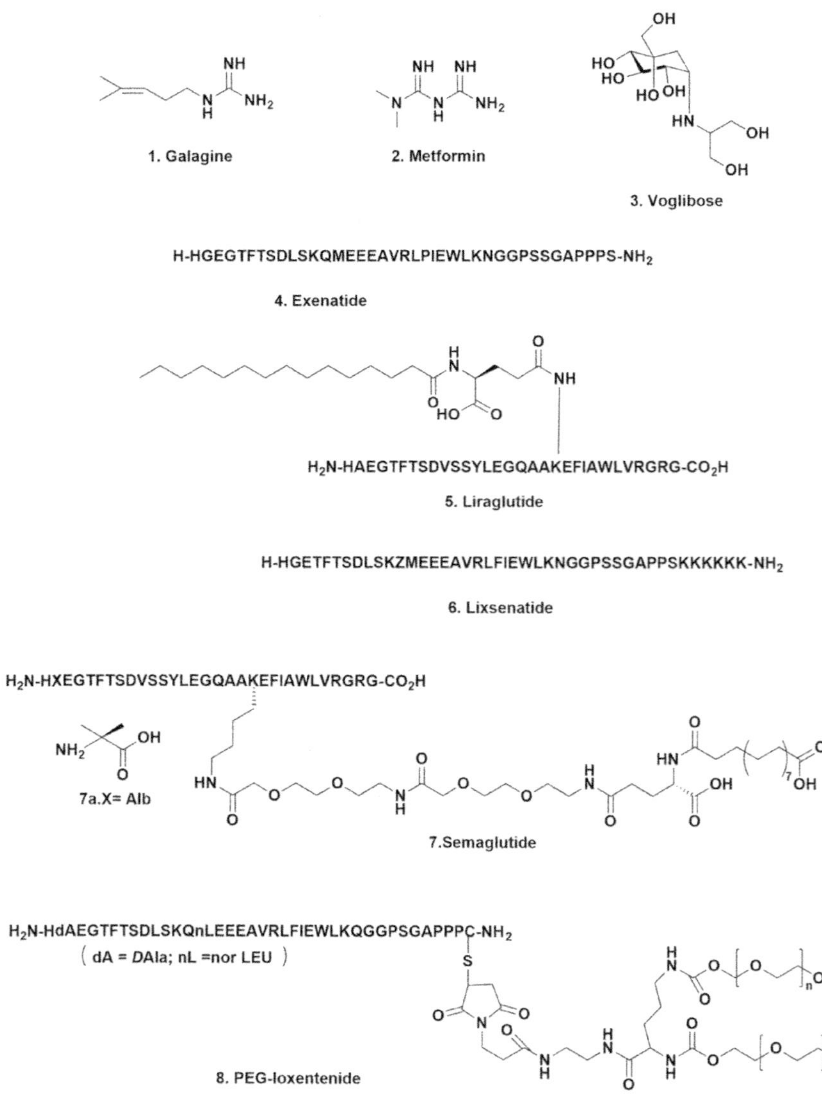

Figure 1.1 Different antidiabetic structures.

then, in 1957, the French physician Jean Steme, first reported its potential as a treatment for adult onset diabetes [18]. Over the next few years, Sterne continued to publish on the utility and mechanism(s) of this agent in a number of publications, with the last one in 1964 before he moved to publishing in books [19].

From these reports and discoveries of the potential of this class of compounds, in 1958, metformin was introduced in the UK and other European countries, but it was not until 36 years later, in 1994, that metformin was approved by the US FDA. It was introduced in 1995, with two reports at that time that were key publications that confirmed the favorable risk/benefit ratio in the management of T2DM [20, 21]. In 2017, Bailey published a review giving the history of metformin and in that article, table 1.1 is a timeline with excellent commentary demonstrating the path from 1772 to 2011, with relevant citations given at each major point in the story [22].

From the perspective of using this compound as an antitumor agent, currently (July, 2022), from a search of the NIH clinical trials database (www.clinicaltrials.gov) using metformin as the drug candidate, there are 72 Phase 3 studies using metformin as a potential antitumor drug that are recruiting or are underway, with over 400 at the same trial level that have been completed.

1.5.2 Drugs Other than Guanides that Have Been Approved from 1997 to 09/2019 against T2DM

In a series of reviews written by the author and colleagues from 1997 to the latest in 2020 [23–28], we analyzed the chemistry behind the drugs that were approved by the US FDA or comparable agencies in other areas of the world, and only counted a drug one time. The first review specialized in antitumor and anti-infective agents, but from the 2003 review we extended the reviews to include most diseases where drugs had been approved since 1981.

These reviews were cumulative, with the latest in 2020 [28] being an open access article in the *Journal of Natural Products*. In the 2003 review [24], we introduced a subset of definitions that further divided the sources when chemical syntheses were used for production as approved drug entities, molecules that required some "chemical forensics" to formally decide if they were based on a natural product, and/or were recognized by the biological system as a mimic. The basic terms are shown in Table 1.3. If more details are desired by the reader, the 2003 review [24] will need to be consulted as detailed reasons with examples were given in that article.

As of the latest published review in 2020, there were 39 non-insulin antidiabetic agents approved by at least one governmental agency. Obviously, if one looks at the pharmaceutical industry data, the numbers will be greater, but as mentioned earlier, only the first approval of a drug is counted irrespective of the number of

Table 1.3 Newman and Cragg Codes.

Code	Brief definition/review year
B	Biological macromolecule/1997
N	Unaltered natural product/1997
NB	Botanical drug (defined mixture)/2012
ND	Natural product derivative/1997
S	Synthetic drug/1997
S*	Synthetic drug (NP pharmacophore)/1997
V	Vaccine/2003
/NM	Mimic of natural product/2003

Table 1.4 Antidiabetic agent sources as of September 2019.

N	ND	S	S/NM	S*	S*/NM
1	8	4	16	1	9

countries that might approve it later on. As of the end of September 2019, their breakdown was as shown in Table 1.4.

The single natural product in Table 1.4 is voglibose (Figure 1.1–3), an α-glucosidase inhibitor first isolated in Japan in 1981 from *Streptomyces hygroscopicus* var *limons* and approved in 1994 in Japan for treatment for T2DM. Two other natural product derivatives (structures not shown) acarbose and miglitol were also approved in 1990 and 1998, respectively, targeting the same enzyme system.

1.5.2.1 Modified Peptides that are Incretin Mimics (GLP-1 Agonists)

In 2005, the peptide known as extenatide or Byetta®, a 39-residue peptide (Figure 1.1–4) was approved for T2DM treatment. This agent is what is known as a glucagon-like peptide 1 (GLP-1) agonist and it was based on one of the peptides in the saliva of the lizard known colloquially as "Gila monster" *Heloderma suspectum*. It is also known as an incretin mimic since GLP-1 is the naturally occurring incretin hormone. Since 2005, a number of other GLP-1 mimics, all based upon modifications of the extenatide base skeleton have been approved, including liraglutide (Figure 1.1–5) in 2009, lixisenatide (Figure 1.1–6) in 2013, semaglutide (Figure 1.1–7) in 2017, and a pegylated version, PEG-loxenatide (Figure 1.1–8) in 2019. With semaglutide, a non-natural amino acid Alb (Figure

1.1–7a. **X=Alb**) was inserted to avoid degradation, and in PEG-loxentenide (Figure 1.1–8) two unusual amino acids were used, the "D" isomer of alanine and *nor*leucine. All of these are used for treatment of T2DM working as GLP-1 agonists and fall under the ND category.

1.5.2.2 Agents against Dipeptidyl Peptidase IV

Another contemporary target in T2DM is DPP-IV (dipeptidyl peptidase IV), and the agents that target this enzyme fall under the S/NM category. Eleven compounds directed against this target were approved between 2006 and 2016, with their structures shown in Figure 1.2 starting with sitagliptin (Figure 1.2–9) in 2006. After this initial agent, over the next ten years other agents were approved as follows: vildagliptin (Figure 1.2–10) in 2007, saxagliptin (Figure 1.2–11) in 2009, alogliptin (Figure 1.2–12) in 2010, linagliptin (Figure 1.2–13) in 2011, teneligliptin (Figure 1.2–14) and anagliptin (Figure 1.2–15) in 2012. Then, in 2015,

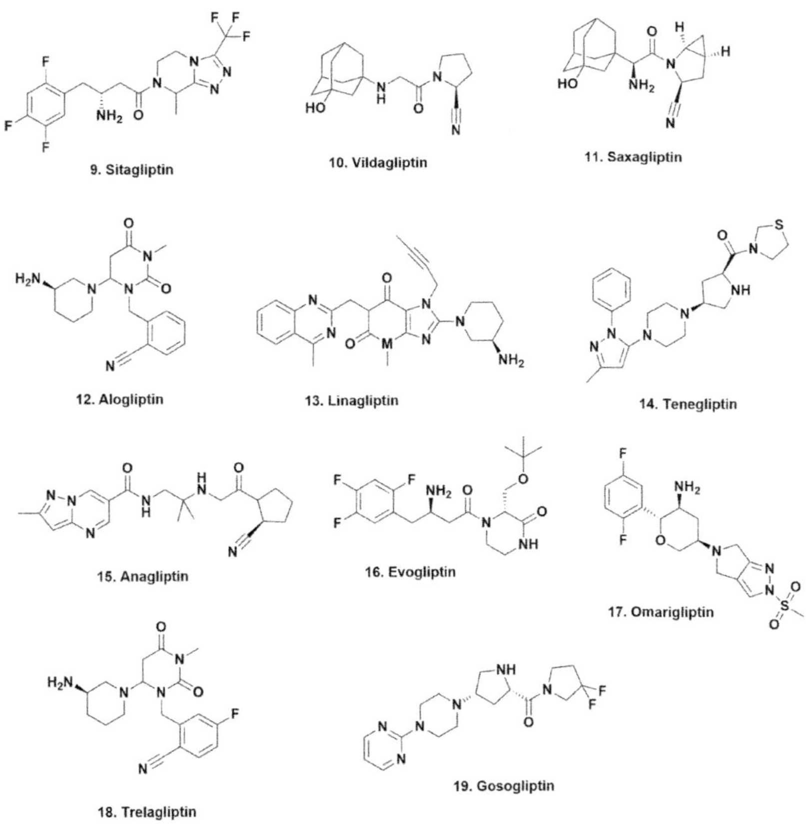

Figure 1.2 Different antidiabetic structures.

three more "gliptins" were launched: evogliptin (Figure 1.2–16), omarigliptin (Figure 1.2–17), and trelagliptin (Figure 1.2–18), and in 2016, the latest to date was gosogliptin (Figure 1.2–19) following approval in Russia.

1.5.2.3 Agents against the Sodium-dependent Glucose Transporter (SGLT-1/2)

These are nine SGLT-1/2 inhibitors whose structures were based upon the non-selective natural product phlorizin (Figure 1.3–20), which led to dapagliflozin (Figure 1.3–21) in 2012, canagliflozin (Figure 1.3–22) in 2013, and then in 2014, four compounds using this base structure were approved, starting with empagliflozin (Figure 1.3–23), continuing with ipragliflozin proline (Figure 1.3–24) and tofogliflozin (Figure 1.3–25), with the last one in 2014 being luseogliflozin (Figure 1.3–26) where a thio-sugar was used. There was then a gap until 2017 when ertugliflozin (Figure 1.3–27) was approved. It should be noted that this compound structure is close to that of the 2012 dapagliflozin (Figure 1.3–21).

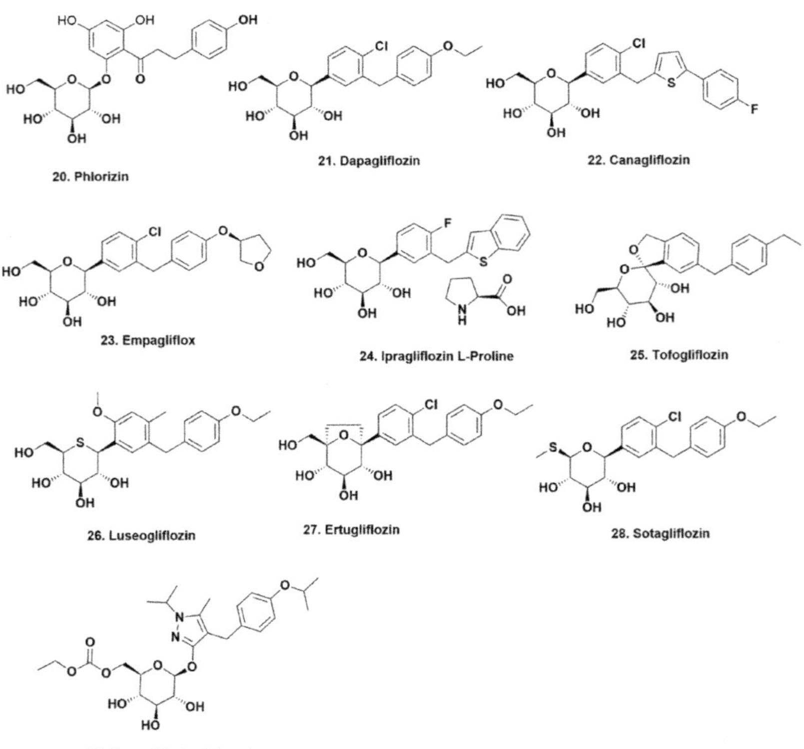

Figure 1.3 Different antidiabetic structures.

In the first nine months of 2019, two more compounds of this class were approved. Sotagliflozin (Figure 1.3–28), which also closely resembles dapagliflozin (Figure 1.3–21), but with a methyl sulfur substitute in the sugar moiety in place of the normal hydroxyl group. In 2019, remogliflozin etabonate (Figure 1.3–29) was approved. The "gliflozins" that are targeted against this protein complex fall under the S*/NM code. In addition, a recent paper by Shaffner et al. [29] covers the pharmacology of these inhibitors in detail and should be consulted in addition to the translational medicine aspect of these agents that was covered by Beitelshees et al. [30].

1.6 Post September 2019 Drug Approvals

Since the publication of the 2020 review [28] as of the middle of June 2022, the following three agents have been approved/launched for the first time (Table 1.5). Please note that their actual dates of approval might differ from their launch date in a specific country and that only the first approval and/or launch is noted.

These three drugs include one small molecule with a novel triazine structure (Imeglimin, Figure 1.4–30) that has activities similar to metformin and can be considered as a cyclic metformin derivative [31]. From a natural product aspect, the other two are peptidic in nature and are direct agonists of specific receptors with one being approved in 2021 and the other in 2022. The details of producing tirzepatide (Figure 1.4–32) on a kilogram scale was published by the Lilly scientists involved in 2021, and demonstrate the methodologies necessary to proceed from a lab-scale of less than a gram to kilogram quantities under cGMP conditions [32].

Wang, in 2022, published an excellent paper on the design parameters that were used to develop tirzepatide and this discussion is an excellent primer on how to design complex peptidic drugs [33]. In that paper, Wang identifies a web site (www.comparediabetesdrugs.com) that compares drugs used in the treatment of both T1DM and T2DM that is an excellent resource for scientists interested in this class of drugs.

Table 1.5 Agents approved 2021–2022.

Name/year	Mechanism	Newman and Cragg codes	Structure #
Imeglimin/2021	GSIS enhancement	S*/NM	Figure 1.4–30
Zegalogue/2021	Glucagon receptor agonist	ND	Figure 1.4–31
Tirzepatide/2022	GIP/GLP-1 agonist	ND	Figure 1.4–32

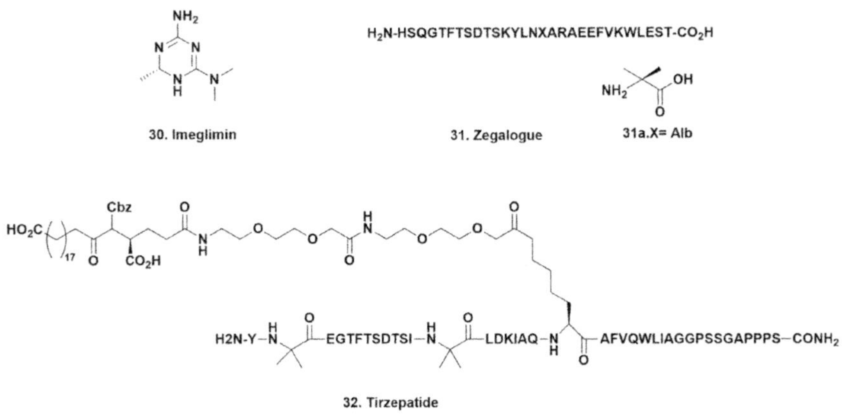

Figure 1.4 Different antidiabetic structures.

1.8 Conclusion

Although insulin was first obtained for human use by extraction and then chemical modification from bovine or porcine sources, once the biotechnological capabilities of recombinant DNA were demonstrated in other areas (though on much smaller/simpler compounds) such as the production of first Humulin® and then modifications of insulin via these techniques, producing insulin no longer was the bottleneck in treatment of T1DM. In addition, as has been mentioned, there are nowadays treatment regimens for T2DM that include insulin treatments [11, 12], in addition to combination with the regular T2DM oral agents.

However, what is extremely interesting from a natural product chemistry perspective, is the progression from mediaeval herbal remedies (Goat's rue aka *Galega officinalis* Linn.) via chemistry to give metformin (Figure 1.1–2) and other derivatives that were developed from another early natural product, phlorizin (Figure 1.3–20), that have been, and in most cases, still are in use as oral treatments for T2DM. What is also of interest is that due to the major advances in peptide synthesis, the two agents referred to in Table 1.5 that were approved for use in 2021 and 2022 under the "ND" code may well be prototypes for molecules in the future directed against specific organelles.

The clinical work around "metabolic syndrome" is ongoing as clinicians learn more about the problems that fall under that term, and it is possible that some of the newer agents may help resolve these problems in due course.

A Note on Structures

All structures were drawn using the program ChemDraw (v.19.1) and then inserted into the text. In the case of the polypeptides, which were greater than 30 amino acids, in certain cases the final structure also contained non-peptidic side-chains such as fatty acids and/or polyethylene glycol (PEG) substituents. Due to their size, the single letter codes corresponding to the international usage for L-amino acids were used within ChemDraw. Details of this nomenclature are shown in the following: https://www.fao.org/3/Y2775E/y2775e0e.htm. If there were modifications of amino acids used then they are shown under the relevant base figure.

References

1 Saklayen, M.G. (2018). The global epidemic of the metabolic syndrome. *Curr. Hyperten. Rpts.* 20: 12.

2 Eknoyan, G. and Nagy, J. (2005). A history of diabetes mellitus or how a disease of the kidneys evolved into a kidney disease. *Adv. Chronic Kidney Dis.* 12: 223–229.

3 Rosenfeld, L. (2002). Insulin: Discovery and controversy. *Clin. Chem.* 48: 2270–2288.

4 Sanger, F. and Tuppy, H. (1951). The amino-acid sequence in the phenylalanyl chain of insulin. 1. The identification of lower peptides from partial hydrolysates. *Biochem. J.* 49: 463–481.

5 Sanger, F. and Tuppy, H. (1951). The amino-acid sequence in the phenylalanyl chain of insulin. 2. The investigation of peptides from enzymic hydrolysates. *Biochem. J.* 49: 481–490.

6 Itakura, K., Hirose, T., Crea, R. et al. (1977). Expression in *Escherichia coli* of a chemically synthesized gene for the hormone somatostatin. *Science.* 198: 1056–1063.

7 Riggs, A.D. (2021). Making, cloning, and the expression of human insulin genes in bacteria: The path to humulin. *Endo. Revs.* 42: 374–380.

8 Goeddel, D.V., Kleid, D.G., Bolivar. F. et al. (1979). Expression in *Escherichia coli* of chemically synthesized genes for human insulin. *Proc. Natl. Acad. Sci. USA.* 76: 106–110.

9 Crea, R., Kraszewski, A., Hirose, T. et al. (1978). Chemical synthesis of genes for human insulin. *Proc. Natl. Acad. Sci. USA.* 75: 5765–5769.

10 Guney, Z. (2019). Insulin and its analogues-what are they for? (pros and cons). *Trends Diabetes Metab.* 2: 1–2.

11 Sebastian, S.N., Co, E.L., Mehendale, M. et al. (2022). Insulin analogs in the treatment of type II diabetes and future perspectives. *Disease-a-Month.* 101417. doi: 10.1016/j.disamonth.2022.101417.

12 Levin, P., Hoogwerf, B.J., Snell-Bergeon, J. et al. (2021). Ultra rapid-acting inhaled insulin improves glucose control in patients with type 2 diabetes mellitus. *Endocrine Pract.* 27: 449–454.

13 Madhusoodanan, J. (2022). Searching for better biomarkers for metabolic sndrome. *ACS Cent. Sci.* 8: 682–685.

14 Culpeper, N. (2015). *The project Gutenberg ebook of the complete herbal* (EBook #49513). Available from: www.gutenberg.org/files/49513/49513-h/49513-h.htm (accessed 30 May 2022).

15 Bailey, C.J. and Day, C. (2004). Metformin: Its botanical background. *Pract. Diabetes Inf.* 21: 115–117.

16 Watanabe, C.K. (1918). Studies in the metabolic changes induced by administration of guanidine bases. Influence of injected guanidine hydrochloride upon blood sugar content. *J. Biol. Chem.* 33: 253–265.

17 Werner, E.A. and Bell. J. (1922). The preparation of methylguanidine, and of ββ-dimethylguanidine by the interaction of dicyandiamide, and methylammonium and dimethylammonium chlorides respectively. *J. Chem. Soc. Trans.* 121: 1790–1794.

18 Sterne, J. (1957). Du nouveau dans les antidiabétiques. La NN dimethylamine guanyl guanidine (N.N.D.G.). *Maroc. Med.* 36: 1295–1296.

19 Sterne, J. (1964). Mechanism of action of antidiabetic biguanides. *Presse Med.* 72: 17–19.

20 Bailey, C.J. and Turner, R.C. (1996). Drug therapy: Metformin. *N. Engl. J. Med.* 334: 574–579.

21 DeFronzo, R.A. and Goodman, A.M. (1995). Efficacy of metformin in patients with non-insulin dependent *diabetes mellitus. N. Engl. J. Med.* 333: 541–549.

22 Bailey, CJ. (2017). Metformin: Historical overview. *Diabetilogia.* 60: 1566–1576.

23 Cragg, G.M., Newman, D.J. and Snader, K.M. (1997). Natural products in drug discovery and development. *J. Nat. Prod.* 60: 52–60.

24 Newman, D.J., Cragg, G.M. and Snader, K.M. (2003). Natural products as sources of new drugs over the period 1981–2002. *J. Nat. Prod.* 66: 1022–1037.

25 Newman, D.J. and Cragg, G.M. (2007). Natural products as sources of new drugs over the last 25 years. *J. Nat. Prod.* 70: 461–477.

26 Newman, D.J. and Cragg, G.M. (2012). Natural products as sources of new drugs over the 30 years from 1981 to 2010. *J. Nat. Prod.* 75: 311–335.

27 Newman, D.J. and Cragg, G.M. (2016). Natural products as sources of new drugs 1981 to 2014. *J. Nat. Prod.* 79: 629–661.

28 Newman, D.J. and Cragg, G.M. (2020). Natural products as sources of new drugs over the nearly four decades from 01/1981 to 09/2019. *J. Nat. Prod.* 83: 770–803.

29 Shaffner, J., Chen, B., Malhotra, D.K. et al. (2021). Therapeutic targeting of SGLT2: A new era in the treatment of diabetes and diabetic kdney disease. *Front. Endocrinol.* 12: 749010.

30 Beitelshees, A.L., Leslie, B.R. and Taylor, S.I. (2019). Sodium-glucose cotransporter 2 inhibitors: A case study in translational research. *Diabetes.* 68: 1109–1120.

31 Raczyńska, E.D., Gal, J.-F., Maria, P.-C. et al. (2019). Biguanide antidiabetic drugs: Imeglimin exhibits higher proton basicity but smaller lithium-cation basicity than metformin *in vacuo*. *ACS Omega*. 3: 17842–17852.

32 Frederick, M.O., Boyse, R.A., Braden, T.M. et al. (2021). *Org. Process. Res. Dev.* 25: 1628–1636.

33 Wang, L. (2022). Designing a dual GLP-1R/GIPR agonist from tirzepatide: Comparing residues between tirzepatide, GLP-1, and GIP. *Drug Design Develop. Ther.* 16: 1547–1559.

2

Marine Natural Products in the Management of Diabetes Mellitus

Murali Krishna Paidi[1,2], *Kanchan Siddaprasad Udata*[1,2], *and Subir Kumar Mandal*[1,2,*]

[1] *Academy of Scientific and Innovative Research (AcSIR), Ghaziabad-201002, India*
[2] *Applied Phycology and Biotechnology Division, CSIR-Central Salt and Marine Chemicals Research Institute, Bhavnagar-364002, India*
[*] *Corresponding author*

2.1 Introduction

Diabetes mellitus (DM) is described as a group of metabolic disorders characterized by elevated levels of blood glucose [1]. Either diabetes occurs when the pancreas does not produce sufficient insulin (a peptide hormone that regulates the blood glucose levels), or when the body (cells) ineffectively uses the available insulin that is sufficiently produced from the pancreas. The prevalence of diabetes has been increasing in developing countries more than in developed countries. The International Diabetes Federation (IDF) has projected the global prevalence is 9.3% in 2019, which will increase by 25% in 2030, and 51% in 2045 if the current diabetes risk management and treatment programs remain unchanged [2]. In many developing countries, people are adapting to high-calorie, low-quality fast foods and other processed food items that are directly linked with the metabolic syndrome, obesity and type 2 diabetes mellitus (T2DM).

2.2 Global Trend and Prevalence of Diabetes

Diabetes is a most common chronic disorder, and the global burden of diabetes has increased significantly since 1990. It has imposed a global burden on public health and socioeconomic development [3]. Recently, the world population was projected to be ~7875 million in 2021, out of this number, about 537 million

Natural Products and their Bioactives in Antidiabetic Drug Discovery, First Edition.
Edited by Kanti Bhooshan Pandey, Maitree Suttajit and Pinar Atukeren.

(6.81%) people living with diabetes will be increased significantly to 693 million by 2045 [1]. In 2019, the IDF estimated that 436 million people live with diabetes and the regional percentage of diabetes is higher (38.4%) in the Western Pacific region followed by the lowest (4.5%) in the African region (Figure 2.1). In the case of low and middle-income countries about 3 in 4 adults are living with diabetes [2]. The prevalence of both type 1 diabetes mellitus (T1DM) and T2DM has been increasing among children and adolescents, and it is estimated that those below age 20 with type 1 diabetes exceed one million [3].

2.3 Risk Factors and Complications of Diabetes

Chronic hyperglycemia during diabetes is associated with long-term microvascular complications affecting the eyes, kidneys, and nerves, and increases the risk of cardiovascular diseases (CVD). The majority of diabetes is categorized into two types: T1DM and T2DM (Table 2.1); although some of the hyperglycemia conditions are difficult to classify. For example, gestational diabetes mellitus (GDM) refers to glucose intolerance, and it was first recognized during the pregnancy stage. Currently, in many peoples (non-pregnant women or mass), glucose intolerance is increasing due to changes in modern diets and sedentary lifestyles. These changes are directly linked to obesity and fat accumulation in the body that causes T2DM and CVD.

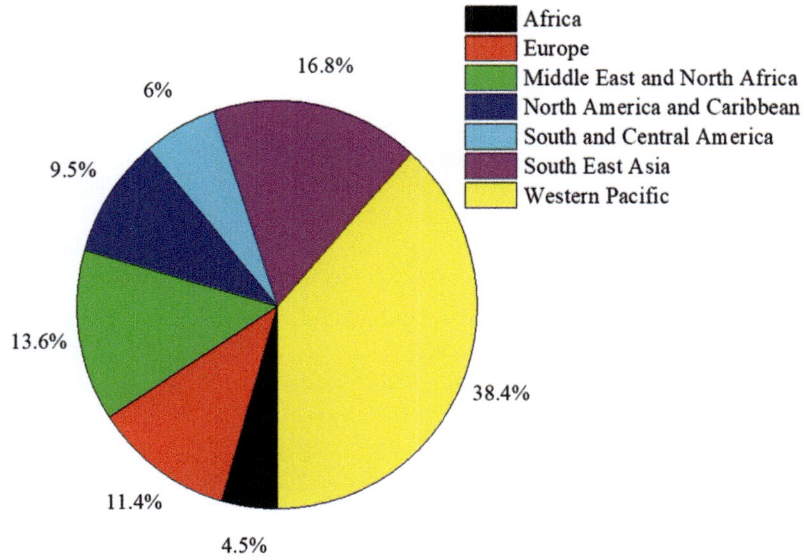

Figure 2.1 Pie chart representing the regional percentage of diabetes in the world in 2021. Adapted from [2] (International Diabetes Federation).

Table 2.1 Diabetes types, major risk factors, and health related problems.

Type	Risk factors	Stage	Frequency	Complications
Type 1 Diabetes Mellitus (T1DM)/ Juvenile DM	Rare type	Childhood and early adulthood	10% of people with diabetes	Increased thirst
	Family history (if parent or sibling has T1DM)			Frequent urination
	Autoimmune disease			Extreme hunger
	Destruction of β-cells			Fatigue, irritations Blurred vision
	Absolute insulin deficiency			Glycated hemoglobin (HbA1c)
				Depression
				Blood vessel damage and other acute symptoms
Type 2 Diabetes (T2DM)	Most common type	~29 to 65 ages	95% of people with diabetes	T1DM symptoms and hyperosmolar, hyperglycemic state (HHS) and other chronic symptoms
	Partial dysfunction of β-cells, Insulin resistance			
	Overweight and obese			
	Family history (if parent or sibling has T2DM)			
	Age			
	Polycystic ovary syndrome			

Table 2.1 (Continued)

Type	Risk factors	Stage	Frequency	Complications
	Abnormal cholesterol and triglyceride levels			
Immune-mediated DM	Rare T1DM slowly evolving in adults	Adults to older ages	—	May develop T1DM by diabetes autoantibodies
	A single glutamic acid decarboxylase antibody (GDA) autoantibody and uncontrolled β-cells function			
Prediabetes	Blood glucose levels are higher than normal	~29 to 65 ages	—	Frequent urination
				Extreme hunger
				Infections of gums and other mouth related infections
Ketosis-prone T2 DM/ Ketosis-prone diabetes (KPD)	Severe dysfunction of β-cells	At any age	—	Hypoglycemia
				Cerebral and pulmonary edema and kidney damage
Monogenic DM:				
i) Monogenic defect of β-cells function	Specific/single gene mutation causes diabetes by making the pancreas less functional	Neonatal diabetes occurs in the first 6 months	4% of people with diabetes	Both Type 1 and 2 related complications

(Continued)

Table 2.1 (Continued)

Type	Risk factors	Stage	Frequency	Complications
ii) Mono-genic defect in insulin action		or		
		Maturity onset diabetes of the young (MODY) in early adulthood		
Cystic fibrosis related DM (CFRD)	Genetic mutation	Specific to people suffering from cystic fibrosis disease.	2% in children, 19% in adolescents and 50% of individuals aged 20– 75 yr in the CF population	Abnormal glucose metabolism
	Dysfunction of both β-cells and islet cells mass and numbers			Microvascular complications
				Hyperglycemia and neuro, gastro and retinopathy symptoms
Drug or chemical induced DM	Impair insulin secretion or action	Organ transplant	—	Hyperglycemia
	Destroy β-cells function	HIV/AIDS treatment		
		Glucocorticoid steroids use		
Gestational DM	Hyperglycemia is due to the production of excess hormones by the placenta that makes body cells more resistant to insulin	Pregnancy stage	—	In baby: excess growth (macrosomia) Low blood glucose T2DM in later life

Table 2.1 (Continued)

Type	Risk factors	Stage	Frequency	Complications
	Women older than 25 years			In mother: preeclampsia and T2DM in the next pregnancy stage
Diabetes Insipidus/ Vasopressin Sensitive Diabetes Insipidus	Central diabetes insipidus	Suddenly at any age	—	Excessive urine production and thirst
	Lack of the hormone vasopressin			Chronic dehydration, low blood temperature
	Brain injury, tuberculosis and other diseases			Weight loss, hypertension, and kidney damage

Generally, diabetes related complications are divided into chronic and acute complications.

Chronic health complications: untreated diabetes in people with high blood glucose levels induces retinopathy [4, 5], nephropathy (kidney damage), neuropathy (nerve damage), damage to blood vessels, receding gum line in the mouth, amputation of feet, and heart attacks and strokes [6, 7]. All of these symptoms develop gradually and can lead to serious organ damage in diabetic persons if they have untreated or uncontrolled conditions.

Acute health complications: These can develop suddenly and lead to chronic or long-term conditions. Hypoglycemia is due to lower blood glucose levels in the body and hyperglycemia to high blood glucose levels [8]. Hyperosmolar Hyperglycemic State (HHS) is characterized by severe dehydration, very high blood sugars, and diabetic ketoacidosis (DKA) conditions [9]. Unchecked or untreated hypoglycemia or hyperglycemia is a dangerous acute complication that leads to diabetic coma. In 2017, diabetes was recognized as the ninth leading cause of mortality. Annually, more than one million deaths can be attributed to diabetes alone [10]. The World Health Organization (WHO) projected in November 2019 estimates of about 1.5 million deaths directly caused by diabetes worldwide [11]. In the last 10 years from 2011 to 2021, deaths attributable to diabetes have risen in almost all countries (Figure 2.2) [2].

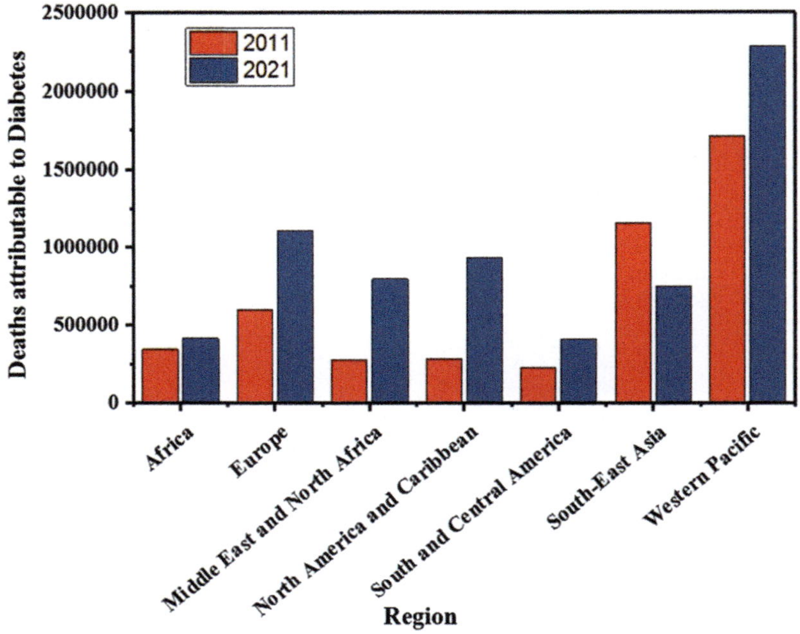

Figure 2.2 Regional projection of the number of deaths attributable to diabetes in the last 10 years. Adapted from [2] (International Diabetes Federation).

The global health care expenditure on diabetes is more than double in 2021 as compared to 2011 and it is increasing continuously. According to the IDF project in 2021, the total health care expenses for diabetes for 2030 and 2045 will be triple in Africa, the Middle East and North Africa, South and Central America, South East Asia, and Western Pacific regional countries (Figure 2.3).

2.3 Diabetes Management

Numerous synthetic antidiabetic drugs are available in the market for the management of diabetes and related complications. However, these antidiabetic drugs are expensive and show adverse effects in patients. For example, the use of acarbose, metformin, glimepiride, and sodium-glucose co-transporters (SGLT2) inhibitors can induce direct excretion of glucose from the body. Whereas, in the case of prediabetes, overusage of these drugs can cause hypoglycemia. Moreover, the concomitant use of amlodipine and telmisartan along with metformin and glimepiride causes hypoglycemia in diabetic patients [12]. Metformin is the first prescribed drug for T2DM; however, its medication has a serious warning for

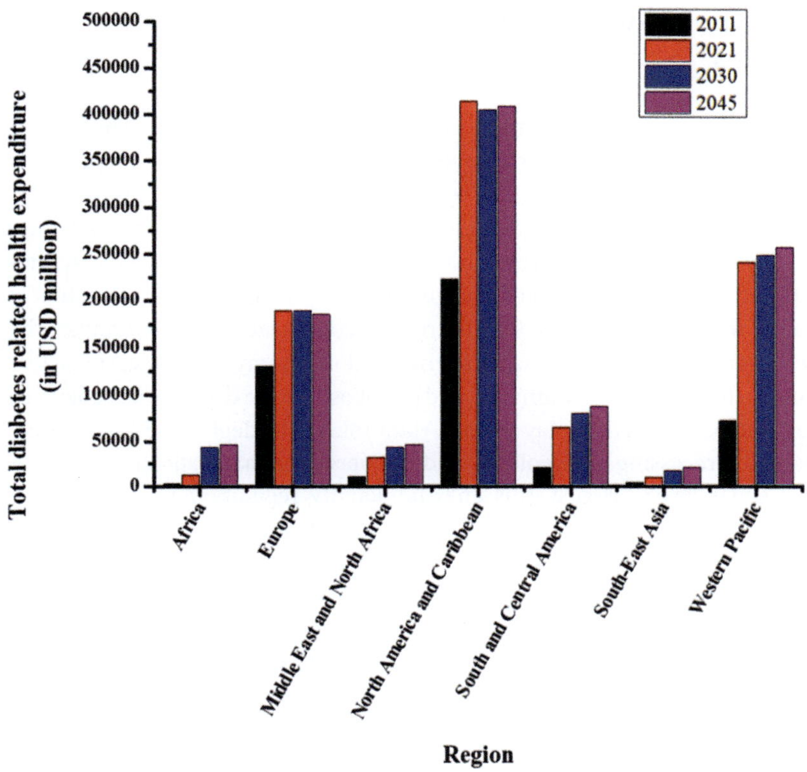

Figure 2.3 Regional-wise projection of the total expenditure for diabetes related health management. Adapted from [2] (International Diabetes Federation).

lactic acidosis. Long-term medication of metformin is also associated with the deficiency of vitamin B12.

Novel antidiabetic drugs like dipeptidyl peptidase 4 (DPP-4) inhibitors are also known as gliptins. Gliptins are a new class of oral antidiabetic agents including sitagliptin, saxagliptin, alogliptin, and vildagliptin, but their consumption also showed some adverse effects such as upper respiratory tract infection, nasopharyngitis, headache, and urinary tract infection in T2DM patients [13]. In addition, gliptins are associated with a risk of hypoglycemia when used in the conjugation of sulfonylureas [14]. In addition, these synthetic antidiabetic drugs are expensive. Due to these reasons, traditional herbal medications [15] and marine based bioactive compounds came into the picture for the management of diabetes related complications [16].

2.4 Marine Natural Bioproducts for the Management of Diabetes

Marine ecosystems are major natural treasures for various products. They play a crucial role in biomedical research and drug development. Marine bioproducts have several inhibitory effects on aldose reductase, α-amylase, α-glucosidase, glycogen synthase kinase 3β (GSK-3β), β-glucosidase, PTP-1B, DPP-4, lipase enzymes, and enhancing the AMPK enzyme phosphorylation. All these enzymes are biocatalysts and are the key components regulating carbohydrate metabolism that is involved in the pathogenesis of diabetes related complications. In particular, the marine organisms cyanobacteria, bacteria, and actinomycetes have bioactive compounds that can inhibit α-amylase and α-glucosidase, and are the key enzymes in sugar metabolism as they play an important role in the degradation of polysaccharides and processing of glycolipids and glycoproteins that leads to an increase of blood sugar levels. Hence, this enzyme activity regulation is good for the treatment of diabetes and obesity [17]. Pyrostatin A and B compounds derived from *Streptomyces* sp. show the specific inhibitory activity of *N*-Acetyl-β-d-glucosaminidase (GlcNAc-ase) and β-d-Galactosidase enzymes [18]. Aquastatin A is a fungal derivative extracted from the marine fungus *Cosmospora* Sp. SF-5060 isolated from inter-tidal sediments. Aquastatin-A has the potential to control sugar levels in T2DM patients by the inhibition of protein tyrosine phosphatase-1B (PTP-1B). PTP-1B is a negative regulator of the insulin signaling pathway and it is considered a primary therapeutic target for the management of T2DM [19].

In addition to cyanobacteria, bacteria, and fungi, marine algae including macro and microalgae have a great potential contribution to the production of various antioxidants, anti-inflammatory, anticancer, antidiabetic, and other bioactive compounds. Microalgae (phytoplankton) are unicellular photosynthetic microorganisms that can thrive in almost all water bodies. They are a rich source of primary and secondary metabolites that have potential bioactivity. Advanced work in the field of aquatic biotechnology has isolated a series of microalgae with promising antidiabetic properties [17]. Recently, marine microalgae cultivation has gained attention and is attracting many natural green chemists and pharmacists due to this being a potential source of new antidiabetic compounds [20]. The diatom *Thalassiosira* species are a very large complex genus and distributed in almost all parts of temperate, tropical, and subtropical regions. They have high productivity of fucoxanthin [21] and are a primary source of omega-3 PUFAs [22]. The microalgae *Isochrysis galbana* and *Nannochloropsis oculata* have a high percentage of water soluble and insoluble polysaccharides and also contain polyunsaturated fatty acids (PUFAs) including docosahexaenoic acid (DHA) and eicosapentaenoic acid (EPA) [23]. Diabetic rats treated with microalgae, either *I.*

galbana or *N. oculata* for 8 weeks, showed significantly declined glucose levels from 600 mg/dl to 100 mg/dl as compared to untreated diabetic rats [24]. Microalgae derived compounds or crude extracts show strong inhibitory effects of α-amylase, α-glucosidase, β-glucosidase, PTP-1B, DPP-4, and lipase enzymes [25]. For example, the ethyl acetate fraction from green algae *Chlorella* and diatom *Nitzscia laevis* showed a negative effect on the formation of total advanced glycation end products (AGEs) [26]. Similarly, the methanolic extracts of *Porphyridium* sp., *Chlorella* sp., *Nannochloropsis* sp., and *Skeletonema* sp. showed the following order of α-glucosidase inhibition: 12.63%, 12.53%, 11.26%, 9.38%, and 7.29%, respectively, at *in vivo* conditions [20].

Fucoxanthin is a xanthophyll carotenoid abundant in marine brown algae. Several reports found that the fucoxanthin is a potential antidiabetic compound that inhibits glucose transport like proteins (GULPs), prevents chronic disease [27], and acts as an anti-obesity agent [28] (Table 2.2). The high fat (HF) diet induced obese mice were fed fucoxanthin-rich wakame lipid (*Undaria pinnatifida*) (HFWLs) for 5 weeks. Thereafter, diabetes related parameters such as blood

Table 2.2 Marine natural antidiabetic bioactive compounds isolated from marine organisms.

Compound	Marine organisms	Biological functions	References
Marine Bacteria/fungus/Sponge			
Aquastastin A	*Cosmospora sp SF-5060*	Inhibition of protein tyrosine phosphatase (PTP1B) enzyme activity	[36]
Callyspongynic acid	*Callyspongia truncata*	Inhibits α-glucosidase	[37]
α-galactosylceramide (α-GalCer),	*Agelas mauritianus*	Protects the pancreatic β-cells	[38]
Dysidine (Terpene)	*Dysidea villosa*	Inhibition of protein tyrosine phosphatase (PTP1B) enzyme activity	[39]
Pyrostatins A and B	*Streptomyces* sp.	*N*-acetyl-glucosaminidase inhibition	[40]
Organic compounds crude (methanolic) extracts	*Microbacterium oleivorans, Planococcus refitoensis, Planomicrobium okeanokoites*	Inhibitory effect of protein tyrosine phosphatase (PTP1B) and β-glucosidase enzymes	[41]

(Continued)

Table 2.2 (Continued)

Compound	Marine organisms	Biological functions	References
Culture supernatants of marine *Actinomycetes*	*Actinomycetes*	Inhibitory effect of β-glucosidase *N*-acetyl-β-d-glucosaminidase Polyglutamyl peptidase α-amylase enzymes	[40]
Marine algal products from micro and macroalgae			
Bromophenols	*Symphyocladia lactiuscula*	Aldose reductase and PTP1B inhibitory activity	[42, 43]
Phenolic extract	*Palmaria* sp., *Ascophyllum* sp. and *Alaria* sp.	Inhibition of α-amylase and α-glucosidase enzyme activity	[44]
Eicosapentaenoic acid (EPA)	*Nitzschia laevis*	Prevent albuminuria progression in T2DM subjects Inhibitory effect on the formation of endogenous *N*-Carboxymethyllysine (CML) formation	[34, 32]
Astaxanthin	*Chlorella zofingiensis*	Antiglycoxidative agent Inhibits AGE formation	[45]
Docosahexaenoic acid (DHA)	*Isochrysis galbana*	Prevents albuminuria progression in T2DM subjects Regulator of glucose and lipid metabolism	[34, 24]
Aqueous/methanolic extract	*Chaetocerose socialis, C. furcellatus, Skeletonema marinoi*	PTP1B inhibitory activity	[46]
Eicosapentaenoic acid (EPA)	*Nannochloropsis oculata*	Regulator of glucose and lipid metabolism	[24]
Carotenoids	*Chlorella pyrenoidosa*	Inhibition of AGE formation	[26]
Lutein	*Chlorella protothecoides*	Inhibition of AGE formation	[47]

Table 2.2 (Continued)

Compound	Marine organisms	Biological functions	References
Fucoxanthin	*Phaeodactylum tricornutum*	Acts as an anti-obesogenic effect by reducing plasma TGA and cholesterol levels	[27]
		Downregulates fatty acid synthase (FAS) proteins by the activation of AMPK signaling pathway in the liver	
	Sargassum angustifolium	Regeneration of pancreatic β-cells	[28–31, 48]
	Undaria pinnatifida (Wakame)	Anti-obesity effect	
		Reduction of MCP-1, TNF-α, IL-6	
		Inhibits the α-amylase and α-glucosidase, sucrose and maltase enzyme activity, as well as enhances the section of insulin	
		Increasing glucose transporter 4 (GLUT4) mRNA in skeletal muscle and anti-obesity effect through UCP1 in white fat and weight loss	
	Undaria pinnatifida	Aldolase inhibition	[48]
3,4-dibromo-5-(2-Bromo-3,4-dihydroxy-6-(ethoxymethyl) benzyl)benzene-1-2-diol	*Rhodomela confervoides*	PTP1B inhibition	[43, 49]
2,4,6-tribromophenol and 2,-4-dibromophenol	*Grateloupia elliptica*	α-glucosidase, sucrose and maltase inhibition	[50]
Phenolic extracts	*Palmaria* sp.	α-amylase inhibition	[44]
Eckol, Dieckol and 7-phoroeckol	*Eisenia bicyclis*	α-amylase inhibition	[51]

(Continued)

Table 2.2 (Continued)

Compound	Marine organisms	Biological functions	References
Raw extracts	*Ulva rigida*	Reduced glucose level in rats	[52]
Phloroglucinol, Diphorethohydroxycarmalol, 6–6-bieckol, Octaphlorethol A, Ishophoroglucin	*Ishige okamurae*	α-amylase and α-glucosidase inhibition	[53]
Aqueous extract	*Padina boergesenii*	Increased activity of glycolytic enzymes and decreased activity of gluconeogenic enzymes in diabetic rats	[54]
Stigmasterol	*Gelidium spinosum*	α-amylase inhibition	[31]
Fucoidan	*S. fusiforme*	Inhibition of oxidative stress in STZ induced diabetic rats	[55]
Phenolic extract	*Alaria and ascophyllum*	α-amylase inhibition	[25]
Butyl isotopephthalate	*Laminaria japonica*	α-glucosidase Inhibition	[56]
Eckol,dioxinodehydroeckol, 7-phloroeckol	*Ecklonia stolonifera*	Inhibition of aldose reductase	[36]
Crude polysaccharides	*Himanthalia elongata*	Blood glucose controller	[57]
Fucoidan	*Himanthalia elongata*	Blood glucose controller	[58]
Natural sodium alginate	*Laminaria angustata*	Blood glucose controller	[59]
DDBT	*Sargassum patens*	Inhibition of α-amylase and α-glucosidase enzyme activity	[59]
Sargahydroquinoic acid, sargaquinoic acid	*Sargassum yezoense*	Decreases insulin resistance by adjusting adubosis	[59]
Thunberol	*Sargassum thunbergii*	PTP1B inhibition	[60]

Table 2.2 (Continued)

Compound	Marine organisms	Biological functions	References
Octaphlorethol A	*Ishige foliacea*	Prevents damage to pancreatic β-cells, inhibition of PEPCK and G6Pase activity, increasing glucose uptake by GLUT4 in skeletal muscle by activating AMPK	[61]
Diphlorethohydroxycarmalol (DPHC)	*Ishige okamurae*	Reduces blood glucose level and absorbs rapidly, reducing glucose and lipid	[62]
		Inhibition of α-amylase and α-glucosidase enzyme activity	

glucose, insulin, leptin concentrations, and obesity were evaluated. The body weights of HFWL treated mice significantly declined, more than the control HF-fed mice (HFc) [29]. From this study, diet administration with a high percentage of fucoxanthin potentially suppressed the obesity problems involved in T2DM. The ethanolic and acetonic extracts of *Sargassum hemiphylum* contain a major amount of polyphenols (36.66 mg/g) and a minor amount of fucoxanthin (15.12 mg/g). These extracts inhibited the α-amylase and α-glucosidase, sucrose and maltase enzyme activity, as well as enhanced the section of insulin [30]. In an *in vivo* experimental study by Oliyaei et al., streptozotocin and nicotinamide induced T2DM mice were gavaged daily with encapsulated fucoxanthin for 3 weeks. Thereafter, the fucoxanthin treated (encapsulated fucoxanthin or free fucoxanthin) diabetic groups showed decreased weight gain and significantly lower fasting blood glucose levels that are similar to metformin (50 mg/kg) treated groups [31].

Microalgae bioproducts, namely astaxanthin, lutine, and eicosapentaenoic acid (EPA), from microalgae *Chlorella zofingiensis* and *Nitzschia laevis*, exhibit a significant inhibitory effect on the formation of endogenous *N*-Carboxymethylene (CML). CML is an advanced glycation end product (AGE). The AGE is a key maker of high blood glucose induced pathophysiological phenomena involved in diabetes related retinopathy in diabetic patients [32]. Omega-3 PUFAs (*n*-3) are the dietary components primarily derived from microalgae (diatoms and other green algae) and deposited at a high percentage in marine fish. Omega-3 PUFAs are good anti-inflammatory bioactive lipids. Inflammation in adipose tissue is a major factor in the development of insulin resistance (IR) in adipose tissues. Dietary supplementation of DHA and EPA enriched oils suppress IR [33].

Supplementation of DHA and EPA acts as angiotensin-converting enzyme inhibitors (ACE-I) or angiotensin receptor blockers (ARBs). These activities reduce albuminuria and coronary artery disease (CAD) in subjects of T2DM [34]. DHA and EPA are effective neurotherapeutic agents, and the dietary supplementation of 1000 mg DHA and 200 mg EPA reduced painful neuropathy in T2DM patients and decreased pro-inflammations and oxidative stress [35].

2.5 Conclusion

Diabetes is a chronic metabolic disorder that imposes a global burden on the public health and socioeconomic development of a country. A person with T1DM needs continuous insulin therapy to live and is not cured with currently available medications, while T2DM can be cured at an early stage diagnosis with proper medications. The progress in advanced gene therapy is still at the laboratory stage only and needs further routine trials on human cells before going through clinical treatments. Every day, the number of diabetic patients increases, which raises the demand and scarcity of insulin and other antidiabetic drugs. To overcome these situations the implementation of medications with natural bioproducts is an effective solution for the management of T2 diabetes related complications. Marine ecosystems are a major source of different antidiabetic bioproducts compared to terrestrial ecosystems. In particular, marine algae contribute more than 90% of global photosynthesis and are leading primary producers. The cultivation of algae (i.e. micro or macroalgae) may improve the source of antidiabetic compounds that can be used for the management of diabetes related complications by the pharmaceutical industries.

Acknowledgments

This is a CSIR-CSMCRI contribution 155/2022. PMK acknowledges CSIR for financial support as SRF (131170/2K19/1). The authors acknowledge CSIR-CSMCRI and AcSIR for providing research facilities.

References

1 Cho, N.H., Shaw, J.E., Karuranga, S. et al. (2018). IDF Diabetes Atlas: Global estimates of diabetes prevalence for 2017 and projections for 2045. *Diabetes Res. Clin. Pract.* 138: 271–281.

2 IDF. (2021). *IDF Diabetes Atlas*. 10th edn.

3 Lin, X., Xu, Y., Pan, X. et al. (2020). Global, regional, and national burden and trend of diabetes in 195 countries and territories: An analysis from 1990 to 2025. *Sci Rep.* 10: 1–11.

4 Bourne, R.R.A., Steinmetz, J.D., Saylan, M. et al. (2021). Causes of blindness and vision impairment in 2020 and trends over 30 years, and prevalence of avoidable blindness in relation to VISION 2020: The right to sight: An analysis for the global burden of disease study. *Lancet Glob. Heal.* 9: e144–160.

5 Nderitu, P., Nunez Do Rio, J.M., Webster, M.L. et al. (2022). Automated image curation in diabetic retinopathy screening using deep learning. *Sci. Rep.* 12: 11196.

6 Lawlor, D.A., Fraser, A., Ebrahim, S. et al. (2007). Independent associations of fasting insulin, glucose, and glycated haemoglobin with stroke and coronary heart disease in older women. *PLoS Med.* 4: 1396–1404.

7 Kaptoge, S., Di Angelantonio, E., Lowe, G. et al. (2010). C-reactive protein concentration and risk of coronary heart disease, stroke, and mortality: An individual participant meta-analysis. *Lancet.* 375: 132–140.

8 Rewers, A. (2018). Acute metabolic complications in diabetes. In: *Diabetes in Americaerica* (ed. C.C. Cowie, S.S. Casagrande, A. Menke, et al,). 3rd edn. Chapter 17. Bethesda (MD): National Institute of Diabetes and Digestive and Kidney Diseases (US). PMID: 33651551

9 Yasuda, T. (2016). Hyperglycemic hyperosmolar state. *Nihon Rinsho.* 74: 92–96.

10 Khan, M.A.B., Hashim, M.J., King, J.K. et al. (2020). Epidemiology of type 2 diabetes - global burden of disease and forecasted trends. *J. Epidemiol. Glob. Health.* 10: 107–111.

11 Gregg, E., Buckley, J., Ali, M. et al. (2021). Improving Health Outcomes of People with Diabetes Mellitus: Target Setting to Reduce the Global Burden of Diabetes Mellitus by 2030. *World Heal. Organ.* 88.

12 Manandhar Shrestha, J.T., Shrestha, H., Prajapati, M. et al.et (2017). Adverse effects of oral hypoglycemic agents and adherence to them among patients with type 2 diabetes mellitus in Nepal. *J. Lumbini Med. Coll.* 5: 34.

13 Pathak, R. and Bridgeman, M.B. (2010). Dipeptidyl peptidase-4 (DPP-4) inhibitors in the management of diabetes. *Pharmacy Ther.* 35: 509–513.

14 Salvo, F., Moore, N., Arnaudl, M. et al. (2016). Addition of dipeptidyl peptidase-4 inhibitors to sulphonylureas and risk of hypoglycaemia: Systematic review and meta-analysis. *BMJ.* 353: i2231.

15 Sudha, P., Zinjarde, S.S., Bhargava, S.Y. and a et al. (2011). Potent α-amylase inhibitory activity of Indian Ayurvedic medicinal plants. *BMC Complement. Altern. Med.* 11: 5.

16 Pathak, K., Gogoi, U., Saikia, R. et al. et al (2022). Marine-derived antidiabetic compounds: An insight into their sources, chemistry, SAR, and molecular mechanisms. *Stud. Nat. Prod. Chem.* 73: 467–504.

17 Lauritano, C. and Ianora, A. (2016). Marine organisms with anti-diabetes properties. *Mar. Drugs*. 14: 1–14.

18 Aoyama, T., Kojima, F., Imada, C. et al. (1995). Pyrostatins A and B, new inhibitors of *N*-Acetyl-β-d-Glucosaminidase, produced by *Streptomyces* sp. SA-3501. *J. Enzyme Inhib*. 8: 223–232.

19 Patel, A.M., Anand, I.S. and Suva, M.A. (2014). Role of protein tyrosine phosphatase-1B inhibitors in type 2 diabetes mellitus. *J. Pharm. Sci. Tech.* 4: 2–6.

20 Priatni, S., Ratnaningrum, D., and and Kosasih, W. (2021). The screening of antidiabetic activity and the cultivation study of local marine microalgae. *IOP Conf. Ser. Mater. Sci. Eng.* 1011: 1–8.

21 Singh, K., Krishna Paidi, M., Kulshrestha, A. et al. et (2022). Deep eutectic solvents based biorefining of value-added chemicals from the diatom Thalassiosira andamanica at room temperature. *Sep. Purif. Technol.* 298: 121636.

22 Krishna, P.M., Polisetti, V., Damarla, K. et al. (2021). Improved biorefinery pathways of marine diatoms using a water miscible ionic liquid and its colloidal solution: Efficient lipid extraction and in situ synthesis of fluorescent carbon dots for bio-imaging applications. *RSC Adv*. 11: 21207–21215.

23 Gilbert-López, B., Mendiola, J.A., Fontecha, J. et al. (2015). Downstream processing of *Isochrysis galbana*: A step towards microalgal biorefinery. *Green Chem*. 17: 4599–4609.

24 Nuño, K., Villarruel-López, A., Puebla-Pérez, A.M. et al. (2013). Effects of the marine microalgae Isochrysis galbana and Nannochloropsis oculata in diabetic rats. *J. Funct. Foods*. 5: 106–115.

25 Nasab, S.B., Homaei, A., Pletschke, B.I. et al. (2020). Marine resources effective in controlling and treating diabetes and its associated complications. *Process. Biochem*. 92: 313–342.

26 Sun, Z., Peng, X., Liu, J. et al.et (2010). Inhibitory effects of microalgal extracts on the formation of advanced glycation endproducts (AGEs). *Food Chem*. 120: 261–267.

27 Bae, M., Kim, M.B., Park, Y.K. et al. (2020). Health benefits of fucoxanthin in the prevention of chronic diseases. *Biochim. Biophys. Acta – Mol. Cell. Biol. Lipids*. 1865: 158618.

28 Maeda, H., Hosokawa, M., Sashima, T. et al. (2005). Fucoxanthin from edible seaweed, Undaria pinnatifida, shows antiobesity effect through UCP1 expression in white adipose tissues. *Biochem. Biophys. Res. Commun*. 332: 392–397.

29 Maeda, H., Hosokawa, M., Sashima, T. et al. (2009). Anti-obesity and anti-diabetic effects of fucoxanthin on diet-induced obesity conditions in a murine model. *Mol. Med. Rep*. 2: 897–902.

30 Hwang, P.A., Hung, Y.L., Tsai, Y.K. et al. (2015). The brown seaweed Sargassum hemiphyllum exhibits α-amylase and α-glucosidase inhibitory activity and enhances insulin release in vitro. *Cytotechnology*. 67: 653–660.

31 Oliyaei, N., Moosavi-Nasab, M., Tamaddon, A.M. et al. (2021). Antidiabetic effect of fucoxanthin extracted from Sargassum angustifolium on streptozotocin-nicotinamide-induced type 2 diabetic mice. *Food Sci. Nutr.* 9: 3521–3529.

32 Sun, Z., Liu, J., Zeng, X. et al. (2011). Protective actions of microalgae against endogenous and exogenous advanced glycation endproducts (AGEs) in human retinal pigment epithelial cells. *Food Funct.* 2: 251–258.

33 Abbott, K.A., Burrows, T.L., Acharya, S. et al. (2020). DHA-enriched fish oil reduces insulin resistance in overweight and obese adults. *Prostaglandins Leukot. Essent. Fat. Acids.* 159: 102154.

34 Elajami, T.K., Alfaddagh, A., Lakshminarayan, D. et al. (2017). Eicosapentaenoic and docosahexaenoic acids attenuate progression of albuminuria in patients with type 2 diabetes mellitus and coronary artery disease. *J. Am. Heart Assoc.* 6: 1–13.

35 Durán, A.M., Beeson, W.L., Firek, A. et al. (2022). Dietary omega-3 polyunsaturated fatty-acid supplementation upregulates protective cellular pathways in patients with type 2 diabetes exhibiting improvement in painful diabetic neuropathy. *Nutrients.* 14: 1–23.

36 Debbab, A., Aly, A.H., Lin, W.H. et al. (2010). Bioactive compounds from marine bacteria and fungi: Minireview. *Microb. Biotechnol.* 3: 544–563.

37 Nakao, Y., Uehara, T., Matunaga, S.et al. (2002). Callyspongynic acid, a polyacetylenic acid which inhibits α-glucosidase, from the marine sponge Callyspongia truncata. *J. Nat. Prod.* 65: 922–924.

38 Van Kaer, L. (2001). Drugs from the sea: A marine sponge-derived compound prevents Type 1 diabetes. *Sci. World J.* 1: 630–632.

39 Zhang, Y., Li, Y., Guo, Y.W. et al. (2009). A sesquiterpene quinone, dysidine, from the sponge Dysidea villosa, activates the insulin pathway through inhibition of PTPases. *Acta Pharmacol. Sin.* 30: 333–345.

40 Imada, C. (2005). Enzyme inhibitors and other bioactive compounds from marine actinomycetes. *Antonie van Leeuwenhoek.* 87: 59–63.

41 Pandey, S., Sree, A., Dash, S.S. et al. (2013). Diversity of marine bacteria producing beta-glucosidase inhibitors. *Microb. Cell Fact.* 12: 1–7.

42 Wang, W., Okada, Y., Shi, H. et al. (2005). Structures and aldose reductase inhibitory effects of bromophenols from the red alga Symphyocladia latiuscula. *J. Nat. Prod.* 68: 620–622.

43 Shi, D., Li, J., Jiang, B., Guo, S. et al. (2012) Bromophenols as inhibitors of protein tyrosine phosphatase 1B with antidiabetic properties. *Bioorganic Med. Chem. Lett.* 22: 2827–2832.

44 Nwosu, F., Morris, J., Lund, V.A. et al. (2011). Anti-proliferative and potential anti-diabetic effects of phenolic-rich extracts from edible marine algae. *Food Chem.* 126: 1006–1012.

45 Sun. Z., Liu, J., Zeng, X. et al. (2011). Astaxanthin is responsible for antiglycoxidative properties of microalga Chlorella zofingiensis. *Food Chem.* 126: 1629–1635.

46 Ingebrigtsen, R.A., Hansen, E., Andersen, J.H. et al. (2016). Light and temperature effects on bioactivity in diatoms. *J. Appl. Phycol.* 28: 939–950.

47 Sun, Z., Liu, J., Zeng, X. et al. (2011). Protective actions of microalgae against endogenous and exogenous advanced glycation endproducts (AGEs) in human retinal pigment epithelial cells. *Food Funct.* 2: 251–258.

48 Peng, J., Yuan, J.P., Wu, C.F. et al. (2011). Fucoxanthin, a marine carotenoid present in brown seaweeds and diatoms: Metabolism and bioactivities relevant to human health. *Mar. Drugs.* 9: 1806–1828.

49 Shi, D., Guo, S., Jiang, B. et al. (2013). HPN, a synthetic analogue of bromophenol from red alga Rhodomela confervoides: Synthesis and anti-diabetic effects in C57BL/KsJ-db/db mice. *Mar. Drugs*. 11: 350–362.

50 Kim, K.Y., Nam, K.A., Kurihara, H. et al. (2008). Potent α-glucosidase inhibitors purified from the red alga Grateloupia elliptica. *Phytochemistry*. 69: 2820–2825.

51 Moon, H.E., Islam, M.N., Ahn, B.R. et al. (2014). Protein tyrosine phosphatase 1B and α-glucosidase inhibitory phlorotannins from edible brown algae, Ecklonia stolonifera and Eisenia bicyclis. *OUP*. 75: 1472–1480.

52 Celikler, S., Tas, S., Vatan, O. et al. (2009). Anti-hyperglycemic and antigenotoxic potential of Ulva rigida ethanolic extract in the experimental diabetes mellitus. *Food Chem. Toxicol*. 47: 1837–1840.

53 Yang, H.W., Fernando, K.H.N., Oh, J.Y. et al. (2019). Anti-obesity and anti-diabetic effects of ishige okamurae. *Mar. Drugs*. 17: 1–11.

54 Senthilkumar, P., Sudha, S. and Prakash, S. (2014). Antidiabetic activity of aqueous extract of Padina boergesenii in streptozotocin-induced diabetic rats. *Int. J. Pharm. Pharm. Sci*. 6: 418–422.

55 Cheng, Y., Sibusiso, L., Hou, L. et al. (2019). Sargassum fusiforme fucoidan modifies the gut microbiota during alleviation of streptozotocin-induced hyperglycemia in mice. *Int. J. Biol. Macromol*. 131: 1162–1170.

56 Bu, T., Liu, M., Zheng, L. et al. (2010). α-glucosidase inhibition and the in vivo hypoglycemic effect of butyl-isobutyl-phthalate derived from the Laminaria japonica rhizoid. *Phyther. Res*. 24: 1588–1591.

57 Lamela, M., Anca, J., Villar, R. et al. (1989). Hypoglycemic activity op several seaweed extracts. *J. Ethnopharmacol*. 27: 35–43.

58 Lamela, A.M., Anca, J.M., Vazquez-Freire, M.J. et al. (1993). Hypoglycemic activity of a fucan from Himanthalia elongata. *Médicaments et Aliments, Approche Ethnopharmacologique: Actes du 2e Colloque Européen D'ethnopharmacologie et de la 11e Conférence Internationale D'ethnomédecine.* Paris, France: Orstom.

59 Kimura, Y., Watanabe, K. and Okuda, H. (1996). Effects of soluble sodium alginate on cholesterol excretion and glucose tolerance in rats. *J. Ethnopharmacol*. 54: 47–54.

60 He, W.F., Yao, L.G., Liu, H.L. et al. (2014). Thunberol, a new sterol from the Chinese brown alga Sargassum thunbergii. *J. Asian Nat. Prod. Res*. 16: 685–689.

61 Lee, S.H., Ko, S.C., Kang, M.C. et al. (2016). Octaphlorethol A, a marine algae product, exhibits antidiabetic effects in type 2 diabetic mice by activating AMP-activated protein kinase and upregulating the expression of glucose transporter 4. *Food Chem. Toxicol.* 91: 58–64.

62 Heo, S.J., Hwang, J.Y., Choi, J.I. et al. (2009). Diphlorethohydroxycarmalol isolated from Ishige okamurae, a brown algae, a potent α-glucosidase and α-amylase inhibitor, alleviates postprandial hyperglycemia in diabetic mice. *Eur. J. Pharmacol.* 615: 252–256.

3

Carbohydrate-based Antidiabetic Agents from Natural Products

Sandeep Singh

Hadassah Biological Psychiatry Laboratory, Hadassah - Hebrew University Medical Center, Jerusalem 91120, Israel

3.1 Introduction

According to the International Diabetes Federation (10th Edition of the IDF Diabetes Atlas 2021), one in ten (10.5%) adults around the world are currently living with diabetes. The total number is predicted to rise to 643 million (11.3%) by 2030 and to 783 million (12.2%) by 2045, and this number is rising rapidly, with many more people at high risk of developing diabetes. Diabetes leads to serious health complications (Figure 3.1) and places enormous demands on health services, as well as the wider economy and social structures.

Diabetes is a chronic, metabolic disease characterized by elevated levels of blood glucose (or blood sugar), which can cause serious damage to the heart, blood vessels, eyes, kidneys, and brain [1, 2]. Since blood glucose is a major driving factor of diabetes, glucose-based molecules have been extensively studied for the treatment of diabetes. Therefore, effective control of blood glucose is the basis of treatment for all patients with diabetes. On account of the seriousness of the diabetes epidemic, pioneering scientists are highly motivated to discover molecules that can target the pathways involved in glucose metabolism. To cure diabetes, we should monitor glucose levels, which are a simple form of polysaccharides, at definite intervals. Polysaccharides are vital to all living beings and this concept immediately attracts carbohydrate-based conjugates that act on particular targets to control glucose levels. The carbohydrate-based antidiabetic natural compounds can be mainly termed antidiabetic glucosidase inhibitors and sodium glucose cotransporter-2 (SGLT2) inhibitors, which depend on the basis of their mode of action. In this chapter, carbohydrate-based antidiabetic agents, especially from natural sources, are explained.

Natural Products and their Bioactives in Antidiabetic Drug Discovery, First Edition.
Edited by Kanti Bhooshan Pandey, Maitree Suttajit and Pinar Atukeren.

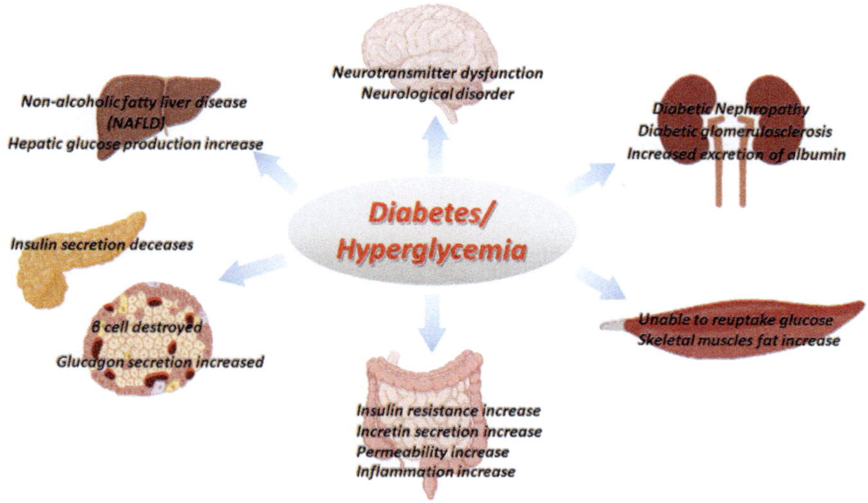

Figure 3.1 Complications associated with diabetes.

3.2 Causes of Diabetes and Types

The major cause of diabetes is either the pancreas not producing enough insulin, or the cells not responding appropriately to insulin [3]. Insulin is a hormone that is responsible for glucose from meals to gain entry to cells, further used for energy. Diabetes is a common metabolic disease that is characterized by an increased level of glucose in the blood, which results in a hyperglycemic condition. This further induces an autoimmune response and destroys the pancreatic β-cells that are solely responsible for insulin secretion. Due to secretion of insulin in deficient amounts, further it affects the glucose metabolism that marks resistance to the action of insulin, known as insulin resistance (IR). Due to this impaired insulin action and/or insulin resistance, organs do not use glucose effectively, which, in turn, triggers a hyperglycemic state.

Diabetes is broadly classified into three types: type 1 diabetes mellitus (T1DM), which results when the pancreas is unable to produce insulin due to a loss of β-cells that are mistakenly destroyed by the body's immune system. Sometimes it is also referred to as juvenile diabetes, which affects children and adults (people can be diagnosed at an early age as well as in adulthood). Type 2 diabetes mellitus (T2DM) is an impairment in a system that regulates and uses sugar (glucose) as a fuel. The long term, i.e. chronic phase, results in too much sugar circulating in the main bloodstream. Ultimately, high blood sugar levels lead to disorders of the circulation, immune systems, and neurodegeneration. Gestational diabetes mellitus (GDM) is a type of diabetes that occurs only during pregnancy. GDM can cause health problems in both mother and baby. It is sometimes related to the hormonal changes of pregnancy that make the body less able to use insulin.

3.3 Importance of Carbohydrate-based Natural Antidiabetic Drugs

As complications start from high blood glucose levels, current antidiabetic therapies have the prime target of reaching normoglycemia. However, most of the oral hypoglycemic drugs [4–6] have several side effects and are inadequate for about 30% of patients [7]. On the other hand, their efficacy decreases over the time, and several concerns exist regarding their safety [8], which limits their use; for example, insulin, sulfonylureas, and glitazones increase IR by facilitating weight gain, and pioglitazone causes edema and cardiovascular problems [9]. So, there is an urgent need for more effective and safer hypoglycemic drugs [10]. Therefore, this has opened a new window for carbohydrate-based drugs or glycoconjugates to combat diabetes.

The importance of carbohydrates to cure diabetes can be understood with recent ongoing research, where natural products have shown huge potential from the beginning. The importance of fructooligosaccharides and isomaltooligosaccharides as dietary components to reduce blood glucose levels has been identified [11]. There is a compound known as type II arabinogalactan found in green tea leaves (*Camellia sinensis*) named 7WA that decreases glucose levels through activation of insulin secretion. Structural analysis showed that it consists of galactopyranosyl moieties with 1,3- and 1,6-linkage branching at the 3rd oxygen of 1,6-linked galactose moiety along with the 4th, 6th oxygen of 1,3-linked galactose moiety [12]. There is another target to reduce secondary complications related to diabetes, i.e. advanced glycation end products (AGEs) inhibition. The AGE inhibitors such as pyridoxamine inhibited retinopathy and neuropathy in diabetic rat models. The astragalosides isolated from the roots of *Astragalus radix* have the potential to inhibit AGEs [13].

3.4 Glycosidase Inhibitors as Antidiabetic Drugs

Glycosidases are ubiquitous intracellular and extracellular enzymes that are responsible for the hydrolysis of glycosidic linkages. For antidiabetic drugs of plant origin, inhibition of α-glucosidase activity has been revealed as a good strategy for regulating glycemia, and glucosidase inhibitors have been used as medicine for diabetes (Figure 3.2). Currently, only three α-glucosidase inhibitors are utilized in clinical practice: acarbose, miglitol, and voglibose, and therefore researchers are searching for novel inhibitors with higher efficacy.

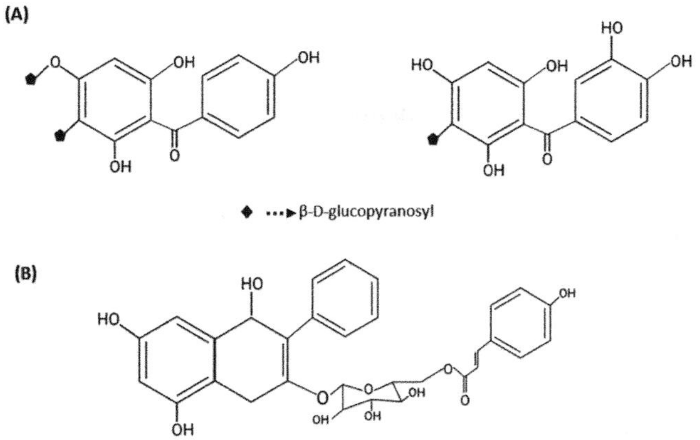

Figure 3.2 Glucosidase inhibitor (A) glycoconjugate (*Cyclopia genistoides*). (B) Glycosylated phenolic (*Edgeworthia gardneri*).

3.4.1 Acarbose

The first α-glucosidase inhibitor approved in Europe and the USA for the treatment of T2DM was acarbose [14]. Acarbose (from *Actinoplanes* sp. found in beetroot) is probably the most widely used digestive enzyme inhibitor for the treatment of T2DM, acting on α-glucosidase, α-amylase, sucrase, and maltase, but without insulinotropic properties [15–17]. Acarbose has been purified from *A. hulinensis* (1094) with a yield of 8.48%, purity of 98%, and which is further identified by LC/MS and NMR ($C_{25}H_{43}NO_{18}$; m/z: 645.6 g/mol). This acarbose has been found to decrease the postprandial blood glucose level in T2DM in a combination therapy with other control drugs [18]. It has the potential to decrease the level of postprandial hyperglycemia and might delay the start of insulin therapy as well as also potentiate the hypoglycemic effects of sulfonylureas or insulin [19].

3.4.2 Miglitol

A reversible α-glucosidase inhibitor, miglitol (derived from 1-deoxynojirimycin, isolated from leaves of the mulberry tree), exhibits glycemic control in the same manner as acarbose but with improved activity. Originally obtained from various *Bacillus* and *Streptomyces* strains, it is now obtained by the oxidation of 1-amino-D-sorbitol to 6-amino-L-sorbose by fermentation using Gluconobacter oxydans. Miglitol is a 1-deoxynojirimycin derivative that reversibly inhibits intestinal α-glucosidase enzymes responsible for the digestion of carbohydrates to absorbable monosaccharides [20, 21]. As miglitol is well absorbed and acarbose is not, it is suggested that miglitol has a systemic hypoglycemic effect, probably related to its close structural similarity to glucose, which warrants further investigation [22].

3.4.3 Voglibose

Voglibose (1, *N*-[2-hydroxy-1-(hydroxymethyl) ethyl] valiolamine, basenR), having more potent disaccharidase inhibitory activity against maltase and sucrase than naturally occurring pseudo-oligosaccharide α-glucosidase inhibitors, has been developed for the treatment of diabetes mellitus as a blood glucose ameliorating agent [23]. Voglibose, one of the most important α-glucosidase inhibitors, delays the digestion and absorption of carbohydrates, thereby inhibiting postprandial hyperglycemia and hyperinsulinemia, and aids in the treatment of diabetes [24].

3.5 Novel Natural Carbohydrate-based α-glucosidase Inhibitors

Extracts of *Cyclopia genistoides* Vent. (Honeybush) are principally renowned for high levels of the tetrahydroxyxanthone C-glucoside mangiferin, a bioactive compound with potent α-glucosidase inhibitory activity [25] (Figure 3.2A). These compounds were tested for α-glucosidase inhibitory activity against α-glucosidase extracted from rat intestinal acetone powder and acarbose was used as control at 65 μM and showed significant effect.

Recently, many phenolics from the flower of *Edgeworthia gardneri* (Tibet Tea), have been reported for α-glucosidase activity to test their antihyperglycemic activity (Figure 3.2B). This, linked with glycosyl, acts a an α-glucosidase inhibitor; it decreases the postprandial glucose level in diabetic-induced and normal mice, as well as in fasting diabetic-induced mice [26]. Flavonoids are abundant polyphenols that are found naturally in daily consumed fruits, vegetables, nuts, etc. and are potential hyperglycemic agents [27, 28]. Flavonoids have been reported as α-glucosidase inhibitors in many reports [29, 30]. There are four structurally-related flavonols (quercetin, kaempferol, quercetagetin, and galangin) that have been evaluated for their ability to inhibit human α-glucosidases (sucrase, maltase, and isomaltase), and compared with the antidiabetic drug acarbose and flavan-3-ol(−)-epigallocatechin-3-gallate (EGCG) [29]. Another flavonoid-based glycoconjugate isolated from the *Spiraea cantoniensis* plant, cultivated in the Japanese temperate region, has potent maltase-inhibiting activity. The enzyme-assay guided fractionation of the extract from the dried flowers of *S. cantoniensis* led to the isolation of three flavonol caffeoyl glycosides (quercetin 3-O-(6-O-caffeoyl)-α-galactoside, kaempferol 3-O-(6-O-caffeoyl)-α-galactoside, and kaempferol 3-O-(6-O-caffeoyl)-α-glucoside), as rat intestinal maltase inhibitors (Figure 3.3A–C). The relatively high content (0.03% each) in this plant compared to that of other plants suggests that it could be physiologically useful for the treatment of diabetes, although *in vivo* experiments are needed [31–33].

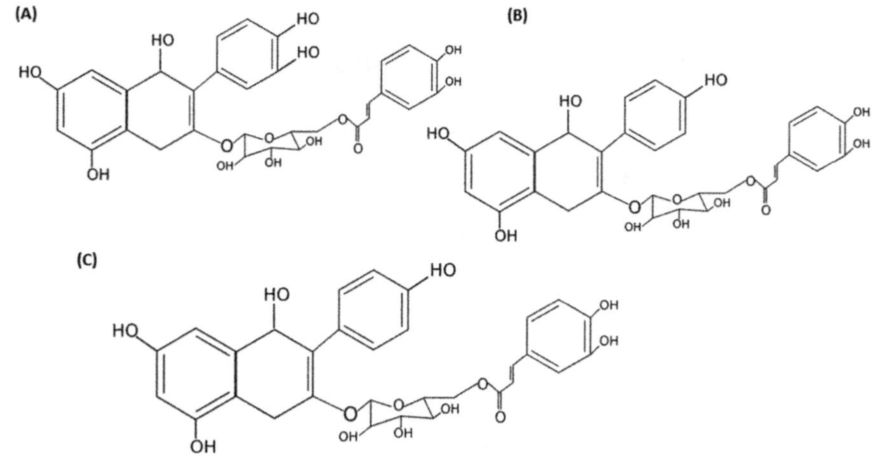

Figure 3.3 Flavanol glycoconjugates (glucosidase inhibitor from *Spiraea cantoniensis*).

3.6 Other Important Antidiabetic Drugs Derived from Plants

The scientific community has done a lot of research from natural sources that are based on carbohydrates, which has opened a new window to antidiabetic therapy. Plants have always been a very rich source of drugs and many of the currently available drugs are derived directly or indirectly from plants. More than 1000 plants have been reported to have antidiabetic activity. A wide range of natural plant-derived chemical compounds have confirmed potential roles [34, 35].

Baicalin is a glycosyloxyflavone, which is the 7-O-glucuronide of baicalein. It is an active ingredient of Chinese herbal medicine *Scutellaria baicalensis* and *Thalictrum baicalense*. Baicalein treatment significantly improved hyperglycemia, glucose tolerance, and blood insulin levels in middle-aged obese diabetic mice [36]. Two steroidal glycoconjugates found in rhizomes of *Anemarrhena asphodeloides*, namely, pseudoprotoimosaponin AIII, and prototimosaponin AIII, showed hypoglycemic effects in a dose-dependent manner in streptozotocin-diabetic mice but no effects on glucose uptake and insulin release, suggesting that the hypoglycemic mechanism may be due to the inhibition of hepatic gluconeogenesis and/or glycogenolysis [37, 38].

A steroidal saponin charantin isolated from *Momordica charantia* mimicked insulin activity by checking the insulin secretion and therefore lowering the glucose levels in blood [39]. Another saponin, isolated from *G. lucidum*, contains various bioactive molecules, such as triterpenoids, polysaccharides, nucleotides, fatty acids, glycoproteins, sterols, steroids, proteins, and peptides. Further, another saponin known as chloragin, isolated from *Chlorophytum nimonii*, showed significant antihyperglycemic potential in streptozotocin-induced diabetic rats [40]. Triterpenoids, polysaccharides, glycoproteins, and proteins decrease glucose levels [41, 42]. Polysaccharides also decrease glucose levels by controlling several enzymes in the glucose metabolism pathway, such as glucose-6-phosphate dehydrogenase, fructose-1,6-bisphosphatase (FBPase), etc. [43, 44].

Yoshikawa and co-worker have isolated three types of elastosides (G, H, and I), which are known as Taranome from the immature shoots of *Aralia elata*, which effectively act as an antihyperglycemic in rat models [45, 46]. A gypenoside saponin isolated from *Gynostemma pentaphyllum* activates the secretion of insulin in rats' islets of Langerhans, resulting in enhanced glucose tolerance [47]. The coagulin L, which has been isolated from *Withania somnifera*, has the potential to decrease the level of postprandial increase in glucose in hyperglycemic streptozotocin diabetic rats [48].

There are so many other compounds that are carbohydrate-based antidiabetic drugs such as icariin, isolated from *Epimedium* (Berberidaceae), and flavonoid glycoside kaempferitrin (*Bauhinia forficata*), found to aid weight loss and myocardial collagen, which are related to diabetes complications and significantly lowered blood glucose, respectively. Hesperidin (flavonoid glycosides) isolated from citrus fruit, is protected against a high level of glucose-induced cell apoptosis by down-regulating caspase-9, caspase-3, and Bax/Bcl-2. Furthermore, it significantly inhibited the phosphorylation of c-Jun N-terminal kinases (JNK) and activated p38 MAPK in high glucose-fed RGC-5 cells [49]. Another glycoside, naringin, also isolated from citrus fruits, has been reported to reduce blood glucose, total cholesterol, triglycerides, and low-density lipoproteins in fructose-fed rats. Naringin restored acetylcholine-mediated vasorelaxation, suggesting its potential influence on fructose-induced metabolic alterations and endothelial dysfunction. Naringin improved serum nitrate/nitrite (NOx), endothelial nitric oxide synthase (eNOS), and phosphorylated eNOS (p-eNOS) protein expression, and preserved endothelium-dependent relaxation in the aortae of fructose-fed rats [50, 51].

3.7 Conclusion

In conclusion, using bioactive molecules from natural sources is an attractive therapeutic process but still requires more understanding by using them in different experimental models. Nowadays, carbohydrate-based compounds have emerged as very promising for the management of diabetes, as biomolecules contain better solubility and pharmacokinetics have fewer side effects. By using a natural and synthetic strategy, glycoconjugates can be a better option for diabetes and its complications. In future therapies for the treatment of diabetes, more thorough research is still needed.

References

1 Tao, Z., Shi, A. and Zhao, J. (2015). Epidemiological perspectives of diabetes. *Cell Biochem. Biophys.* 73(1): 181–185.

2 Boles, A., Kandimalla, R. and Reddy, P.H. (2017). Dynamics of diabetes and obesity: Epidemiological perspective. *Biochim. Biophys. Acta Mol. Basis Dis.* 1863(5): 1026–1036.

3 Masharani, U. and German, M.S. (2011). Pancreatic hormones and diabetes mellitus. In: *Greenspan's Basic & Clinical Endocrinology [Internet].* (ed. D.G. Gardner and D. Shoback), 9th edn., New York: McGraw-Hill. [cited 2022 Aug 11]. Available from: https://accessmedicine.mhmedical.com/content.aspx?aid=8407307.

4 Cheng, A.Y.Y. and Fantus, I.G. (2005). Oral antihyperglycemic therapy for type 2 diabetes mellitus. *CMAJ.* 172(2): 213–226.

5 Mizuno, C.S., Chittiboyina, A.G., Kurtz, T.W. et al. (2008). Type 2 diabetes and oral antihyperglycemic drugs. *Curr. Med. Chem.* 15(1): 61–74.

6 Uwaifo, G.I. and Ratner, R.E. (2005). Novel pharmacologic agents for type 2 diabetes. *Endocrinol. Metab. Clin. North Am.* 34(1): 155–197.

7 Wagman, A.S. and Nuss, J.M. (2001). Current therapies and emerging targets for the treatment of diabetes. *Curr. Pharm. Des.* 7(6): 417–450.

8 Israili, Z.H. (2011). Advances in the treatment of type 2 diabetes mellitus. *Am. J. Ther.* 18(2): 117–152.

9 Clar, C., Gill, J.A., Court, R. et al. (2012). Systematic review of SGLT2 receptor inhibitors in dual or triple therapy in type 2 diabetes. *BMJ Open.* 2(5): e001007.

10 Aguilar, R.B. (2011). Evaluating treatment algorithms for the management of patients with type 2 diabetes mellitus: A perspective on the definition of treatment success. *Clin. Ther.* 33(4): 408–424.

11 Bharti, S.K., Krishnan, S., Kumar, A. et al. (2015). Mechanism-based antidiabetic activity of fructo- and isomalto-oligosaccharides: Validation by in vivo, in silico and in vitro interaction potential. *Process Biochem.* 2(50): 317–327.

12 Wang, H., Shi, S., Bao, B. et al. (2015). Structure characterization of an arabinogalactan from green tea and its anti-diabetic effect. *Carbohydr. Polym.* 124: 98–108.

13 Motomura, K., Fujiwara, Y., Kiyota, N. et al. (2009). Astragalosides isolated from the root of astragalus radix inhibit the formation of advanced glycation end products. *J. Agric. Food Chem.* 57(17): 7666–7672.

14 Saeedi, P., Petersohn, I., Salpea, P. et al. (2019). Global and regional diabetes prevalence estimates for 2019 and projections for 2030 and 2045: Results from the International Diabetes Federation Diabetes Atlas, 9th edition. *Diabetes Res. Clin. Pract.* 157: 107843.

15 Bedekar, A., Shah, K. and Koffas, M. (2010). Natural products for type II diabetes treatment. *Adv. Appl. Microbiol.* 71: 21–73.

16 Wehmeier, U.F. (2003). The biosynthesis and metabolism of acarbose in Actinoplanes sp. SE 50/110: A progress report. *Biocatal. Biotransformation.* 21(4–5): 279–284.

17 Wehmeier, U.F. and Piepersberg, W. (2004). Biotechnology and molecular biology of the alpha-glucosidase inhibitor acarbose. *Appl. Microbiol. Biotechnol.* 63(6): 613–625.

18 Tuyen, D.T., Yew, G.Y., Cuong, N.T. et al. (2021). Selection, purification, and evaluation of acarbose—an α-glucosidase inhibitor from *Actinoplanes sp. Chemosphere.* 265: 129167.

19 Yee, H.S. and Fong, N.T. (1996). A review of the safety and efficacy of acarbose in diabetes mellitus. *Pharmacotherapy.* 16(5): 792–805.

20 Scott, L.J. and Spencer, C.M. (2000). Miglitol: A review of its therapeutic potential in type 2 diabetes mellitus. *Drugs.* 59(3): 521–549.

21 Sels, J.P., Huijberts, M.S. and Wolffenbuttel, B.H. (1999). Miglitol, a new alpha-glucosidase inhibitor. *Expert Opin. Pharmacother.* 1(1): 149–156.

22 Joubert, P., Venter, H. and Foukaridis, G. (1990). The effect of miglitol and acarbose after an oral glucose load: A novel hypoglycaemic mechanism? *Br. J. Clin. Pharmacol.* 30(3): 391–396.

23 Chen, X., Zheng, Y. and Shen, Y. (2006). Voglibose (Basen®, AO-128), one of the most important α-Glucosidase inhibitors. *CMC.* 13(1): 109–116.

24 Kaku, K. (2014). Efficacy of voglibose in type 2 diabetes. *Expert Opin. Pharmacother.* 15(8): 1181–1190.

25 Beelders, T., Brand, D.J., de Beer, D. et al. (2014). Benzophenone *C* - and *O* -glucosides from *Cyclopia genistoides* (Honeybush) inhibit mammalian α-Glucosidase. *J. Nat. Prod.* 77(12): 2694–2699.

26 Ma, Y.Y., Zhao, D.G., Zhou, A.Y. et al. (2015). α-Glucosidase inhibition and antihyperglycemic activity of phenolics from the flowers of Edgeworthia gardneri. *J. Agric. Food Chem.* 63(37): 8162–8169.

27 Xiao, J. (2017). Dietary flavonoid aglycones and their glycosides: Which show better biological significance? *Crit. Rev. Food Sci. Nutr.* 57(9): 1874–1905.

28 Hussain, T., Tan, B., Murtaza, G. et al. (2020). Flavonoids and type 2 diabetes: Evidence of efficacy in clinical and animal studies and delivery strategies to enhance their therapeutic efficacy. *Pharmacological Res.* 152: 104629.

29 Barber, E., Houghton, M.J. and Williamson, G. (2021). Flavonoids as human intestinal α-Glucosidase inhibitors. *Foods* 10(8): 1939.

30 Şöhretoğlu, D. and Sari, S. (2020). Flavonoids as alpha-glucosidase inhibitors: Mechanistic approaches merged with enzyme kinetics and molecular modelling. *Phytochem Rev.* 19(5): 1081–1092.

31 Yoshida, K., Hishida, A., Iida, O. et al. (2008). Flavonol caffeoylglycosides as α-Glucosidase inhibitors from Spiraea cantoniensis flower [Internet]. *ACS Pub.* American Chemical Society; [cited 2022 Sep 16]. Available from: https://pubs.acs.org/doi/pdf/10.1021/jf8007579.

32 Isaza, J.H., Ito, H. and Yoshida, T. (2001). A flavonol glycoside-lignan ester and accompanying acylated glucosides from Monochaetum multiflorum. *Phytochemistry*. 58(2): 321–327.

33 Cho, M.J., Howard, L.R., Prior, R.L. et al. (2005). Flavonol glycosides and antioxidant capacity of various blackberry and blueberry genotypes determined by high-performance liquid chromatography/mass spectrometry. *J. Sci. Food Agric.* 85(13): 2149–2158.

34 Rizvi, S.I. and Mishra, N. (2013). Traditional Indian medicines used for the management of diabetes mellitus. *J. Diabetes Res.* 2013: 1–11.

35 Salehi, B., Ata, A., Anil Kumar, V. et al. (2019). Antidiabetic potential of medicinal plants and their active components. *Biomolecules*. 9(10): 551.

36 Fu, Y., Luo, J., Jia, Z. et al. (2014). Baicalein protects against type 2 diabetes via promoting islet β-cell function in obese diabetic mice. *Int. J. Endocrinol.* 2014: 1–13.

37 Nakashima, N., Kimura, I., Kimura, M. et al. (1993). Isolation of pseudoprototimosaponin AIII from rhizomes of Anemarrhena asphodeloides and its hypoglycemic activity in streptozotocin-induced diabetic mice. *J. Nat. Prod.* 56(3): 345–350.

38 Miura, T., Ichiki, H., Iwamoto, N. et al. (2001). Antidiabetic activity of the rhizoma of Anemarrhena asphodeloides and active components, mangiferin and its glucoside. *Biol. Pharm. Bull.* 24(9): 1009–1011.

39 Ng, T.B., Wong, C.M., Li, W.W. et al. (1986). Insulin-like molecules in Momordica charantia seeds. *J. Ethnopharmacol.* 15(1): 107–117.

40 Lakshmi, V., Kumar, R., Pandey, K. et al. (2009). Structure and activities of a steroidal saponin from Chlorophytum nimonii (Grah) Dalz. *Nat. Prod. Res.* 23(10): 963–972.

41 Fatmawati, S., Kurashiki, K., Takeno, S. et al. (2009). The inhibitory effect on aldose reductase by an extract of Ganoderma lucidum. *Phytother. Res.* 23(1): 28–32.

42 Teng, B.S., Wang, C.D., Zhang, D. et al. (2012). Hypoglycemic effect and mechanism of a proteoglycan from ganoderma lucidum on streptozotocin-induced type 2 diabetic rats. *Eur. Rev. Med. Pharmacol. Sci.* 16(2): 166–175.

43 Agius, L. (2007). New hepatic targets for glycaemic control in diabetes. *Best Pract. Res. Clin. Endocrinol. Metab.* 21(4): 587–605.

44 Xiao, C., Wu, Q.P., Cai, W. et al. (2012). Hypoglycemic effects of Ganoderma lucidum polysaccharides in type 2 diabetic mice. *Arch. Pharm. Res.* 35(10): 1793–1801.

45 Yoshikawa, M., Matsuda, H., Harada, E. et al. (1994). Elatoside E, a new hypoglycemic principle from the root cortex of Aralia elata Seem: Structure-related hypoglycemic activity of oleanolic acid glycosides. *Chem. Pharm. Bull. (Tokyo)*. 42(6): 1354–1356.

46 Yoshikawa, M., Yoshizumi, S., Ueno, T. et al. (1995). Medicinal foodstuffs. I. Hypoglycemic constituents from a garnish foodstuff "taranome," the young shoot of Aralia elata SEEM.: Elatosides G, H, I, J, and K. *Chem. Pharm. Bull. (Tokyo).* 43(11): 1878–1882.

47 Norberg, A., Hoa, N.K., Liepinsh, E. et al. (2004). A novel insulin-releasing substance, phanoside, from the plant Gynostemma pentaphyllum. *J. Biol. Chem.* 279(40): 41361–41367.

48 Maurya, R., Akanksha, Jayendra, et al. (2008). Coagulanolide, a withanolide from Withania coagulans fruits and antihyperglycemic activity. *Bioorg. Med. Chem. Lett.* 18(24): 6534–6537.

49 Liu, W.Y., Liou, S.S., Hong, T.Y. et al. (2017). Protective effects of Hesperidin (Citrus Flavonone) on high glucose induced oxidative stress and apoptosis in a cellular model for diabetic retinopathy. *Nutrients* 9(12): E1312.

50 Malakul, W., Pengnet, S., Kumchoom, C. et al. (2018). Naringin ameliorates endothelial dysfunction in fructose-fed rats. *Exp. Ther. Med.* 15(3): 3140–3146.

51 Blahova, J., Martiniakova, M., Babikova, M. et al. (2021). Pharmaceutical drugs and natural therapeutic products for the treatment of type 2 diabetes mellitus. *Pharmaceuticals (Basel).* 14(8): 806.

4

Functional Foods in Clinical Trials in the Intervention of Diabetes Complications

*Bavani Arumugam and Umah Rani Kuppusamy**

Department of Biomedical Science, Faculty of Medicine, Universiti Malaya, 50603 Kuala Lumpur, Malaysia
** Corresponding author*

4.1 Introduction

Diabetes Mellitus (DM) is a metabolic disorder marked by persistent hyperglycemia (high blood sugar levels) and glucose intolerance. The World Health Organization (WHO) anticipates that by the year 2030, more than 19% of the world's adult population will have DM [1]. DM is typically brought on by the body's inability to secrete adequate amounts of insulin, failure of insulin action, or both, which culminates in dysregulation of carbohydrate, fat, and protein metabolism [2]. The current broadly used, simplified classification of diabetes is Type 1 Diabetes Mellitus (T1DM), Type 2 Diabetes Mellitus (T2DM), Gestational Diabetes Mellitus (GDM), and other types of diabetes caused by a variety of factors [3]. T1DM, which accounts for 5–10% of all diabetes cases, is caused by autoimmune destruction of the β-cells in the pancreatic islets, ensuing in an absolute lack of insulin action. These patients require lifelong insulin therapy. Meanwhile, T2DM, which accounts for 90–95% of all diabetes cases, is primarily caused by insulin resistance due to decreased cellular response to insulin and impaired insulin production due to β-cell dysfunction [3]. Poor glycemic control is known to trigger oxidative stress and inflammation that eventually lead to microvascular complications such as diabetic retinopathy, nephropathy, neuropathy, and macrovascular complications such as cardiovascular disease [4]. Obesity, which is strongly associated with metabolic syndrome (MetS), is considered to be one of the primary causes of insulin resistance, which manifests in T2DM. Weight control through increased physical activity and calorie restriction may be useful in the prevention and management of obesity-mediated T2DM. Drugs such as

Natural Products and their Bioactives in Antidiabetic Drug Discovery, First Edition.
Edited by Kanti Bhooshan Pandey, Maitree Suttajit and Pinar Atukeren.

thiazolidinediones (TZDs), sulfonylurea derivatives, biguanides, meglitinides, bile acid sequestrant, dipeptidyl peptidase IV (DPPIV) inhibitors, sodium-glucose transport protein (SGLT2), and α-glucosidase inhibitors are among the oral hypoglycemic medications that are frequently used to treat T2DM [5]. Owing to the undesirable side effects and lack of efficacy in some patients, as well as the high cost of existing diabetes medications, the quest for novel alternative treatments continues.

Recently, there has been a surge of interest in healthy eating habits, in line with Hippocrates' tenet "Let food be thy medicine, and medicine be thy food." The use of functional foods in disease treatment is becoming increasingly popular, particularly as an adjuvant or replacement for drug therapy. Although these foods may have been used for centuries often the proof is anecdotal, and the knowledge is passed on from generation to generation. Extensive research into the benefits and medicinal properties of food and its ingredients has paved the way for the use of functional food for the treatment of diseases including T2DM and related complications. A number of these functional foods and or food ingredients have made it to clinical trials.

This chapter highlights selected functional foods or food ingredients claimed to be useful in the management of diabetes and related complications that have been subjected to or are undergoing clinical trials. In addition, the challenges of clinical trials and future perspectives on the use of functional foods for the treatment/management of diabetes and related complications are also discussed.

4.2 How Does Chronic Hyperglycemia Lead to Diabetes-related Complications?

Chronic hyperglycemia is associated with microvascular (nephropathy, neuropathy, and retinopathy), as well as macrovascular (coronary heart disease, peripheral vascular disease, and stroke) complications that increase mortality and morbidity [6]. Acute fluctuations of hyperglycemia can be even more detrimental in causing diabetic complications [7]. The etiology of hyperglycemia-mediated complications is well established [1, 8, 9], and summarized in Figure 4.1.

4.3 Classification/Categories of Functional Foods

The term "functional food" was first coined in Japan in 1981 and was given a formal legislative term, FOSHU ("Foods for Specific Health Uses"). Three important criteria that include (i) effectiveness in clinical trial studies, (ii) safety in clinical and non-clinical studies, and (iii) determination of active/effective components must

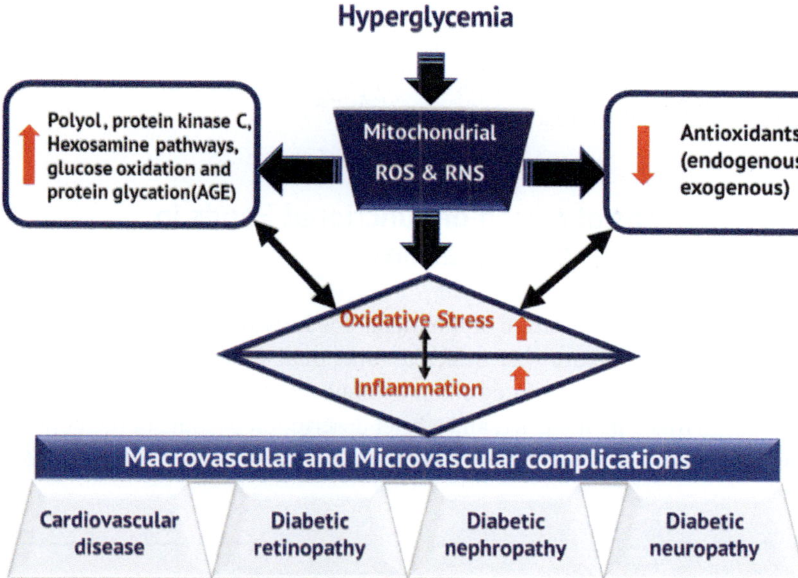

Figure 4.1 Hyperglycemia-induced biochemical alterations that ultimately exacerbate diabetic complications (microvascular and macrovascular). Hyperglycemia causes superoxide (reactive oxygen species, ROS) overproduction in the mitochondria, accompanied by increased nitric oxide (reactive nitrogen species, RNS) generation that lead to more potent free radicals such as the hydroxyl radicals and peroxynitrite radicals that can amplify macromolecular damages. The glycosylation of the antioxidant enzyme proteins could reduce their activity. When the body's antioxidant system is unable to cope with the removal of the excess free radicals, oxidative stress prevails. This leads to further damage and metabolic dysregulation through activation of protein kinase C, increased polyol and hexosamine pathway flux and increased advanced glycation end-product (AGE) formation as well as endothelial injury and dysfunction. Oxidative stress also stimulates the production of inflammatory mediators such as proinflammatory cytokines, which in turn increases the synthesis of free radicals, hence exacerbating diabetic complications (microvascular and macrovascular) [4, 7, 10, 11].

be fulfilled by the food to qualify for FOSHU [12]. The American Functional Food Centre defines functional foods as "Natural or processed foods that contain biologically-active compounds which, in defined, effective, non-toxic amounts, provide a clinically proven and documented health benefit utilizing specific biomarkers, to promote optimal health and reduce the risk of chronic/viral diseases and manage their symptoms" [13], and this is a widely accepted definition currently. According to the American Association of Dietetics (ADA), functional foods can be categorized based on how these foods are prepared; *conventional foods* are unmodified, whole foods that contain bioactive compounds with health benefits; *modified foods* are enhanced or fortified with food ingredients or bioactive compounds that have health benefits. Some of these food ingredients are also developed as nutraceuticals. Meanwhile, *synthetic food ingredients* are ingredients synthesized in the laboratory that may confer health benefits. This chapter focuses primarily on conventional and modified foods.

4.4 Mechanisms of Action of Functional Foods in Preventing Diabetes Complications

The general mechanisms of action of functional foods as antidiabetic agents have been summarized in Figure 4.2. Selected functional foods beneficial for the management of diabetes and related complications that have made it into clinical trials are summarized in Table 4.1, and a brief description of some of the popular foods is provided in the following subsections.

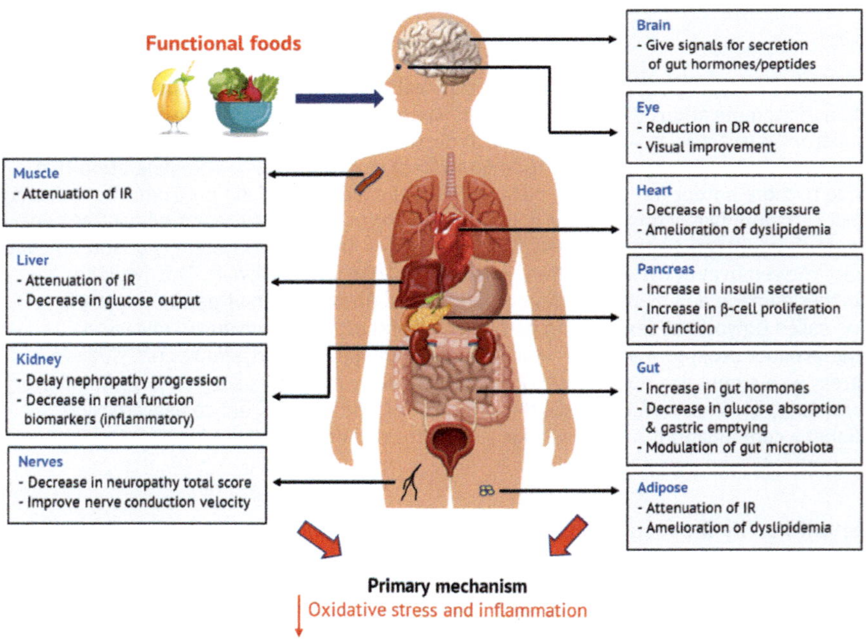

Figure 4.2 General antidiabetic effects (mechanisms of action) of functional foods in a human body. Attenuation of insulin resistance (IR) in adipose, liver, and muscle could be achieved through the increase in glucose uptake and/or insulin sensitivity. Amelioration of dyslipidemia in adipose and liver could be attributed to the decrease in fatty acid synthesis/low-density lipoprotein (LDL) cholesterol or improvement in the regulation of lipid metabolism. Information presented was partially adapted from [14]. (Source of human body vector and functional foods: Designed by Freepik).

Table 4.1 Selected functional foods that have been tested in clinical trials for the treatment of diabetes complications.

Functional food	Disease/Clinical trial code/Study design (SD)/Status/Phase (if any)	Biomarkers	Intervention/Significant findings	References
Conventional food Kiwifruit	*Prediabetes* **Code:** ACTRN12618000329268 **SD:** Randomized-parallel **Status:** Completed	HbA1c, Vit C, BP, LP, UA, glycemic, inflammatory, OS markers, AMs	**Intervention:** 2 kiwifruits (200g) for 12 weeks **Findings:** Increase in Vit C and status in prediabetics with no metabolic disturbance	[15]
Conventional food Taro, taro and probiotic, taro and beet juice	*Prediabetes* **Ethics approval number:** KET329/UN2.F1/ETIK/PPM.00.02/2019 **SD:** Randomized-crossover-DB **Status:** Unknown	Gut microbiota, AMs	**Intervention:** (i) 50% taro flour + 50% wheat flour, (ii) products in (i) + probiotic *L. plantarum* IS-10506, (iii) products in (i) beetroot **Findings:** Interindividual variations affected evaluation of gut microbiota modulation in prediabetics. There was a taro consumption–*Butyricimonas* association	[16]
Conventional food Gold kiwifruit	*Prediabetes* **Code:** CTRN12616000858493 **SD:** Non-randomized **Status:** Active	Glycemic indices, plasma [Vit C], hormones, LP and high-sensitivity CRP, fecal microbiota composition	**Intervention:** 2 SunGold kiwifruit/day for 12 weeks **Findings:** Increase in *Coriobacteriaceae* in prediabetic individuals	[17]
Conventional food Cocoa	*T2DM* **Code:** NCT01886989 **SD:** Randomized-crossover **Status:** Completed	Postprandial Apo-A1 and Apo-B, non-esterified FA, NMR-derived lipoprotein subclass profiles, inflammation biomarkers	**Intervention:** 20g cocoa powder with 960 mg polyphenols, 201 mg proanthocyanidins, 480 mg flavanols, 40 mg epicatechin **Findings:** Attenuation of inflammation and postprandial dyslipidemia upon a high-fat diet in T2DM adults	[18]

(Continued)

Table 4.1 (Continued)

Functional food	Disease/Clinical trial code/Study design (SD)/Status/Phase (if any)	Biomarkers	Intervention/Significant findings	References
Conventional food Raw yellow onion	*T2DM* **Code:** IRCT2012103111335N1 **SD:** Randomized-parallel-TB **Status:** Completed	Hormonal and metabolic indices, liver-related functional enzymes	**Intervention:** 100 to 160 g/day onion; 30 to 40 g/day small onion for 8 weeks **Findings:** Attenuation of IR and hyperglycemia in breast cancer patients during doxorubicin-based chemotherapy	[19]
Conventional food Dark chocolate	*Diabetes* **Code:** NCT04847999 **SD:** Randomized-crossover-DB **Status:** Completed	BG (iAUC), Peak [BG], BG total AUC	**Intervention:** 1 bar (34 g) sugar-free dark chocolate sweetened with stevia, inulin and erythritol **Findings:** Reduction in lower BG iAUC in diabetics.	[20]
Conventional food Peanut	*Diabetes* **Code:** NCT03194152 **SD:** Randomized-parallel **Status:** Completed	BG, inflammatory, LP, microbiome, urinary polyphenol metabolites, appetite and other hunger sensations	**Intervention:** 2 packs of peanuts/day (28 g/packet) for 12 weeks **Findings:** Improved responsiveness of some metabolic and clinical parameters and MetS risk. Personalized nutrition is important due to differences in microbiota composition	[21]
Conventional food Almond	*T2DM* **Code:** NCT01656850 **SD:** Randomized-crossover **Status:** Completed	Biomarkers of glucoregulation, inflammation, OS, LP	**Intervention:** ~60 g/d almonds (ALM) (contains OA and α-tocopherol) added to NCEP step II diet **Findings:** Better glycemic control and improvement in glycemic status in T2DM patients.	[22]

Conventional food White sesame oil	*T2DM* **Code:** Not available **SD:** Randomized **Status:** Unknown	FBG, insulin, HbA1c, hepatic AO enzymes, serum biochemistry, cardiac, hepatic and renal functions	**Intervention:** 30 ml of white sesame oil for 90 days **Findings:** Reversed oxidative damage to pancreatic β-cells. Improved hepatic AO enzymes, reduced OS and regulated glucose/myocardial function in T2DM adults.	[23]
Conventional food Orange juice concentrate	*T2DM* **Code:** NCT04412798 **SD:** Randomized-crossover-DB **Status:** Recruiting	Glycemic (venous blood and capillary) and insulin responses, HbA1c, AMs. AD related markers	**Intervention:** 250 ml of 100% orange juice (concentrate) (single serving) to T2DM adults (male) **Findings:** Pending publication	—
Modified food Diet rich in polyphenol, fiber and vegetable-protein functional foods	*T2DM* **Code:** NCT03421301 **SD:** Randomized-DB **Status:** Completed	Fecal microbiota composition, biochemical parameters, HbA1c and free FA, branched-chain AA, lipopolysaccharide, trimethylamine-*N*-oxide	**Intervention:** Reduced-energy diet with polyphenol-rich, high-fiber and vegetable-protein functional foods for 3 months **Findings:** Increased *Faecalibacterium prausnitzii* and *Akkermansia muciniphila* (species known to have anti-inflammatory effects), AO activity; May improve glycemic control, inflammation and dyslipidemia in T2DM patients.	[24]
Modified food Lupin biscuits (Skinnybik™ Lupin) containing 20% lupin flour	*T2DM* **Code:**ACTRN12620000353998 **SD:** Non-randomized-crossover pragmatic study **Status:** Completed	Interstitial glucose level, satiety and palatability, bowel motions	**Intervention:** 2 biscuits for morning and afternoon tea for 3 consecutive days **Findings:** Significant reduction in 90 min postprandial glucose and satiety in hospitalized T2DM patients.	[25]

(Continued)

Table 4.1 (Continued)

Functional food	Disease/Clinical trial code/Study design (SD)/Status/Phase (if any)	Biomarkers	Intervention/Significant findings	References
Modified food Kothala Himbutu (*Salacia reticulata*) biscuit	*Diabetes* **Code:** NCT02290925 **SD:** Randomized-crossover-quadrupole **Status:** Completed (Phase 3)	HbA1c, liver and renal adverse effects	**Intervention:** 4 biscuits twice/day snack for 3 months; Two arms: 1. Placebo biscuit then Kothala Himbutu biscuit and 2. Kothala Himbutu biscuit then placebo biscuit to be given to T2DM adults. **Findings:** Pending publication	—
Modified food Pomegranate peel added to bread	*T2DM* **Code:** NCT05232305 **SD:** Randomized-parallel-DB **Status:** Not yet recruiting	HbA1c, insulin, LP, TBARS, AMs	**Intervention:** Bread containing 500 mg pomegranate peel powder to be given to T2DM adults for 8 weeks **Findings:** Pending publication	—
Conventional food Peanut	*Prediabetes/T2DM and CVD* **Code:** NCT03654651 **SD:** Randomized-crossover **Status:** Completed (Phase 2)	FBG, fructosamine, fasting insulin, LP/lipoprotein, central and peripheral BP, pulse wave	**Intervention:** 28 g/day of peanuts (dry unroasted, unsalted) as a night-time snack for 6 weeks **Findings:** No significant effect observed with FBG and cardiometabolic endpoints in adults with elevated FBG	[26]
Conventional food Walnut	*Obesity and T2DM* **Code:** NCT02673281 **SD:** Randomized-crossover-DB **Status:** Active	Resting metabolic rate, body composition, neurocognitive testing, fasting MRI, visual analog scales	**Intervention:** 8 g walnut (walnut shake)/day for 5 days **Findings:** May improve cognitive control and increase salience; control highly desirable food cues in obese adults and promote healthy weight; reduce the risk of developing diabetes	[27]

Conventional food Cashew nut	*T2DM and CVD* **Code:** CTRI/2017/07/009022 **SD:** Randomized-parallel **Status:** Closed for recruitment	Glycemic, LP, AMs, general health clinical indicators	**Intervention:** 30 g cashew nuts/day for 12 weeks **Findings:** Increased [HDL cholesterol] and reduced BP (systolic) in T2DM patients. No adverse effects on glycemia, LP and body weight	[28]
Modified food Low fat yogurt enriched with flaxseed	*Diabetes and CVD* **Code:** NCT02436369 **SD:** Randomized-parallel **Status:** Completed	AMs, FBG, HbA1c, LP	**Intervention:** 200 g of 2.5% fat yogurt containing 30 g flaxseed daily for 8 weeks **Findings:** Decreased HbA1c; significant changes in TC and TG in T2DM individuals	[29]
Modified food Pecan (*Carya illinoensis* (Wangenh.) K.Koch)-rich diet	*Obesity/ T2DM and CVD* **Code:** NCT01950806 **SD:** Randomized-crossover **Status:** Completed	Fasting insulin and glucose, IR, β-cell function, OS, AO markers, inflammation, endothelial function, BP, LP	**Intervention:** Control diet with whole pecans (~42.5 g) for 4 weeks followed by control diet for another 4 weeks after 2 weeks washout period **Findings:** Improved fasting insulin, IR, and β-cell function in mildly obese and overweight adults with central adiposity. Lowered risk of cardiometabolic disease	[30]
Modified food Low fat-spreads with added plant sterols (PS)	*T2DM and CVD* **Code:** NCT02288585 **SD:** Randomized-parallel-double blind **Status:** Completed	Fasting serum/plasma LP, glucose and insulin	**Intervention:** Low-fat spreads with added PS (2 g/day) for 6 weeks **Findings:** Reduced elevated LDL-cholesterol and TG in individuals with and at risk of T2DM. No significant changes in glycemic and postprandial lipid	[31]

(Continued)

Table 4.1 (Continued)

Functional food	Disease/Clinical trial code/Study design (SD)/Status/Phase (if any)	Biomarkers	Intervention/Significant findings	References
Modified food Chlorogenic acid (CGA) enriched coffee	*T2DM, chronic renal insufficiency and CVD* **Code:** NCT02524938 **SD:** Randomized-parallel-DB **Status:** Unknown (Phase I and II)	HbA1c, eGFR, fasting glucose, fructosamine, AGEs, sRAGE, MDA, urinary albumin excretion, creatinine, BP, cholesterol, inflammation and OS markers	**Intervention:** CGA 400 mg/day (CGA-enriched coffee) for 12 weeks for T2DM adults **Findings:** Pending publication	—
Conventional food Fish	*Diabetic retinopathy (DR)* **Code:** ISRCTN35739639 **SD:** Randomized **Status:** Completed	DR cases that require vitrectomy, laser photocoagulation, and/or antiangiogenic therapy	**Intervention:** At least 500 mg/day of LCω3PUFAs **Findings:** Reduced risk of incident sight-threatening DR among T2DM adults	[32]
Conventional food Cocoa powder	*Diabetic neuropathy* **Code:** NCT05247034 **SD:** Randomized-parallel-TB **Status:** Recruiting	Rate-dependent depression on the frequency of stimulation of the H reflex, TCNS, BSFS, BP, glucose, TG, LP, HbA1c, HOMA, serum insulin, Inflammatory markers, AMs	**Intervention:** Diet + 4 capsules of cocoa powder (500 mg each)/daily for 12 weeks. Each capsules contain 12.5 mg of flavonoids for T2DM and diabetic neuropathy adults **Findings:** Pending publication	—

Conventional food Fish oil concentrate	*Diabetic neuropathy* **Code:** NCT05169060 **SD:** Randomized-parallel **Status:** Not yet recruiting (early phase I)	Omega-3 index, serum resolvin D1, resolvin E1, neuroprotectin D1, omega-3 PUFA metabolites, inflammatory markers, serum omega-6 and omega-3 FA ratio, MNSI, ankle reflex, vibration perception, FBG, fasting insulin, LP, HbA1c, urine microalbuminuria	**Intervention:** 2 or 4 fish oil (concentrate enriched in omega-3 PUFA) capsules (1000 mg)/day for 16 weeks for T2DM and diabetic peripheral neuropathy adults **Findings:** Pending publication	—

Abbreviations: AA, amino acid; AD, Alzheimer's disease; AGEs, advanced glycation end products; AMD, Age-related macular degeneration; AMs, anthropometric measurements; AO, antioxidant; Apo, apolipoproteins; AUC, area under the curve; BP, blood pressure; BSFS, Bristol stool form scale; Carb., carbohydrates; CGA, chlorogenic acid; CRP, C-reactive protein; CVD, cardiovascular disease; DB, double blind; eGFR, estimated glomerular filtration rate; FA, fatty acids; FBG, fating blood glucose; GI, glycemic index; HbA1c, glycated hemolglobin; iAUC, incremental area under the curve; IR, insulin resistance; LCω3PUFAs, long-chain ω−3 polyunsaturated fatty acids; LP, lipid profile; MDA, malondialdehyde; MetS, metabolic syndrome; MNSI, Michigan neuropathy screening instrument; NCEP, National cholesterol education program; NMR, nuclear magnetic resonance; OA, oleic acid; OS, oxidative stress; PUFA, polyunsaturated fatty acids; Ref., reference; SD, study design; sRAGE, soluble receptor for AGE; T2DM, Type 2 diabetes mellitus; TB, triple blind; TBARS, thiobarbituric acid reactive substances; TC, total cholesterol; TCNS, Toronto Clinical Scoring System; TG, triglycerides; UA, uric acid; Vit, vitamin.

4.4.1 Turmeric

Turmeric (*Curcuma longa*) is a popular culinary spice in many cultures. The primary bio-ingredient found in the root (rhizome) responsible for the majority of the health benefits is curcumin (a polyphenol, and sub-component of curcuminoids) that gives an intense yellow color to food. Curcumin has been well investigated in many preclinical and clinical studies, which revealed its effectiveness against cardiometabolic diseases due to its anti-inflammatory and antioxidant properties [33]. It is currently available in the market as supplements (turmeric extract rich in curcumin or pure curcumin with improved bioaccessibility such as Theracurmin® and Cucuwin®) or included in food products such as snacks (e.g. Nature's Heart's mango turmeric cashew and pistachio), ready-to-eat-meals (e.g. Bio-Zentrale Naturprodukte's Lunch 'n Go Vegetable Paella with Curcumin), beverages (e.g. Clearspring's Japanese matcha turmeric tea) that promote health benefits [33, 34].

Several diabetes intervention studies involving turmeric and its active component, curcumin, are still ongoing. Supplementation of 80 mg nano-curcumin/day for 8 weeks to T2DM adults showed a reduction in fasting blood glucose (FBG) level, glycated hemoglobin (HbA1c), and neuropathy total score [35]. An increased dosage of 1320 mg curcuminoids/day for 10 weeks showed a reduction in FBG level and waist circumference in T2DM adults in a separate study [36]. Turmeric (500 mg) with piperine (5 mg) when given for 120 days showed a reduction in HbA1c, homeostatic model assessment for insulin resistance (HOMA-IR), and serum triglycerides levels in individuals with T2DM [37]. Meanwhile, consuming powdered rhizomes of turmeric (2100 mg)/day for 8 weeks decreased body weight and improved the lipid profile in hyperlipidemic T2DM patients [38]. In a separate trial, patients with impaired glucose tolerance who received 180 mg turmeric extract daily for 6 months had a lower level of oxidized low-density lipoprotein (LDL) cholesterol [39].

4.4.2 Berries

Dietary berries such as blueberries, cranberries, raspberries, strawberries, and many others, can be consumed as fresh (whole fruits), processed (frozen and dried fruits), dried powder, unsweetened juices, and extract forms. The berries are low in calories, contain micronutrients such as vitamin C, E, and folic acid, and are rich in polyphenols such as flavonoids (primarily anthocyanins followed by flavan-3-ols and flavonols) and tannins [40]. Heat treatment generally reduces the polyphenol content [41] while freeze-drying preserves it [42], therefore, they are commonly used in feeding trials [43].

The efficacy of the different berries in attenuating T2DM and related complications is well established [44]. The berries on their own or in combination with another berry or functional food could improve glycemic control, lipid profiles, and markers for cardiometabolic risks. A delayed postprandial phase of glucose response was observed among healthy young adults ($n = 17$) when they consumed a blueberry powder beverage (34 g; 310 mg anthocyanin or 80 g; 724 mg anthocyanin) in both sugar-matched and/or no-added-sugar conditions in a double-blind crossover study [45]. In a randomized postprandial meal study, a breakfast meal supplemented with frozen whole red raspberries (125 g) reduced the 2 hours postprandial glucose and insulin (area under the curve), peak insulin and glucose, and triglycerides in adults with prediabetes, insulin resistance, and obesity/overweight subjects ($n = 32$) [46]. Raspberry supplementation in adults with T2DM on medications showed a reduction in 2 and 4 hours blood glucose in a Phase I study, and a decrease in inflammatory markers (interleukin-6, IL-6; high sensitivity tumor necrosis factor-alpha, hs-TNF-α) in a Phase 2 study [47]. Meanwhile, freeze-dried blueberry supplementation for 8 weeks improved HbA1c, fructosamine, and triglycerides in men with T2DM [48]. Anthocyanins, which are abundant in blueberries, may interfere with carbohydrate digestion, delay intestinal glucose absorption by inhibiting glucose transporter-2 (GLUT-2) [49], increase GLUT-4 expression in white adipose tissue and skeletal muscle by stimulating AMP-activated protein kinase (AMPK), and decrease glucose production in the liver [50].

4.4.3 Soy

The soybean is a legume and edible seed with many identified health benefits such as anti-inflammatory, anti-obesity, and antioxidant activities [51, 52]. The bio-ingredients in soybeans include isoflavones (daidzein and genistein), phytoestrogen, and peptides [53]. The isoflavones stimulate β-cells, enhance glucose uptake, improve insulin secretion, promote GLP-1 secretion, and inhibit carbohydrate hydrolyzing enzymes [51], and regulate insulin signaling pathway-extracellular regulated kinase (ERK) 1/2, AMPK, and protein kinase B (Akt) [53]. Consumption of soy proteins has long been linked with a lower incidence of diabetes and improvement in diabetes-related complications [54]. Administration of soy protein without isoflavones in patients with diet- or metformin-controlled T2DM showed intrinsic glycemic control (reduced HbA1c) and, in the presence of isoflavones, it improved insulin resistance and LDL cholesterol [55]. Short-term consumption of roasted soy-nut and textured soy protein (TSP) caused a significant reduction in total cholesterol, LDL cholesterol, very-low-density lipoprotein (VLDL) cholesterol, FBG, insulin, HOMA-IR, malondialdehyde, and apolipoprotein B100, and increase in total antioxidant capacity in older women with borderline MetS [56]. Consumption of soy milk alone did not improve inflammation and oxidative stress in diabetic nephropathy patients in a randomized crossover clinical trial [57]. However, probiotic soymilk that was prepared through a fermentation process improved oxidative stress indices in diabetic kidney disease (DKD) patients [58]. Fermentation improves the absorption of isoflavones by converting the glycosidic to the aglycone form. In addition, the presence of probiotics favors the regulation of oxidative stress pathways by boosting beneficial gut microbiota [59].

4.4.4 Nuts

Nuts such as walnuts, almonds, pistachios, hazelnuts, macadamias, cashews, and pecans are highly nutritious and are rich in unsaturated fatty acids, lignans, fibers, vitamins, minerals and other bio-ingredients. The Food and Drug Administration (FDA) recognizes the beneficial effects of nuts in reducing cardiovascular risks [60]. Diabetes and cardiovascular disease organizations, such as the Canadian Cardiovascular Society, recommend increased nut consumption [61]. The replacement of a carbohydrate diet with a full-dose nut diet (monounsaturated fats) has been proven to improve glycemic control and lipid markers (LDL cholesterol) in T2DM patients [62]. A recent systematic review and meta-analysis of randomized controlled clinical trials (16 studies involving a total of 1041 patients) showed that consumption of peanuts and tree nuts reduced concentrations of total cholesterol and triglycerides significantly in T2DM patients, therefore, reducing the risk of developing cardiovascular disease [63]. A separate meta-analysis and systematic review of 15 randomized controlled trials involving 899 subjects revealed that consumption of peanuts and tree nuts does not affect the adiposity and weight of T2DM patients [64]. A meta-analysis of 12 clinical trials on walnut diets revealed that walnut reduced LDL cholesterol in individuals with abnormal glucose homeostasis [65]. Walnut in diets of T2DM adults was shown to improve the intake of dietary nutrients such as calcium, magnesium, thiamin, fat, total saturated fatty acids, and mono- and poly-unsaturated fatty acids, known for T2DM and its related cardiometabolic risk factor reduction [66]. Antidiabetic activities of almonds, peanuts, cashews, and pecans that are actively studied in clinical trials, have been included in Table 4.1.

4.4.5 Selected Vegetables

Okra (*Abelmoschus culentesus*), or commonly known as lady's finger, is a vegetable with high fiber content, traditionally used to treat diabetes. The juice of overnight soaked okra pieces or its roasted seeds are consumed to treat diabetes. The okra pods are commonly boiled, fried, or used to make soups and stews. Consumption of 10 g okra powder blended in 150 g conventional yogurt for 8 weeks by T2DM patients resulted in a significant decrease in HOMA-IR, FBG, total cholesterol, LDL-cholesterol, triacylglycerol, and insulin sensitivity check index [67]. In a separate study, oral consumption of 1000 mg of okra (powdered whole fruit in capsules) every 6 hours for 8 weeks reduced fasting and random blood glucose, and HbA1c levels in T2DM individuals [68]. Different parts of the okra plant contain various polyphenolic compounds, for instance, the fruits contain isoquercitrin, quercetin, quercetin-3-O-gentiobioside, protocatechuic acid, rutin, and catechin [69], while the seeds contain flavonol derivatives [70] and other bio-active ingredients responsible for its antioxidant and antidiabetic properties.

Bitter melon or bitter gourd (*Momocardia charantia*) is another popular vegetable known for its hypoglycemic activity. Its bioactive compounds such as vicine, polypeptide-p, charantin, and saponins present in the plant [71] exert their antidiabetic effect through several mechanisms that include protection of pancreatic β-cells, modulation of peroxisome proliferator-activated receptor gamma (PPAR-γ) expression, attenuation of mitogen-activated protein kinases (MAPK) activation, inhibition of intestinal glucose uptake, insulin secretagogue-like effect, and many others [71]. A systematic review and meta-analysis of 10 clinical studies on T2DM proved that the *M. charantia* formulation significantly reduced FBG, postprandial blood glucose, and HbA1c compared to placebo [72]. Powdered bitter gourd supplementation (2.5 g)/daily for 8 weeks showed the potential to attenuate fasting plasma glucose in prediabetic individuals [73].

Bell pepper (capsicum; *Capsicum annum* Group) has gained interest among the diabetic population due to its effectiveness in reducing blood glucose levels and risks of related complications. The presence of anthocyanins that inhibit the enzymes lipase and α-glucosidase slows down the digestion of lipids and carbohydrates [74] while capsaicin mediates fat burning [75]. Bell pepper is rich in antioxidants such as phenolic compounds (quercetin, luteolin), lycopene, β-carotene, and vitamin C [74] that prevent oxidative damage and the development of associated chronic diseases including diabetes. A randomized controlled study that was conducted among T2DM patients ($n = 50$) showed that consumption of 100 ml bell pepper juice (twice/day) along with yoga therapy for four consecutive days was beneficial in diabetes control and cardiovascular functions as it reduced postprandial blood glucose, pulse rate, and systolic blood pressure compared to yoga therapy alone [76].

In general, consumption of at least 400 g of vegetables and fruits excluding starchy roots and cassava is recommended by the WHO for the prevention of chronic diseases such as diabetes, cancer, heart disease, and stroke [77]. A recent study showed that consumption of bitter and strong-tasting (BST) *Brassica* (curly kale, red cabbage, pointed cabbage, white cabbage, celeriac) and some root vegetables (beetroots, carrot) (500 g) for 12 weeks improved glycemic control, insulin sensitivity, lipid profile, blood pressure, and body mass compared to regular mild and sweet tasting (MST) vegetables (500 g) (Nordic diet) in T2DM individuals [78]. These activities have been attributed to the high dietary fiber content, nitrate-accumulating capacity, oxidative capacity as well as changes in β-carotene plasma concentration and potentially other phytochemicals such as polyacetylenes, phenolics, and glucosinolates in the vegetables [78]. A randomized crossover trial on various vegetables (mustard, cabbage, turnip, and radish) and beans (kidney beans and gram) powder enriched chapati (wheat flour premix) showed a significant reduction in postprandial glucose and postprandial insulin in healthy young adults [79]. The vegetables and beans have soluble fibers which are responsible for delayed gastric emptying and hence, improvement in metabolic control in diabetic patients.

4.5 Functional Food Clinical Trial Challenges

Functional foods would receive wider acceptance as the intervention of choice against diseases to replace or as an adjuvant to allopathic medicine if validated through clinical trials. However, functional food clinical trials (especially if pertaining to a specific disease such as diabetes) present a plethora of challenges at every stage of the process, from the trial organization and trial execution to data dissemination once the study is completed.

4.5.1 Funding, Technical Support, and Patient Recruitment Issues

Funding is an important issue as sponsors are reluctant to invest huge amounts of money in functional food trials since the active ingredients in foods cannot be patented and the prospect of financial gain through commercialization is limited [80]. Technical support concerning documentation preparation such as ethics and other related applications is lacking in disease-based clinical trials using functional foods compared to drugs.

Usually, a pilot study in healthy subjects is required to ensure that the bioactives present in the functional food are safe and well tolerated. This will pave the way for the appropriate recruitment of subjects for major interventional clinical trials. However, the current requirement by the regulatory bodies for more stringent controls of trials such as including double-blind, randomized, placebo-controlled, longer trials, large numbers of patients enrolled in the trial, and more complex inclusion and exclusion criteria make patient recruitment difficult and time-consuming, in addition to incurring a huge cost [80].

Recruitment of an appropriate number of age-matched patients for the placebo group could pose difficulties as most diabetic patients may already be on medication and specific dietary control [81]. Meanwhile, crossover trials allow individuals to serve as their control and participate in two or more interventions with appropriate washout periods, which saves the effort required for subject recruitment. However, this prolongs the trial period with escalated costs and thus adds to the difficulties in achieving successful completion of the trial [82].

The prolonged period of studies may lead to a lack of compliance, high dropout rates, escalated cost, and logistic issues as well as potential issues in maintaining the stability of test material (functional food) [83].

4.5.2 Clinical Trial Execution-related Issues

Unlike drug trials where the placebo group is given pills without the active component (which is easily accomplished), preparation of functional food placebo can be very tricky as it is not easy to prepare food with the same flavor, taste, texture, and appearance without the active component [84].

The free-living diet provides food variety and convenience for subjects, but it has compliance issues, diet changes, and difficulties in assessing the efficacy of test material. The food variety may influence the gut microbiota, adding another layer of complexity to establishing functional food efficacy [83]. A metabolically controlled diet is an ideal design to test the efficacy of test materials, but it requires high commitment from subjects apart from being labor-intensive, expensive, and lacking in food variety [83]. Sometimes the patient on a clinical trial may share the food (meant for their consumption) with others (non-participating individuals), hence reducing the appropriate intake of the active ingredient, which could result in misleading results. Preparation of food without compromising its active ingredient properties needs special attention as food harvesting, preparation (including cooking methods), storage, and contaminants (pesticides, heavy metals, and mycotoxins) may influence the food quality and the concentration of the active ingredients. These factors along with variation in the individual's genetic make-up could also influence the bioaccessibility of the food, gut microbiota, and bioavailability of the active ingredient(s) [85–87]. This ultimately poses a problem in estimating and ensuring a consistent, efficacious dose, which is imperative for obtaining a credible and reliable result.

Compliance in the study will gradually decline if the food provided lacks suitable flavor, texture, and variety of tastes, thereby jeopardizing the trial results. Easy access to the functional food that may be available in the supermarket may compromise the control of the quantity consumed by the recruited patients, jeopardizing the authenticity of the trial results as it affects the dose of the active ingredient consumed. The placebo group may also have access to the functional food in a clinical trial and, if they consume it, the clinical trial results will be compromised. Compliance of the patient on functional food clinical trials can be estimated through the measurement of an appropriate biomarker level in plasma.

However, the consistency of the results and the biomarker levels may be influenced by factors affecting the individual's digestion, assimilation, and absorption of the food [80].

4.5.3 Dissemination of Trial Findings to the Public

Journal publications and the media play a significant role in the dissemination of clinical trial findings. Positive results of functional food trials may receive large media attention to the extent of sensationalization. In most instances, public acceptance is attributed to the fact that the findings are instantly decipherable by the layman. However, a biased reporting of the findings through publications or media may lead to a premature acceptance of findings of undeserved (poorly designed), compromised trials. It is also important that the credibility, reliability, and reproducibility of the findings must be acceptable by the medical community as they have a great influence on patient care and patient acceptance of functional food as a replacement or adjuvant to the currently available allopathic medicines. However, despite the numerous impressive clinical trial findings, most medical practitioners are skeptical about the use of functional food to manage diseases [80]. Patients would inevitably be confronted with the dilemma of whether or not to trust the usage of functional foods to manage their ailment.

4.6 Future Perspectives and Conclusion

The current surge in interest in functional foods for the management of diseases such as diabetes and its complications has led to a parallel rise in functional food clinical trials, to improve the credibility of efficacy claims. However, this raises numerous challenges and issues that need to be addressed for more concrete, undisputed acceptance of the trial findings. Increasing consumer demand for functional foods, on the other hand, might improve revenue and growth in the food industry, as well as drive research and food-product innovation. There must be nationwide as well as international concerted efforts between academics/researchers, the medical community, funding agencies, regulatory bodies, and journals/media to ensure that the end-user receives credible, reproducible, and reliable information about the safety and efficacy of functional foods. Reliable information on functional foods may also be shared digitally through well-regulated apps. It is also vital to educate the public that, while functional foods can help with disease prevention and management, other health practices such as exercise, caloric control, appropriate stress management, and adequate sleep are equally important in ensuring successful disease management and delaying the onset of complications.

References

1 Ohiagu, F.O., Chikezie, P.C. and Chikezie, C.M. (2021). Pathophysiology of diabetes mellitus complications: Metabolic events and control. *Biomed. Res. Ther.* 8(3): 4243–4257.

2 World Health Organization (WHO). (2021). Diabetes [Internet]. WHO. [cited 2022 Sep 5]. Available from: https://www.who.int/news-room/fact-sheets/detail/diabetes#:~:text=Type%202%20diabetes%20(formerly%20called,body%20weight%20and%20physical%20inactivity).

3 American Diabetes Association. (2021). Classification and diagnosis of diabetes: Standards of medical care in diabetes. *Diabetes Care.* 44(Suppl.1): S15–S33.

4 Oguntibeju, O.O. (2019). Type 2 diabetes mellitus, oxidative stress and inflammation: Examining the links. *Int. J. Physiol. Pathophysiol. Pharmacol.* 11(3): 45–63.

5 Feingold, K.R. (2000). Oral and injectable (non-insulin) pharmacological agents for the treatment of type 2 diabetes. In: *Endotext [Internet].* (ed. K.R. Feingold, B. Anawalt, A. Boyce, et al.). South Dartmouth (MA): MDText.com, Inc. Available from https://www.ncbi.nlm.nih.gov/books/NBK279141.

6 Forbes, J.M. and Cooper, M.E. (2013). Mechanisms of diabetic complications. *Physiol. Rev.* 93(1): 137–188.

7 Zhang, Z.-Y., Miao, L.-F., Qian, L.-L. et al. (2019). Molecular mechanisms of glucose fluctuations on diabetic complications. *Front. Endocrinol.* 10.

8 Wong, F.N., Chua, K.H., Tan, J.A. et al. (2018). Glycaemic control in type 2 diabetic patients with chronic kidney disease: The impacts on enzymatic antioxidants and soluble rage. *Peer J.* 2018: e4421.

9 Ng, Z.X., Chua, K.H., Tajunisah, I. et al. (2013). Attenuated levels of pro-inflammatory markers in diabetic retinopathy patients undergoing treatment with antihyperglycemic and antihypertensive drugs. *Clinics.* 68(2): 185–191.

10 Ceriello, A. (2003). New insights on oxidative stress and diabetic complications may lead to a "causal" antioxidant therapy. *Diabetes Care* 26(5): 1589–1596.

11 Kuppusamy, U.R., Indran, M., Ahmad, T. et al. (2005). Comparison of oxidative damage in Malaysian end-stage renal disease patients with or without non-insulin-dependent diabetes mellitus. *Clin. Chim. Acta.* 351(1–2): 197–201.

12 Martirosyan, D.M. and Singh, J. (2015). A new definition of functional food by FFC: What makes a new definition unique? *Funct. Foods Health Dis.* 5(6): 209.

13 Martirosyan, D. and Ekblad, M. (2022). Functional foods classification system: Exemplifying through analysis of bioactive compounds. *Funct. Food Sci.* 2(4): 94.

14 Venkatakrishnan, K., Chiu, H.-F. and Wang, C.-K. (2019). Popular functional foods and herbs for the management of type-2-diabetes mellitus: A comprehensive review with special reference to clinical trials and its proposed mechanism. *J. Funct. Foods.* 57: 425–438.

15 Mishra, S., Bentley-Hewitt, K., Lubransky, A. et al. (2022). Metabolic, anthropometric and blood pressure effects of adding two kiwifruit or bottled water into the diets of people with pre-diabetes: A randomised, parallel group, intervention study. *Recent Prog. Nutr.* 2(1): 006; doi:10.21926/rpn.2201006.

16 Surono, I.S., Simatupang, A., Kusumo, P.D. et al. (2022). Effect of different functional food supplements on the gut microbiota of prediabetic Indonesian individuals during weight loss. *Nutrients.* 14(4): 781.

17 Wilson, R., Willis, J., Gearry, R. et al. (2018) Sungold kiwifruit supplementation of individuals with prediabetes alters gut microbiota and improves vitamin C status, anthropometric and clinical markers. *Nutrients.* 10(7): 895.

18 Davis, D.W., Tallent, R., Navalta, J.W. et al. (2020). Effects of acute cocoa supplementation on postprandial apolipoproteins, lipoprotein subclasses, and inflammatory biomarkers in adults with type 2 diabetes after a high-fat meal. *Nutrients.* 12(7): 1902.

19 Jafarpour-Sadegh, F., Montazeri, V., Adili, A. et al. (2016). Consumption of fresh yellow onion ameliorates hyperglycemia and insulin resistance in breast cancer patients during doxorubicin-based chemotherapy: A randomized controlled clinical trial. *Integr. Cancer Ther.* 16(3): 276–289.

20 Oliveira, B., Falkenhain, K. and Little, J.P. (2022). Sugar-free dark chocolate consumption results in lower blood glucose in adults with diabetes. *Nutr. Metab. Insights.* 15: 117863882210769.

21 Wang, S., Zhang, L., Wang, D. et al. (2021). Gut microbiota composition is associated with responses to peanut intervention in multiple parameters among adults with metabolic syndrome risk. *Mol. Nutr. Food Res.* 65(18): 2001051.

22 Chen, C.-M., Liu, J.-F., Li, S.-C. et al. (2017). Almonds ameliorate glycemic control in Chinese patients with better controlled type 2 diabetes: A randomized, crossover, controlled feeding trial. *Nutr. Metab.* 14(1).

23 Aslam, F., Iqbal, S., Nasir, M. et al. (2018). White sesame seed oil mitigates blood glucose level, reduces oxidative stress, and improves biomarkers of hepatic and renal function in participants with type 2 diabetes mellitus. *J. Am. Coll. Nutr.* 38(3): 235–246.

24 Medina-Vera, I., Sanchez-Tapia, M., Noriega-López, L. et al. (2019). A dietary intervention with functional foods reduces metabolic endotoxaemia and attenuates biochemical abnormalities by modifying faecal microbiota in people with type 2 diabetes. *Diabetes Metab.* 45(2): 122–131.

25 Skalkos, S., Moschonis, G., Thomas, C.J. et al. (2020). Effect of lupin-enriched biscuits as substitute mid-meal snacks on post-prandial interstitial glucose excursions in post-surgical hospital patients with type 2 diabetes. *Nutrients.* 12(5): 1239.

26 Sapp, P.A., Kris-Etherton, P.M. and Petersen, K.S. (2021). Peanuts or an isocaloric lower fat, higher carbohydrate night-time snack have similar effects on fasting glucose in adults with elevated fasting glucose concentrations: A 6-week randomized crossover trial. *J. Nutr.* 152(1): 153–162.

27 Farr, O.M., Tuccinardi, D., Upadhyay, J. et al. (2018). Walnut consumption increases activation of the insula to highly desirable food cues: A randomized, double-blind, placebo-controlled, cross-over fMRI study. *Diabetes Obes. Metab.* 20(1): 173–177.

28 Mohan, V., Gayathri, R., Jaacks, L.M. et al. (2018). Cashew nut consumption increases HDL cholesterol and reduces systolic blood pressure in Asian Indians with type 2 diabetes: A 12-week randomized controlled trial. *J. Nutr.* 148(1): 63–69.

29 Hasaniani, N., Rahimlou, M., Ramezani Ahmadi, A. et al. (2019). The effect of flaxseed enriched yogurt on the glycemic status and cardiovascular risk factors in patients with type 2 diabetes mellitus: Randomized, open-labeled, controlled study. *Clin. Nutr. Res.* 8(4): 284.

30 McKay, D., Eliasziw, M., Chen, C. et al. (2018). A pecan-rich diet improves cardiometabolic risk factors in overweight and obese adults: A randomized controlled trial. *Nutrients.* 10(3): 339.

31 Trautwein, E.A., Koppenol, W.P., de Jong, A. et al. (2018). Plant sterols lower LDL-cholesterol and triglycerides in dyslipidemic individuals with or at risk of developing type 2 diabetes; a randomized, double-blind, placebo-controlled study. *Nutr. Diabetes.* 8(1): 30.

32 Chew, E.Y. (2017). Dietary intake of omega-3 fatty acids from fish and risk of diabetic retinopathy. *JAMA.* 317(21): 2226.

33 Munekata, P.E.S., Pateiro, M., Zhang, W. et al. (2021). Health benefits, extraction and development of functional foods with curcuminoids. *J. Funct. Foods.* 79: 104392.

34 Straus, M. (2020). Turmeric food, drink, and dietary supplement opportunities [Internet]. Nutritional Outlook. Nutritional Outlook; 2020 [cited 2022 Sep 08]. Available from: https://www.nutritionaloutlook.com/view/which-are-turmeric-s-biggest-product-opportunities.

35 Asadi, S., Gholami, M.S., Siassi, F. et al. (2019). Nano curcumin supplementation reduced the severity of diabetic sensorimotor polyneuropathy in patients with type 2 diabetes mellitus: A randomized double-blind placebo-controlled clinical trial. *Complement. Ther. Med.* 43: 253–260.

36 Hodaei, H., Adibian, M., Nikpayam, O. et al. (2019). The effect of curcumin supplementation on anthropometric indices, insulin resistance and oxidative stress in patients with type 2 diabetes: A randomized, double-blind clinical trial. *Diabetol. Metab. Syndr.* 11:1.

37 Sousa, D.F., Araújo, M.F., de Mello, V.D. et al. (2020). Cost-effectiveness of passion fruit albedo versus turmeric in the glycemic and lipaemic control of people with type 2 diabetes: Randomized clinical trial. *J. Am. Coll. Nutr.* 40(8): 679–688.

38 Adab, Z., Eghtesadi, S., Vafa, M.R. et al. (2019). Effect of turmeric on glycemic status, lipid profile, HS-CRP, and total antioxidant capacity in hyperlipidemic type 2 diabetes mellitus patients. *Phytother. Res.* 33(4): 1173–1181.
39 Funamoto, M., Shimizu, K., Sunagawa, Y. et al. (2019). Effects of highly absorbable curcumin in patients with impaired glucose tolerance and non-insulin-dependent diabetes mellitus. *J. Diabetes Res.* 2019: 1–7.
40 Basu, A., Rhone, M. and Lyons, T.J. (2010). Berries: Emerging impact on cardiovascular health. *Nutr. Rev.* 68(3): 168–177.
41 Wilkes, K., Howard, L.R., Brownmiller, C. et al. (2013). Changes in Chokeberry (*aronia melanocarpa* L.) polyphenols during juice processing and storage. *J. Agric. Food Chem.* 62(18): 4018–4025.
42 Wojdyło, A., Figiel, A. and Oszmiański, J. (2009). Effect of drying methods with the application of vacuum microwaves on the bioactive compounds, color, and antioxidant activity of strawberry fruits. *J. Agric. Food Chem.* 57(4): 1337–1343.
43 Park, E., Edirisinghe, I., Wei, H. et al. (2016). A dose-response evaluation of freeze-dried strawberries independent of fiber content on metabolic indices in abdominally obese individuals with insulin resistance in a randomized, single-blinded, diet-controlled crossover trial. *Mol. Nutr. Food Res.* 60(5): 1099–1109.
44 Calvano, A., Izuora, K., Oh, E.C. et al. (2019). Dietary berries, insulin resistance and type 2 diabetes: An overview of human feeding trials. *Food & Funct.* 10(10): 6227–6243.
45 Bell, L., Lamport, D.J., Butler, L.T. et al. (2017). A study of glycaemic effects following acute anthocyanin-rich blueberry supplementation in healthy young adults. *Food Funct.* 8(9): 3104–3110.
46 Xiao, D., Zhu, L., Edirisinghe, I. et al. (2019). Attenuation of postmeal metabolic indices with red raspberries in individuals at risk for diabetes: A randomized controlled trial. *Obesity.* 27(4): 542–550.
47 Schell, J., Betts, N.M., Lyons, T.J. et al. (2019). Raspberries improve postprandial glucose and acute and chronic inflammation in adults with type 2 diabetes. *Ann. Nutr. Metab.* 74(2): 165–174.
48 Stote, K.S., Wilson, M.M., Hallenbeck, D. et al. (2020). Effect of blueberry consumption on cardiometabolic health parameters in men with type 2 diabetes: An 8-week, double-blind, randomized, placebo-controlled trial. *Curr. Dev. Nutr.* 4(4): nzaa030.
49 Baron, G., Altomare, A., Regazzoni, L. et al. (2017). Pharmacokinetic profile of bilberry anthocyanins in rats and the role of glucose transporters: LC–MS/MS and computational studies. *J. Pharm. Biomed. Anal.* 144: 112–121.
50 Takikawa, M., Inoue, S., Horio, F. et al. (2010). Dietary anthocyanin-rich bilberry extract ameliorates hyperglycemia and insulin sensitivity via activation of AMP-activated protein kinase in diabetic mice. *J. Nutr.* 140(3): 527–533.
51 Gilbert, E.R. and Liu, D. (2013). Anti-diabetic functions of soy isoflavone genistein: Mechanisms underlying its effects on pancreatic β-cell function. *Food Funct.* 4(2): 200–212.

52 Tidke, S.A., Ramakrishn, D., Kiran, S. et al. (2015). Nutraceutical potential of soybean: Review. *Asian J. Clin. Nutr.* 7(2): 22–32.

53 Chang, C.L., Lin, Y., Bartolome, A.P. et al. (2013). Herbal therapies for type 2 diabetes mellitus: Chemistry, biology, and potential application of selected plants and compounds. *Evid. Based Complement. Alternat. Med.* 2013: 1–33.

54 Hermansen, K., Søndergaard, M., Høie, L. et al. (2001). Beneficial effects of a soy-based dietary supplement on lipid levels and cardiovascular risk markers in type 2 diabetic subjects. *Diabetes Care.* 24(2): 228–233.

55 Konya, J., Sathyapalan, T., Kilpatrick, E.S. et al. (2019). The effects of soy protein and cocoa with or without isoflavones on glycemic control in type 2 diabetes. A double-blind, randomized, placebo-controlled study. *Front. Endocrinol.* 10: 296.

56 Bakhtiari, A., Hajian-Tilaki, K., Omidvar, S. et al. (2019). Clinical and metabolic response to soy administration in older women with metabolic syndrome: A randomized controlled trial. *Diabetol. Metab. Syndr.* 11(1): 47.

57 Miraghajani, M.S., Esmaillzadeh, A., Najafabadi, M.M. et al. (2012). Soy milk consumption, inflammation, coagulation, and oxidative stress among type 2 diabetic patients with nephropathy. *Diabetes Care.* 35(10): 1981–1985.

58 Miraghajani, M., Zaghian, N., Mirlohi, M. et al. (2017). The impact of probiotic soy milk consumption on oxidative stress among type 2 diabetic kidney disease patients: A randomized controlled clinical trial. *J. Ren. Nutr.* 27(5): 317–324.

59 Homayouni-Rad, A., Soroush, A.-R., Khalili, L. et al. (2016). Diabetes management by probiotics: Current knowledge and future pespective. *Int. J. Vitam. Nutr. Res.* 86(5–6): 216–228.

60 Center for Food Safety and Applied Nutrition. Labeling & Nutrition - Qualified Health Claims: Letter of enforcement discretion - nuts and coronary heart disease (docket no 02P-0505) [Internet]. *Archive.* Center for Food Safety and Applied Nutrition; [cited 2022 Sep 11]. Available from: https://wayback.archive-it.org/7993/20171114183724/https:/www.fda.gov/Food/IngredientsPackagingLabeling/LabelingNutrition/ucm072926.htm.

61 Anderson, T.J., Grégoire, J., Hegele, R.A. et al. (2013). 2012 update of the Canadian Cardiovascular Society guidelines for the diagnosis and treatment of dyslipidemia for the prevention of cardiovascular disease in the adult. *Can. J. Cardiol.* 29(2): 151–167.

62 Jenkins, D.J., Kendall, C.W., Lamarche, B. et al. (2019). Correction to: Nuts as a replacement for carbohydrates in the diabetic diet: A reanalysis of a randomised controlled trial. *Diabetologia.* 62(3): 549–552.

63 Xia, J.-Y., Yu, J.-H., Xu, D.-F. et al. (2021). The effects of peanuts and tree nuts on lipid profile in type 2 diabetic patients: A systematic review and meta-analysis of randomized, controlled-feeding clinical studies. *Front. Nutr.* 8: 765571.

64 Fernández-Rodríguez, R., Martínez-Vizcaíno, V., Garrido-Miguel, M. et al. (2021). Nut consumption, body weight, and adiposity in patients with type 2 diabetes: A systematic review and meta-analysis of randomized controlled trials. *Nutr. Rev.* 80(4): 645–655.

65 Malmir, H., Larijani, B. and Esmaillzadeh, A. (2021). The effect of walnut consumption on cardiometabolic profiles of individuals with abnormal glucose homoeostasis: A systematic review and meta-analysis of clinical trials. *Br. J. Nutr.* 128(6): 1037–1049.

66 Njike, V.Y., Costales, V.C., Petraro, P. et al. (2018). The resulting variation in nutrient intake with the inclusion of walnuts in the diets of adults at risk for type 2 diabetes: A randomized, controlled, crossover trial. *Am. J. Health Promot.* 33(3):430–438.

67 Moradi, A., Tarrahi, M.J., Ghasempour, S. et al. (2020). The effect of okra (*abelmoschus esculentus*) on lipid profiles and glycemic indices in type 2 diabetic adults: Randomized double blinded trials. *Phytother. Res.* 34(12): 3325–3332.

68 Saatchi, A., Aghamohammadzadeh, N., Beheshtirouy, S. et al. (2022). Anti-hyperglycemic effect of *abelmoschus culentesus* (OKRA) on patients with diabetes type 2: A randomized clinical trial. *Phytother. Res.* 36(4): 1644–1651.

69 Wu, D.-T., Nie, X.-R., Shen, D.-D. et al. (2020). Phenolic compounds, antioxidant activities, and inhibitory effects on digestive enzymes of different cultivars of okra (abelmoschus esculentus). *Molecules.* 25(6): 1276.

70 Hu, L., Yu, W., Li, Y. et al. (2014). Antioxidant activity of extract and its major constituents from okra seed on rat hepatocytes injured by carbon tetrachloride. *BioMed. Res. Int.* 2014: 1–9.

71 Joseph, B. and Jini, D.(2013). Antidiabetic effects of Momordica charantia (bitter melon) and its medicinal potency. *Asian Pac. J. Trop. Dis.* 3(2): 93–102.

72 Peter, E.L., Kasali, F.M., Deyno, S. et al. (2019). Momordica charantia L. Lowers elevated glycaemia in type 2 diabetes mellitus patients: Systematic review and meta-analysis. *J. Ethnopharmacol.* 231: 311–324.

73 Krawinkel, M.B., Ludwig, C., Swai, M.E. et al. (2018). Bitter gourd reduces elevated fasting plasma glucose levels in an intervention study among prediabetics in Tanzania. *J. Ethnopharmacol.* 216: 1–7.

74 Chávez-Mendoza, C., Sanchez, E., Muñoz-Marquez, E. et al. (2015). Bioactive compounds and antioxidant activity in different grafted varieties of bell pepper. *Antioxidants.* 4(2): 427–446.

75 Mukherjee, A., Mukherjee, S., Biswas, J. et al. (2015). Phytochemicals in obesity control. *Int. J. Curr. Microbiol. App. Sci.* 4(4): 558–567.

76 Nagasukeerthi, P., Mooventhan, A. and Manjunath, N.K. (2017). Short-term effect of add on bell pepper (*Capsicum annuum* var. Grossum) juice with integrated approach of yoga therapy on blood glucose levels and cardiovascular functions in patients with type 2 diabetes mellitus: A randomized controlled study. *Complement. Ther. Med.* 34: 42–45.

77 World Health Organization (WHO). Healthy diet [Internet]. WHO; [cited 2022 Sep 11]. Available from: https://www.who.int/news-room/fact-sheets/detail/healthy-diet#:~:text=For%20adults&text=At%20least%20400%20g%20(i.e.,cassava%20and%20other%20starchy%20roots).

78 Thorup, A.C., Kristensen, H.L., Kidmose, U. et al. (2021). Strong and bitter vegetables from traditional cultivars and cropping methods improve the health status of type 2 diabetics: A randomized control trial. *Nutrients*. 13(6): 1813.

79 Akhtar, S., Layla, A., Sestili, P. et al. (2019). Glycemic and insulinemic responses of vegetables and beans powders supplemented chapattis in healthy humans: A randomized, crossover trial. *BioMed. Res. Int*. 2019: 1–7.

80 Brown, L., Caligiuri, S.P.B., Brown, D. et al. (2018). Clinical trials using functional foods provide unique challenges. *J. Funct. Foods*. 45: 233–238.

81 Garcia, R., Benet, M., Arnau C. et al. (2004). Efficiency of the cross-over design: An empirical estimation. *Stat. Med*. 23(24): 3773–3780.

82 Dennis, B.H., Ershow, A.G., Obarzanek, E. et al. (1999). Well-controlled diet studies in humans. In: *A Practical Guide to Design and Management*. (ed. B.H. Dennis). 3rd edn. Chicago: Am. Diet. Assoc. 10–18.

83 AbuMweis, S.S., Jew, S. and Jones, P.J.H. (2010). Optimizing clinical trial design for assessing the efficacy of functional foods. *Nutr. Rev*. 68(8): 485–499.

84 Orubu, S.E., Hobson, N.J., Basit, A.W. et al. (2016). The milky way: Paediatric milk-based dispersible tablets prepared by direct compression – A proof-of-concept study. *J. Pharm. Pharmacol*. 69(4): 417–431.

85 Ng, Z.X., Chai, J.W. and Kuppusamy, U.R. (2011). Customized cooking method improves total antioxidant activity in selected vegetables. *Int. J. Food Sci. Nutr*. 62(2): 158–163.

86 Wilmanski, T., Diener, C., Rappaport, N. et al. (2021). Gut microbiome pattern reflects healthy ageing and predicts survival in humans. *Nat. Metabol*. 3(2): 274–286.

87 Abdulla, N.M., Adam, B., Blair, I. et al. (2019). Heavy metal content of herbal health supplement products in Dubai – UAE: A cross-sectional study. *BMC Complement. Altern. Med*. 19(1): 276.

5

Role of Nanotechnology in Refining the Antidiabetic Activities of Plant Derived Bioactives

Sonam Mishra[1], Ravindra Dhar[1], Maitree Suttajit[2], and Bishnu Kumar Pandey[3,]*

[1] *Centre of Materials Science, University of Allahabad, Prayagraj - 211002, India*
[2] *Thai Vegetarian Association, Changpuak, Chiang Mai, Thailand*
[3] *Department of Physics, SPM College, University of Allahabad, Prayagraj - 211002, India*
* *Corresponding author*

5.1 Introduction

Diabetes is a chronic disease that occurs when either the pancreas does not produce enough insulin or when our body cannot effectively use the insulin produced by the pancreas. In general, diabetes mellitus can be classified in many categories including type-1 (T1DM), type-2 (T2DM), gestational diabetes mellitus (GDM), and other specific types [1]. There are several anti-diabetic drugs available in the market that are prescribed by diabetologists according to the situation of patients. However, these pharmaceutical drugs are associated with one or multiple side effects. On the other hand, on maintaining optimal blood glucose in the body there remains the risk of the onset of other side effects including hypoglycemia, lipohypertrophy, hypersensitivity, and renal and gastrointestinal distress [2]. Research on effective drugs for the treatment of diabetes mellitus with minimal or no side effects is still continuing.

Since ancient times, Asian literature and Ayurveda advocate the use of medicinal plants in the treatment of many human diseases. Studies reveal that phyto-bioactive compounds such as polyphenols, flavonoids, curcumins, and quercetin have shown much potential in anti-diabetic activity without complications. Moreover, these bioactive compounds are also very effective in inflammation and other complications associated with diabetes. In addition, several Phytochemicals have been found to suppress postprandial hyperglycemia by interrupting carbohydrate

Natural Products and their Bioactives in Antidiabetic Drug Discovery, First Edition.
Edited by Kanti Bhooshan Pandey, Maitree Suttajit and Pinar Atukeren.

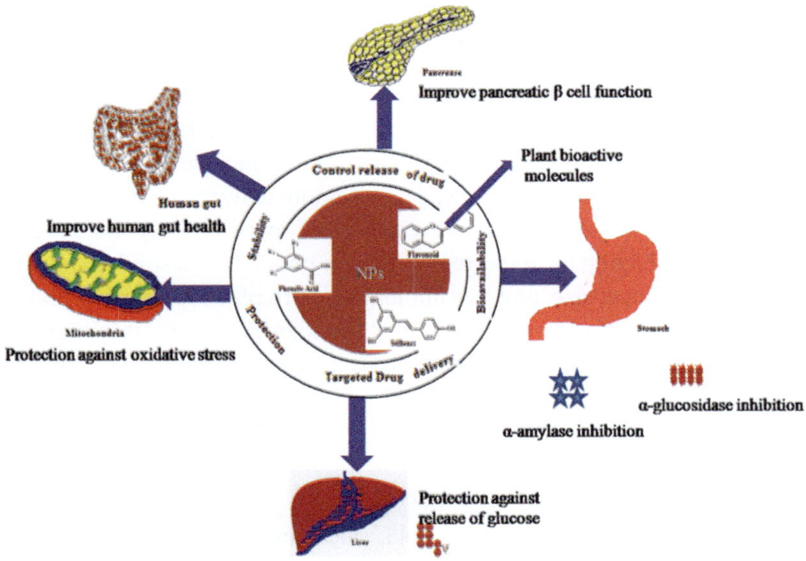

Figure 5.1 The diagram shows the mechanism of action of plant bioactive compounds against anti-diabetic activities.

digestion and retarding glucose absorption through inhibition of intestinal carbohydrate digesting enzymes, such as α-amylase, α-glucosidase, and β-glucosidase, β-cell stimulation, β-cell regeneration, improvement of insulin expression, reversal of insulin resistance, promotion of glucose utilization, and regulation of carbohydrate and lipid metabolism (Figure 5.1) [3].

However, many phytochemicals and food grade bioactive compounds consumed through oral means undergo a substantial loss in bioactivity, and thereby express a compromised anti-diabetic potential. Further, poor water solubility, rapid metabolism, poor bioavailability, and high P-glycoprotein (P-gp) efflux have been found to be accountable for their ineffectiveness [4–6]. Over the years, several pieces of research work have been performed for the development of nano-carriers such as polymeric nanoparticles, liposomes, dendrimers, stimuli-responsive nanoparticles, nanostructured lipid carriers, niosomes, nanomicelles, metallic nanoparticles, and nanofabricated devices, which have been found to produce great success over conventional drug delivery systems in terms of effectiveness, stability, bioavailability, biodistribution, and drug release. In the present chapter, nanotechnology based improvement in anti-diabetic potentials of some plant derived phytochemicals has been discussed.

5.2 Nano-carrier Based Approaches in Anti-diabetic Treatment

Nano-carriers are nanoparticles having particle sizes ranging from 10^{-9}m (i.e. 1 nm) to 10^{-7} m (i.e. 100 nm) with special structure and morphology. The main strategy to design nanoparticles (NPs) is successful delivery of anti-diabetic drugs with the aim to suitably target areas for improved efficacy and safety. However, it is challenging for nano-researchers to obtain the desired nano-carriers with specific shape, size, and morphology that are reproducible and maintain characteristics during *in vivo* and *in vitro* experiments. In this section, various nano-carriers with anti-diabetic bioactives compound delivery are discussed.

5.2.1 Inorganic Nano-carriers

Inorganic nano-carriers including silver, gold, silica, and magnetic nanoparticles have been studied for the purpose of drug delivery and well established results *in vivo* as well as *in vitro* show successful delivery of bioactive compounds for treating various diseases [7]. There are several synthesis mechanisms of nanoparticles including top to bottom and bottom to top approaches. However, the green synthesis method has gained several advantages over other methods. The green synthesis method is environmentally-friendly and free of chemical contamination. Further, during green synthesis, specific biomolecules can be attached on the surface of nanoparticles. Moreover, Ag and Au NPs have shown great potential in clinical purposes due to specific properties such as surface plasmon resonance (SPR), which does not possess liposomes, micelles, dendrimers, etc. Also, Ag, Au, and magnetic NPs have also shown very good bio-compatibility towards cellular metabolism.

Metal NPs are highly stable, hydrophilic, and biocompatible. These NPs allow controlled drug release at specific targeted sites of the body. Besides this, inorganic NPs also protect loaded drugs against enzyme degradation and stomach gastric degradation. Moreover, magnetic NPs offer a great advantage in drug delivery owing to controlled drug release at targeted specific organs inside the body using an external magnetic field.

5.2.2 Polymer Based Nano-carriers

The polymer based nano-carriers with diameters ranging from 10–1000 nm are ideal for delivering plant derived bioactive molecules [8]. Natural polymers such as alienate and chitosan as well as synthetic polymers polyvinyl alcohol, poly ethylene glycol, are widely used in the synthesis of polymer based nanoparticles. Moreover, natural polymer chitosan based nano-carriers are found in most abundant nano-carriers and studied for the management of T2DM. Polymer based nano-carriers can be either nanocapsules or nanospheres depending on the polymer used. Polymer based nanocapsules contain a lipid core surrounded by a polymeric membrane. The bioactive compound can be either attached on the surface membrane or dissolved in the lipid core.

5.2.3 Lipid Based Nano-carriers

Solid lipid NPs are synthesized by fatty acid mono, di, or triglycerides and surrounded by a phospholipids monolayer. They are arranged in a manner such that solid lipids reside in the core whereas phospholipids are in the surface. Further, biocompatible surfactants are used to stabilize these NPs. Before drug delivery, drugs are dispersed in the solid liquid core of solid lipid NPs. Moreover, when these NPs are dispersed in aqueous medium it forms a solid lipid NPs suspension. The main benefits of these NPs are controlled drug release and protection of incorporated drugs against chemical degradation inside the body [9, 10].

5.2.4 Nanoliposomes

Nanoliposomes are synthesized by a phospholipids bilayer and stabilized by sterols. Phosphatidylcholine is a common phospholipid derived from egg yolk, and soybean is used for liposomes' production. Liposomes are usually biocompatible and biodegradable structures used for bioactive compound delivery. The nanoliposomes offer several benefits such as increasing drug solubility bioavailability and specific cell targeting ability [11, 12].

5.2.5 Protein and Polysaccharide Based Nano-carriers

Protein based NPs can be extracted from natural biological sources such as plants, animals, microorganisms, and marine sources [13, 14]. These NPs can be easily functionalized with specific drugs and targeting ligands. Similarly, polysaccharide NPs are formed by monosaccharides linked through *O*-glycosidic bonds and make specific physical chemical properties. However, these NPs degrade at high temperatures and are soluble in water, which limits their application in nanomedicines [14].

5.2.6 Quantum Dots

Quantum dots are specific NPs with diameter in the range of 2–8 nm, and their properties of quantum dots very much depend on their size and shape. The quantum dots can be used for targeted drug delivery, biosensors, and bioimaging. An antibody attached with quantum dots has been used for selective separation of HepG2 and Hepa-2 cellular carcinoma cells in blood samples [15–17]. Quantum dots could also be used in sustained and controlled release of drugs, or bioactive compounds at a specific site.

5.2.7 Micelle Based Nano-carriers

Micelles, shown in Figure 5.2, have been widely studied as nano-carriers for hydrophobic drug delivery applications. These nano-carriers are formed by self-assembled amphiphilic co-polymers via aggregation or agglomeration in aqueous media above the critical micellar concentration (CMC) and form a pore cell micellar structure [18–24]. Usually, the particle size of micelles ranges between 5–50 nm. In nano-carrier micelles the outer layer is hydrophilic, which constitutes the micelle shell, and confers stability and aqueous solubility to the micellar structure. On the other hand, the hydrophobic portion of amphiphilic molecules

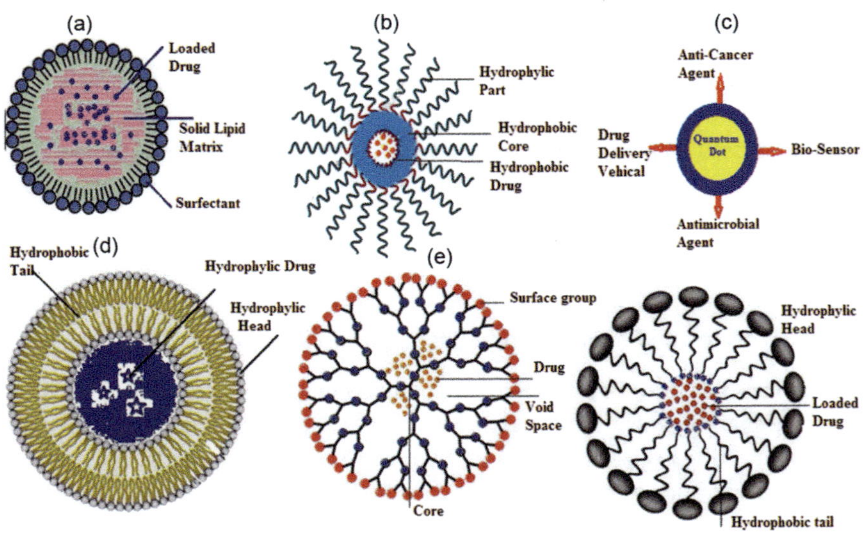

Figure 5.2 Different types of nano-carriers: (a) Solid lipid nanoparticles; (b) polymeric micelles; (c) quantum dot; (d) liposome; (e) dendrimers; (f) micelle.

constitutes the core of the micelle that serves as a reservoir and protects the agent. Figure 5.2 displays various types of nano-carriers.

5.3 Nano-formulated Bioactives for Anti-diabetic Activity

5.3.1 Nano-curcumin

Curcumin is a promising natural compound of turmeric and possesses a variety of properties including anti-oxidant, anti-inflammatory, and anti-bacterial activities. In a recent study, it was shown that curcumin plays a very therapeutic role in managing diabetic complications [25–27]. However, oral delivery of curcumin shows low anti-diabetic potential due to low water solubility and stability in gastrointestinal environments [28–32]. Nano-curcumin offers an excellent opportunity to improve the stability and safety of bioactive compounds with improved bioavailability of curcumin [33–46]. Curcumin-loaded polyglyco acid NPs with mean particle size 281 nm have been studied in a diabetic rat model; the result revealed a higher bioavailability through oral delivery along with an enhanced anti-diabetic effect [47]. Moreover, self-emulsifying curcumin ~213 NPs have shown improved results in treating diabetic neuropathy in Sprague Dawley rats through oral delivery as compared to conventional curcumin [48]. In another study it was found that nano-carrier formulated curcumin has shown 22 times enhanced bioavailability as compared to conventional curcumin [49]. Gauda at al. have shown that nano- curcumin can decrease glucose concentration and increase activity of gene expression in insulin receptors [50]. Several other studies related to nano-curcumin delivery have been performed that confirmed the improved bioavailability and anti-diabetic performance of nano-curcumin.

5.3.2 Nano-silibinin

Silibinin is a natural polyphenolic flavonoid isolated from the seed extract of the herb milk thistle (*Silybum marianum*) and has recently shown great anti-diabetic activity in many animal and cell models [51–55]. However, bioavailability and absorption in the stomach and intestine are relatively low, which limits their broad application [56–58]. Recently, various nano delivery systems have been developed to improve the bioavailability and stability in diabetic animal models. Silibinin-loaded NPs with particle size 230 nm have shown improved bioavailability in systematic circulation with higher restoration of pancreatic cells. These studies confirm higher anti-diabetic activity of silibinin-loaded NPs due to passive transport and restoration of anti-oxidative status [59]. Moreover, silibinin-loaded nanoliposomes showed improved stability and bioactivity in other chronic disease models [60]. These studies show that a nano delivery system with silibinin can improve stability, bioavailability, as well as the therapeutic nature of this bioactive compound to manage diabetes.

5.3.3 Nano-quercetin

Quercetin is a polyphenolic flavonoid found in fruits such as apples, grapes, blueberries, and cherries. Studies have shown that despite its anti-oxidant properties quercetin also exhibits anti-inflammatory, anti-diabetic, and anti-hypertensive activities [61–64]. However, the efficacy of quercetin through an oral delivery system is very low due to gastrointestinal and post-absorptive metabolism [65– 68]. Several approaches have been developed to improve the bioavailability and stability of quercetin in order to improve the bioavailability and reaction mechanism of quercetin in a diabetic cell model. Further, the nano delivery system is a more promising approach in the delivery of quercetin through oral intake to improve anti-diabetic activities. Polyglycolic acid-loaded quercetin NPs of particle size 139 nm have shown improved bioavailability in a diabetic rat model. This study confirmed that quercetin dose and dosing time can be reduced and also the anti-diabetic potential of quercetin can be improved by using a polyglycilic acid nano delivery system. Recently, quercetin nano rods have been developed as nano-carriers, which have shown improved anti-diabetic activities with restored organ function.

5.3.4 Nano-baicalin

Baicalin is a novel anti-diabetic bioactive compound found in some specific plants such as scutellaria. This compound has shown potential bioactivity against T2DM. However, these bioactive compounds are highly hydrophobic, which limits their bioavailability through oral delivery, hence becoming bio-inactive. Meanwhile, many novel approaches have been developed for delivery and improved bioavailability [69, 70].

Baicalin-loaded lipid nano-carriers with particles size 92 nm have shown higher anti-diabetic activity than conventional baicalin. Further, baicalin-loaded lipid nano-carriers also reduced drug doses and duration to improve diabetic conditions [71]. Baicalin-loaded nano-emulsion has shown a seven times increase in bioavailability. Further, baicalin-loaded nano-emulsion stability has also been studied up to six months and the obtained results confirm that nano-emulsion-loaded baicalin improves bioavailability through oral delivery in the treatment of type 2 diabetes [72].

Baicalin-loaded nanoliposomes with an average particle size 375 nm have been studied in the area of bioavailability of targeted organs including lever, kidney, and pancreas. The obtained results suggest improved bioavailability. The novel nano delivery system shows successful delivery of baicalin for further treatment of chronic diseases including T2DM [73].

5.3.5 Nano-naringenin

Naringenin is a bioactive compound found in citrus fruits such as lemons, oranges, and related beverages. Studies have shown that naringenin has high anti-diabetic potential in cellular and animal models [74–80]. However, owing to limited bioavailability and stability against gastric acid, noringenin's anti-diabetic application is limited. Naringenin-loaded nano-emulsion has shown the enhanced bioavailability and stability of naringenin in oral delivery systems. In another study, naringenin-loaded NPs with a mean particle size 66 nm showed higher bioavailability and improved hepatoprotective activity in a rat model. Further, 60–80 nm solid lipid NPs have been prepared to load naringenin, and this system showed a better cellular uptake [81].

5.3.6 Nano-morin

Morin is a bioactive compound derived from bioflavonoids found in many fruits, vegetables, and herbs. Morin has shown great potential activities against diabetes and associated diseases [82–85]. It has been reported that morin works against lipogenesis, oxidative stress, glycogenesis, inflammation, and other complications. Morin also works as a potential hepatoprotective agent, which confirms that it can reduce hyperlipidemia. A natural antibiotic can be also derived from morin [86, 87]. It possesses limited bioavailability and oral solubility, and a higher dose of morin has shown increased toxicity. To overcome these limitations, a morin nano delivery system has been studied that has shown greater potential including stability as well as improved bioavailability through oral delivery. The results also advocated that nano-emulsified morin exhibited improved bioavailability, which may be an additional feature in enhancing the bioactivity of this natural compound.

5.3.7 Nano-emodin

Emodin is found in many plants including rhubarb, Japanese knotweed, and buckthorn. Many studies have shown that emodin is an effective bioactive compound against diabetic associated diseases [88–94]. It plays an active role in the treatment of nephropathy and neuropathy at earlier stages. However, due to higher hydrophobicity, it cannot be delivered orally in an efficient manner, which limits its bioavailability and bioaccessibility. Recently novel nano-carrier based delivery of emodin has made it possible to overcome the hurdles and enhanced the efficacy of this natural bioactive compound [95–101]. Emodin-loaded magnesium hollow silicate nano-carriers have shown efficacy in treating angiogenesis associated diabetic retinopathy. An emodin-loaded nano-transfersome sized 292 nm through oral delivery has shown anti-obesity activity [102], which is an alternate approach to reduce T2DM risk. Another study proposed that nano-emulsion-loaded emodin oral delivery has shown higher bioavailability in a rat model [103].

5.3.8 Nano-genistein

Genistein is an isoflavonoid bioactive compound derived from soy and was recently discovered as a potential anti-oxidant and anti-diabetic molecule [104– 110]. However, low oral delivery of genistein leads to lower solubility and loss of bioavailability, which inhibits its broad application against diabetes and other related chronic diseases. Moreover, a higher dose of genistein leads to cellular toxicity and other complications [111, 112]. In recent years, the nano delivery system has facilitated to overcome the higher dose toxicity, and other complications of genistein administration. Genistein-loaded polymeric micelles, with a mean particle size 27 nm, showed higher bioavailability through oral delivery due to higher solubility of the compound and improved permeability [113].

5.3.9 Nano-hesperidins

Hesperidin is a flavonoid derived from lemons and other citrus food and has shown multiple anti-diabetic activities such as glucose control lipid level along with anti-oxidant action with reduced cellular damage and insulin resistance [114–118]. However, low bioavailability and stability through oral delivery show lower efficacy against diabetes and other chronic diseases [119–121]. Although several nano-carrier delivery techniques have shown improved bioavailability and beneficial anti-diabetic effects, studies have reported that hesperidin-loaded solid-lipid NPs through oral delivery possess improved stability and are useful to treat various chronic diseases through their functional activities [121].

5.3.10 Nano-resveratrol

Resveratrol is a polyphenolic bioactive compound that can be derived from fruit such as grapes and seeds. Resveratrol contains three hydroxyl groups that play an important role in scavenging free radicals [122, 123]. Studies show that resveratrol works as a direct anti-oxidant as well as also regulating anti-oxidant enzymes. Resveratrol works as a gene regulator and stimulates mitochondrial biogenesis; moreover, it works as an anti-diabetic, neuro-protective, as well as an anti-cancerous agent [124–126]. Recent studies have confirmed that nano-encapsulation of resveratrol within lipid nano-carriers, nano-emulsions, micelles, polymeric particles, solid dispersions, as well as nanoparticles, has demonstrated to be beneficial over other approaches to achieve better stability, improved bioavailability, specific targeting, advanced therapeutic efficacy, and improved patient compliance [127, 128]. Resveratrol-assembled Au NPs have shown to be potential therapeutic agents in diabetic retinopathy. Resveratrol-loaded nanoliposomes and their PEG-amalgamated modification have been reported to improve glycemic status, redox status, and insulin levels in glucose or streptozotocin-treated β-TC cells [129].

There are several other bioactive compounds reported in the literature that show potential anti-diabetic activities such as ferulic acid, thymoquinone, pelargonidin, catechins, gallic acid, silybum, flavonolignans, scutellarin, eleostearic acid, glycyrrhizin, asiatic acid, berberine, gymnemic acid, mangiferin, luteolin, myricitrin, and apigenin. However, nano-formulation of these bioactive compounds has not yet been much explored, but intense research activities are in progress. Table 5.1 describes the nano-applied plant derived biomolecules showing improved activities in providing protection against diabetes.

5.4 Conclusion and Future Prospective

Nanotechnology has demonstrated a significant role in diabetic mellitus management. Several nano-carriers have been developed as drug vehicles to release bioactive compounds at particular sites including specific organs inside the body, with improved clinical efficacy and fewer side effects. Nano-carriers formulated with anti-diabetic bioactive molecules have reduced the frequency as well as dose of the drugs. Many bioactive compounds extracted from medicinal plants have been nano-formulated with the help of inorganic nanoparticles such as Au and Ag, and ZnO NPs are being frequently used in the management of T2DM. Moreover, several natural polymeric nanomaterials including chitosan, micelles, as well as synthetic polymer nanomaterials, including liposomes and dendrimers, are also used as nano delivery systems for bioactive compounds including polyphenols, flavonoids, etc. Organic based NPs can be easily modified in a very systematic way with the aim of delivery of plant derived bioactive compounds in a targeted and controlled manner. Consequently, these nano-formulated bioactive molecules have shown greater stability and higher resistance against gastric fluid and enzymes, and therefore more bioavailability. Bioavailability screening of these nano-formulations have been performed *in vitro* using cellular and non-cellular modes, and *in vivo* in animal models. Nano-formulated bioactive molecules have shown improved results including glycemic control, improved insulin level, and regeneration of pancreatic β-cells; however, detailed studies on humans are needed. Intensive research is also required to understand the toxicity related effects after long-time consumption.

Table 5.1 A description of nano-applied plant derived biomolecules showing improved activities in providing protection against diabetes.

S.No.	Bioactive compound	Therapeutic limitations	Nano-formulation	Result	References
1	Curcumin	Low water solubility and stability, poor absorption, rapid metabolism	Curcumin-loaded polyglyco-acid NPs, self emulsifying curcumin, self-nanophospholipid, curcumin-ZnO	Improved bioavailability and anti-diabetic performance such as reduction of blood glucose, improvement in serum insulin, activation of GLUT2, and glucokinase genes in pancreas and liver	[27–35]
2	Silibinin	Low absorption in stomach and intestine, low stability	Nanoliposomes	Enhanced bioavailability in systematic circulation with higher restoration of pancreatic cells	[55–58]
3	Quercetin	Lower gastrointestinal and post-absorptive metabolism	Polyglycolic acid NPs	Improved bioavailability and stability	[63–66]
4	Baicalin	Highly hydrophobic, which limits its bioavailability	Lipid nano-carriers, nano-emulsion, nanoliposomes	Improved bioavailability and anti-diabetic effect	[67–71]
5	Naringenin	Limited oral bioavailability and stability against gastric acid	Nano-emulsion, solid lipid NPs	Higher bioavailability and improved hepatoprotective activity in a rat model	[79]
6	Morin	Limited bioavailability as well as oral solubility	Nano-emulsion	Improved bioavailability	[84]
7	Emodin	Higher hydrophobicity, limited bioavailability and bioaccessibility	Magnesium hollow silicate nano-carriers, nano-transfersomes, nano-emulsion	Higher anti-obesity activity, higher bioavailability	[93–100]

8	Genistein	Lower solubility and loss of bioavailability	Polymeric micelles with mean particle size 27 nm	Higher bioavailability, solubility and improved permeability	[108–111]
9	Hesperidins	Lower bioavailability and stability	Solid-lipid NPs	Improved stability and bioavailability	[117–119]
10	Resveratrol	Lower stability, bioavailability	Nano-carriers, nano-emulsions, micelles, polymeric particles; solid dispersions, Au NPs	Improved bioavailability, specific targeting diabetic retinopathy	[126]
11	Apigenin	Lower stability and bioavailability	Nanomixed micelles, carbon nanopowder, nanoliposomes	Improved oral bioavailability, and gastrointestinal absorption	[128–129]
12	Luteolin	Lower stability and bioavailability	Solid lipid nanoparticles	Enhanced solubility, biological half-life and bioavailability	[130]
13	Mangiferin	Lower stability and bioavailability	Phospholipidicnano-micelles, lactoglobulin nanoparticles	Enhanced intestinal permeability and oral bioavailability	[131–132]
14	Gymnemic Acid	Lower gastrointestinal absorption,	AuNPs, nano-emulsion	Improved anti-hyperglycemic activity	[133]
15	Rosmarinic Acid	Lower stability and bioavailability	Chitosan and solid lipid nanoparticles,	Improved wound healing and anti-oxidant effect	[134, 135]
16	Catechins	Low stability, gastrointestinal absorption, and bioavailability	Chitosan nanoparticles, solid lipid nanoparticles	Enhanced anti-diabetic potential in terms of α-glucosidase and α-amylase inhibitory effects	[136]

References

1 Petersmann, A., Müller-Wieland, D., Müller, U.A. et al. (2019). Definition, classification and diagnosis of diabetes mellitus. *Exp. Clin. Endocrinol. Diabetes.* 127: 1–7.

2 Chaudhury, A., Duvoor, C., Reddy Dendi, V.S. et al. (2017). Clinical review of antidiabetic drugs: Implications for type 2 diabetes mellitus management. *Front. Endocrinol. (Lausanne).* 8: 1–12.

3 Dewanjee, S., Chakraborty, P., Mukherjee, B. et al. (2020). Plant-based antidiabetic nanoformulations: The emerging paradigm for effective therapy. *Int. J. Mol. Sci.* 21(6): 2217.

4 Aryaeian, N., Sedehi, S.K. and Arablou, T. (2017). Polyphenols and their effects on diabetes management: A review. *Med. J. Islam. Repub. Iran.* 31: 134.

5 Mojzer, E.B., Hrnčič, M.K., Škerget, M. et al. (2016). Polyphenols: Extraction methods, antioxidative action, bioavailability and anticarcinogenic effects. *Molecules.* 21: 901.

6 Khurana, R.K., Gaspar, B.L., Welsby, G. et al. (2018). Improving the biopharmaceutical attributes of mangiferin using vitamin E-TPGS co-loaded self-assembled phosholipidicnano-mixed micellar systems. *Drug Deliv. Transl. Res.* 8: 617–632.

7 Shah, F., Hasnain, J., Abdullah, Alam, I. et al. (2022). In vivo analgesic, anti-inflammatory, and anti-diabetic screening of Bacopa monnieri-synthesized copper oxide nanoparticles. *ACS Omega.* 7: 4071–4082.

8 Wickramasinghe, A.S.D., Kalansuriya, P. and Attanayake, A.P. (2022). Nanoformulation of plant-based natural products for type 2 diabetes mellitus: From formulation design to therapeutic applications. *Curr. Therapeutic Res.* 96: 100672.

9 Kamaly, N., Yameen, B., Wu, J. et al. (2016). Degradable controlled-release polymers and polymeric nanoparticles: Mechanisms of controlling drug release. *Chem. Rev.* 116(4): 2602–2663.

10 Paliwal, R., Paliwal, S.R., Kenwat, R. et al. (2020). Solid lipid nanoparticles: A review on recent perspectives and patents. *Expert Opin. Ther. Pat.* 30: 179–194.

11 Li, J., Wang, X., Zhang, T. et al. (2015). A review on phospho-lipids and their main applications in drug delivery systems. *Asian J. Pharm. Sci.* 10: 81–98.

12 Mozafari, M.R. (2010). Nanoliposomes: Preparation and analysis. In: *Liposomes: Methods and Protocols, Volume 1: Pharmaceutical Nanocarriers.* (ed. V. Weissig), vol 65, 29–50. Totowa, NJ: Humana Press

13 Balaji, A.B., Pakalapati, H., Khalid, M. et al. (2017). Natural and synthetic biocompatible and biodegradable polymers. In: *Biodegradable and Biocompatible Polymer Composites: Processing, Properties and Applications. Woodhead Publishing Series in Composites Science and Engineering* (ed. N.G. Shimpi), 3–32. Duxford: Woodhead Publishing.

14 Bassas-Galia, M., Follonier, S., Pusnik, M. et al. (2017). Natural polymers: A source of inspiration. In: *Bioresorbable Polymers for Biomedical Applications*. (ed. G. Perale). 31–64. New York: Elsevier.

15 Fatima, I., Rahdar, A., Sargazi, S. et al. (2021). Quantum dots: Synthesis, antibody conjugation, and HER2-Receptor targeting for breast cancer therapy. *J. Funct. Biomater.* 12(4): 75.

16 Shi, Y., Pramanik, A., Tchounwou, C. et al. (2015). Multifunctional biocompatible graphene oxide quantum dots decorated magnetic nanoplatform for efficient capture and two-photon imaging of rare tumor cells. *ACS Appl. Mater. Interfaces.* 7(20): 10935–10943.

17 Abumrad, N.A., Nassir, F. and Marcus, A. (2016). Digestion and absorption of dietary fat, carbohydrate, and protein-clinical key. In: Sleisenger and Fordtran's Gastrointestinal and Liver Disease. (ed. J.T. DiPiro). 1736–1764. Philadelphia, PA: Saunders Elsevier.

18 Huang, Y. and Li, S. (2014). Polymeric micelles, nanocarriers for cancer-targeted drug delivery. *AAPS PharmSciTech.* 15(4): 862–871.

19 Ahmad, Z., Shah, A., Siddiq, M. et al. (2014). Polymeric micelles as drug delivery vehicles. *RSC Adv.* 4: 17028.

20 Kesharwani, S.S., Kaur, S., Tummala, H. et al. (2018). Overcoming multiple drug resistance in cancer using polymeric micelles. *Expert Opin. Drug Del.* 15(11): 1127–1142.

21 Liao, C., Chen, Y., Yao, Y. et al. (2016). Cross-linked small-molecule micelle-based drug delivery system: Concept, synthesis, and biological evaluation. *Chem. Mater.* 28: 7757–7764.

22 Chou, W.H., Galaz, A., Jara, M.O. et al. (2020). Drug-loaded lipid-core micelles in mucoadhesive films as a novel dosage form for Buccal administration of poorly water-soluble and biological drugs. *Pharmaceutics.* 12(12): 1168.

23 Zhang, L., Gu, F., Chan, J. et al. (2008). Nanoparticles in medicine: Therapeutic applications and developments. *Clin. Pharmacol. Ther.* 83: 761–769.

24 Bonechi, C., Martini, S., Ciani, L. et al. (2012). Using liposomes as carriers for polyphenolic compounds: The case of trans-resveratrol. *PLoS ONE.* 7(8): 41438.

25 Rahmani, A.H., Alsahli, M.A., Aly, S.M. et al. (2018). Role of curcumin in disease prevention and treatment. *Adv. Biomed. Res.* 7: 38.

26 Nasri, H., Sahinfard, N., Rafieian, M. et al. (2014). Turmeric: A spice with multifunctional medicinal properties. *J. Herb. Med. Plarmacol.* 3(1): 5–8.

27 Nabavi, S.F., Thiagarajan, R., Rastrelli, L. et al. (2015). Curcumin: A natural product for diabetes and its complications. *Curr. Top. Med. Chem.* 15(23): 2445–2455.

28 Yuan, X.H., Li, H.Z., Bai, H. et al. (2014). Synthesis of novel curcumin analogues for inhibition of 11 beta-hydroxysteroid dehydrogenase type 1 with anti-diabetic properties. *Eur. J. Med. Chem.* 77: 223–230.

29 Raslan, M.M.,Mohamed, S., Maksoud, M.D. et al. (2018). Role of curcumin-zinc oxide composite nanoparticles on streptozotocin-induced diabetic rats. *J. Biotechnol. Biomater.* 8: 55.

30 Sivabalan, S. and Anuradha, C.V. (2010). A comparative study on the antioxidant and glucose-lowering effects of curcumin and bisdemethoxycurcumin analog through in vitro assays. *Int. J. Pharmacol.* 6(5): 664–669.

31 Nahar, P.P., Slitt, A.L. and Seeram, N.P. (2015). Anti-inflammatory effects of novel standardized solid lipid curcumin formulations. *J. Med. Food.* 18(7): 786–792.

32 Prasad, S., Tyagi, A.K. and Aggarwal, B.B. (2014). Recent developments in delivery, bioavailability, absorption and metabolism of curcumin: The golden pigment from golden spice. *Cancer Res. Treat.* 46(1): 2–18.

33 Ghosh, D., Choudhury, S.T., Ghosh, S. et al. (2012). Nanocapsulated curcumin: Oral chemopreventive formulation against diethylnitrosamine induced hepatocellular carcinoma in rat. *Chem. Biol. Interact.* 195(3): 206–214.

34 Hani, U. and Shivakumar, H.G. (2014). Solubility enhancement and delivery systems of curcumin a herbal medicine: A review. *Curr. Drug Deliv.* 11(6): 792–804.

35 Huang, Y., Zhao, Y.L., Liu, F. and Liu, S.Q. (2015). Nano traditional Chinese medicine: Current progresses and future challenges. *Curr. Drug Targets.* 16(13): 1548–1562.

36 Nguyen, M.H., Yu, H., Kiew, T.Y. et al. (2015). Cost-effective alternative to nano-encapsulation: Amorphous curcumin-chitosan nanoparticles complex exhibiting high payload and supersaturation generation. *Eur. J. Pharm. Biopharm.* 96: 1–10.

37 Beloqui, A., Coco, R., Memvanga, P.B. et al. (2014). pH-sensitive nanoparticles for colonic delivery of curcumin in inflammatory bowel disease. *Int. J. Pharmaceut.* 473(1–2): 203–212.

38 Kumar, A., Ahuja, A., Ali, J. et al. (2012). Curcumin loaded nano globules for solubility enhancement: Preparation, characterization and ex vivo release study. *J. Nanosci. Nanotechnol.* 12(11): 8293–8302.

39 Sharma, M., Sharma, S., Sharma, V. et al. (2016). Design of folic acid conjugated chitosan nano-cur-bioenhancers to attenuate the hormone-refractory metastatic prostate carcinoma by augmenting oral bioavailability. *RSC Adv.* 6(30): 25137–25148.

40 Kumar, A., Ahuja, A., Ali, J. et al. (2016). Curcumin-loaded lipid nanocarrier for improving bioavailability, stability and cytotoxicity against malignant glioma cells. *Drug Deliv.* 23(1): 214–229.

41 Zhang, Z.P., Zhang, R.J., Zou, L.Q. et al. (2016). Encapsulation of curcumin in polysaccharide-based hydrogel beads: Impact of bead type on lipid digestion and curcumin bioaccessibility. *Food Hydrocolloid.* 58: 160–170.

42 Ramalingam, P., Yoo, S.W. and Ko, Y.T. (2016). Nanodelivery systems based on mucoadhesive polymer coated solid lipid nanoparticles to improve the oral intake of food curcumin. *Food Res. Int.* 84: 113–119.

43 Molina-Bolivar, J.A. and Galisteo-Gonzalez, F. (2015). Olive-oil nanocapsules stabilized by HSA: Influence of processing variables on particle properties. *J. Nanopart. Res.* 17: 391.

44 Ramalingam, P. and Ko, Y.T. (2015). Enhanced oral delivery of curcumin from N-trimethyl chitosan surface-modified solid lipid nanoparticles: Pharmacokinetic and brain distribution evaluations. *Pharm. Res.* 32(2): 389–402.

45 Grama, C.N., Suryanarayana, P., Patil, M.A. et al. (2013). Efficacy of biodegradable curcumin nanoparticles in delaying cataract in diabetic rat model. *PLoS One*. 8(10): 78217.

46 Joshi, R.P., Negi, G., Kumar, A. et al. (2013). SNEDDS curcumin formulation leads to enhanced protection from pain and functional deficits associated with diabetic neuropathy: An insight into its mechanism for neuroprotection. *Nanomedicine*. 9(6): 776–785.

47 Tsai, Y.M., Jan, W.C., Chien, C.F. et al. (2011). Optimised nano-formulation on the bioavailability of hydrophobic polyphenol,curcumin, in freely-moving rats. *Food Chem.* 127(3): 918–925.

48 Weaam, G., Naglaa, A., Hafiz Lamiaa, M. et al. (2019). Effects of nano-curcumin on gene expression of insulin and insulin receptor. *Bull. Natl. Res. Cent.* 43: 128.

49 Tong, W.W., Zhang, C., Hong, T. et al. (2018). Silibinin alleviates inflammation and induces apoptosis in human rheumatoid arthritis fibroblast-like synoviocytes and has a therapeutic effect on arthritis in rats. *Sci. Rep.* 8: 3241.

50 Fallahzadeh, M.K., Dormanesh, B., Sagheb, M.M. et al. (2012). Effect of addition of silymarin to renin-angiotensin system on proteinuria in type 2 diabetic patients with overt nephropathy: A randomized, double-blind, placebo-controlled trial. *Am. J. Kidney Dis.* 60(6): 896–903.

51 Khazim, K., Gorin, Y., Cavaglieri, R.C. et al. (2013). The antioxidant silybin prevents high glucose-induced oxidative stress and podocyte injury in vitro and in vivo. *Am. J. Physiol. Renal Physiol.* 305(5): 691–700.

52 Bouderba, S. (2013). Protective effect of silibinin against liver oxidative stress in Psammomysobesus, model of insulin resistance and type- 2 diabetes. *Fund. Clin. Pharmacol.* 27: 107–108.

53 Bouderba, S., Sanchez-Martin, C., Villanueva, G.R. et al. (2014). Beneficial effects of silibinin against the progression of metabolicsyndrome, increased oxidative stress, and liver steatosis in Psammomysobesus, a relevant animal model of human obesity and diabetes. *J. Diabetes*. 6(2): 184–192.

54 Lambert, K., Hokayem, M., Thomas, C. et al. (2018). Combination of nutritional polyphenols supplementation with exercise training counteracts insulin resistance and improves endurance in high-fat diet-induced obese rats. *Sci. Rep.* 8: 2885.

55 Cheng, B.A., Gong, H., Li, X.C. et al. (2012). Silibinin inhibits the toxic aggregation of human islet amyloid polypeptide. *Biochem. Bioph. Res. Commun.* 419(3): 495–499.

56 Haddad, Y., Vallerand, D., Brault, A. et al. (2011). Antioxidant and hepatoprotective effects of silibinin in a rat model of nonalcoholic steatohepatitis. *Evid. Based Complement. Alternat. Med.* 2011: 1–10.

57 Milic, N., Milosevic, N., Suvajdzic, L. et al. (2013). New therapeutic potentials of milk thistle (Silybum marianum). *Nat. Prod. Commun.* 8(12): 1801–1810.

58 Das, S., Roy, P., Pal, R. et al. (2014). Engineered silybin nanoparticles educe efficient control in experimental diabetes. *PLoS One.* 9(7): e101818.

59 Borges, G., Degeneve, A., Mullen, W. et al. (2010). Identification of flavonoid and phenolic antioxidants in black currants, blueberries, raspberries, red currants, and cranberries. *J. Agric. Food Chem.* 58: 3901–3909.

60 Di Lorenzo, C., Colombo, F., Biella, S. et al. (2021). Polyphenols and human health: The role of bioavailability. *Nutrients.* 13: 273.

61 Pocernich, C.B., Lange, M.L., Sultana, R. et al. (2011). Nutritional approaches to modulate oxidative stress in Alzheimer's disease. *Curr. Alzheimer Res.* 8: 452–469.

62 Anand David, A.V., Arulmoli, R. and Parasuraman, S. (2016). Overviews of biological importance of quercetin: A bioactive flavonoid. *Pharmacogn. Rev.* 10: 84–89.

63 Zhang, Y.J., Gan, R.Y., Li, S. et al. (2015). Antioxidant phytochemicals for the prevention and treatment of chronic diseases. *Molecules.* 20(12): 21138–21156.

64 Clark, J.L., Zahradka, P. and Taylor, C.G. (2015). Efficacy of flavonoids in the management of high blood pressure. *Nutr. Rev.* 73(12): 799–822.

65 Alam, M.M., Abdullah, K., Singh, B.R. et al. (2016). Ameliorative effect of quercetin nanorods on diabetic mice: Mechanistic and therapeutic strategies. *RSC Adv.* 6(60): 55092–55103.

66 Shao, Y., Yu, Y., Li, C. et al. (2016). Synergistic effect of quercetin and 6-gingerol treatment in streptozotocin induced type 2 diabetic rats and poloxamer P-407 induced hyperlipidemia. *RSC Adv.* 6(15): 12235–12242.

67 Yi, Q., He, X.E., Luo, K.F. et al. Protection of long-term treatment with huang-lian-jie-du-tang on vascular endothelium in rats with type 2 diabetes mellitus. *Curr. Ther. Res. Clin. E.* 73(6): 174–185.

68 Xie, W. and Du, L. (2011). Diabetes is an inflammatory disease: Evidence from traditional Chinese medicines. *Diabetes Obes. Metab.* 13(4): 289–301.

69 Shi, F., Wei, Z., Zhao, Y.et al. (2016). Nanostructured lipid carriers loaded with baicalin: An efficient carrier for enhanced antidiabeticeffects. *Pharmacogn. Mag.* 12(47): 198–202.

70 Dinda, B., SilSarma, I., Dinda, M. et al. (2015). Oroxylumindicum (L.) Kurz an important Asian traditional medicine: From traditional usesto scientific data for its commercial exploitation. *J. Ethnopharmacol.* 161: 255–278.

71 Wei, Y., Guo, J., Zheng, X. et al. (2014). Preparation, pharmacokinetics and biodistribution of baicalin-loaded liposomes. *Int. J. Nanomedicine.* 9: 3623–3630.

72 Al-Rejaie, S.S., Aleisa, A.M., Abuohashish, H.M. et al. (2015). Naringenin neutralizes oxidative stress and nerve growth factor discrepancy in experimental diabetic neuropathy. *Neurol. Res.* 37(10): 924–933.

73 Nie, X., Chen, Z., Pang, L. et al. (2020). Oral nano drug delivery systems for the treatment of type 2 diabetes mellitus: An available administration strategy for antidiabetic phytocompounds. *Int. J. Nanomed.* 15: 10215–10240.

74 Bhattacharya, S., Oksbjerg, N., Young, J.F. et al. (2014). Caffeic acid, naringenin and quercetin enhance glucose-stimulated insulin secretion and glucose sensitivity in INS-1E cells. *Diabetes Obes. Metab.* 16(7): 602–612.

75 Hasanein, P. and Fazeli, F. (2014). Role of naringenin in protection against diabetic hyperalgesia and tactile allodynia in male Wistar rats. *J. Physiol. Biochem.* 70(4): 997–1006.

76 Keshari, A.K., Kumar, G., Kushwaha, P.S. et al. (2016). Isolated flavonoids from Ficusracemosa stem bark possess antidiabetic, hypolipidemic andprotective effects in albino Wistar rats. *J. Ethnopharmacol.* 181: 252–262.

77 Koch, C.E., Ganjam, G.K., Steger, J. et al. (2013). The dietary flavonoids naringenin and quercetin acutely impair glucose metabolism in rodents possibly via inhibition of hypothalamic insulin signalling. *Br. J. Nutr.* 109(6): 1040–1051.

78 Priscilla, D.H., Roy, D., Suresh, A. et al. (2014). Naringenin inhibits alpha-glucosidase activity: A promising strategy for the regulation of postprandial hyperglycemia in high fat diet fed streptozotocin induced diabetic rats. *Chem. Biol. Interact.* 210: 77–85.

79 Zeng, B., Chen, K., Du, P. et al. (2016). Phenolic compounds from clinopodiumchinense (Benth.) O. Kuntze and their inhibitory effects on alpha-glucosidase and vascular endothelial cells injury. *Chem. Biodivers.* 13(5): 596–601.

80 Ji, P., Yu, T., Liu, Y. et al. (2016). Naringenin-loaded solid lipid nanoparticles: Preparation, controlled delivery, cellular uptake, and pulmonary pharmacokinetics. *Drug Des. Dev. Ther.* 10: 911–925.

81 Abuohashish, H.M.,, Al-Rejaie S.S., Al-Hosaini, K.A. et al. (2013). Alleviating effects of morin against experimentally induceddiabetic osteopenia. *Diabetol. Metab. Syndr.* 5(1): 5.

82 AlSharari, S.D., Al-Rejaie, S.S., Abuohashish, H.M. et al. (2014). Ameliorative potential of morin in streptozotocininduced neuropathic pain in rats. *Trop. J. Pharm. Res.* 13(9): 1429–1436.

83 Komirishetty, P., Areti, A., Sistla, R. et al. (2016). Morin mitigates chronic constriction injury (CCI)-induced peripheral neuropathy by inhibiting oxidative stress induced PARP over-activation and neuroinflammation. *Neurochem. Res.* 41(8): 2029–2042.

84 Naowaboot, J., Wannasiri, S. and Pannangpetch, P. (2016). Morin attenuates hepatic insulin resistance in high-fat-diet-induced obese mice. *J. Physiol. Biochem.* 72(2): 269–280.

85 Ikeuchi-Takahashi, Y., Murata, S., Murata, W. et al. (2020). Development of Morin-loaded nanoemulsions containing various polymers; role of polymers in formulation properties and bioavailability. *AAPS Pharm. Sci. Tech.* 21: 150.

86 Paoli, P., Cirri, P., Caselli, A. et al. (2013). The insulin-mimetic effect of morin: A promising molecule in diabetes treatment. *Biochim. Biophys. Acta*. 1830(4): 3102–3111.

87 Lu, Y., Yang, J.H., Li, X. et al. (2011). Emodin, a naturally occurring anthraquinone derivative, suppresses IgE-mediated anaphylactic reaction and mast cell activation. *Biochem. Pharmacol.* 82(11): 1700–1708.

88 Cao, Y.N., Chang, S.F., Dong, J. et al. (2016). Emodin ameliorates high-fat-diet induced insulin resistance in rats by reducing lipid accumulation in skeletal muscle. *Eur. J. Pharmacol.* 780: 194–201.

89 Xue, J.F., Ding, W.J. and Liu, Y. (2010). Anti-diabetic effects of emodin involved in the activation of PPAR gamma on high-fat diet-fed and low dose of streptozotocin - induced diabetic mice. *Fitoterapia*. 81(3): 173–177.

90 Arvindekar, A., More, T., Payghan, P.V. et al. (2015). Evaluation of anti-diabetic and alpha glucosidase inhibitory action of anthraquinones from Rheum emodi. *Food Funct.* 6(8): 2693–2700.

91 Bae, U.J., Song, M.Y., Jang, H.Y. et al. (2015). Emodin isolated from Rheum palmatum prevents cytokine-induced beta-cell damage and the development of type-1 diabetes. *J. Funct. Foods*. 16: 9–19.

92 El Sayed, A.M., Ezzat, S.M., El Naggar, M.M. et al. (2016). In vivo diabetic wound healing effect and HPLC-DAD-ESI-MS/MS profiling of the methanol extracts of eight aloe species. *Rev. Bras. Farmacogn.* 26(3): 352–362.

93 Abu Eid, S., Adams, M., Scherer, T. et al. (2015). Emodin, a putative anti-diabetic active compound, impaired glucose tolerance in an obese mouse. *Wien Klin. Wochenschr.* 127(17–18): 735–735.

94 Lin, L.F., Lin, H.M., Yin, X.B. et al. (2015). Characterization of the constituents in rat plasma after oral administration of radix polygonimultiflori extracts by ultra-performance liquid chromatography/quadrupole time-of-flight mass spectrometry. *Biomed. Chromatogr.* 29(10): 1541–1547.

95 Zhang, L., Chang, J.H., Zhang, B.Q. et al. The pharmacokinetic study on the mechanism of toxicity attenuation of rhubarb total free anthraquinone oral colon-specific drug delivery system. *Fitoterapia*. 104: 86–96.

96 Wang, S.P., Chen, T.K., Chen, R.E. et al. (2012). Emodin loaded solid lipid nanoparticles: Preparation, characterization and antitumor activity studies. *Int. J. Pharmaceut.* 430(1–2): 238–246.

97 Zhang, L., Ma, W.F., Li, J, et al. (2013). Influence of processing on pharmacokinetic of typical constituents in radix polygonimultiflori after oral administration by LC-ESI-MS/MS. *J. Ethnopharmacol.* 148(1): 246–253.

98 Chang, K.C., Li, L.F., Sanborn, T.M. et al. (2016). Characterization of emodin as a therapeutic agent for diabetic cataract. *J. Nat. Prod.* 79(5): 1439–1444.

99 Cai, J.Z, Chen, S.H., Zhang, Q.W. et al. (2013). Determination of emodin in rat plasma by gradient elution LC-ESI-MS and its application to pharmacokinetics. *Lat. Am. J. Pharm.* 32(2): 269–274.

100 Zhang, Y.X., Li, J.S., Peng, W.W. et al. (2013). Comparative pharmacokinetics of aloe-emodin, rhein and emodin determined by liquid chromatographymass spectrometry after oral administration of a rhubarb peony decoction and rhubarb extract to rats. *Pharmazie*. 68(5): 333–339.

101 Lu, K., Xie, S., Han, S. et al. (2014). Preparation of a nano emodin transfersome and study on its anti-obesity mechanism in adipose tissue of diet induced obese rats. *J. Transl. Med.* 12: 72.

102 Shi, Y.B., Li, H.L., Li, J.C. et al. (2015). Development, optimization and evaluation of emodin loaded nanoemulsion prepared by ultrasonic emulsification. *J. Drug Deliv. Sci. Tec.* 27: 46–55.

103 Fu, Z., Zhang, W., Zhen, W. et al. (2010). Genistein induces pancreatic beta-cell proliferation through activation of multiple signaling pathways and prevents insulin-deficient diabetes in mice. *Endocrinology*. 151(7): 3026–3037.

104 Getek, M., Czech, N., Muc-Wierzgon, M. et al. (2014). The active role of leguminous plant components in type 2 diabetes. *Evid. Based Complement. Alternat. Med.* 2014: 293961.

105 Gilbert, E.R. and Liu, D. (2013). Anti-diabetic functions of soy isoflavone genistein: Mechanisms underlying its effects on pancreatic beta-cell function. *Food Funct.* 4(2): 200–212.

106 Kwon, D.Y., Hong, S.M., Ahn, I.S. et al. (2011). Isoflavonoids and peptides from meju, long-term fermented soybeans, increase insulin sensitivity and exert insulinotropic effects in vitro. *Nutrition*. 27(2): 244–252.

107 Mezei, O., Banz, W.J., Steger, R.W. et al. (2003). Soy isoflavones exert antidiabetic and hypolipidemic effects through the PPAR pathways in obese Zucker rats and murine RAW 264.7 cells. *J. Nutr.* 133(5): 1238–1243.

108 Rauter, A.P., Martins, A., Borges, C. et al. (2010). Antihyperglycaemic and protective effects of flavonoids on streptozotocin-induced diabetic rats. *Phytother. Res.* 24(Suppl. 2): S133–S138.

109 Saha, S., Sadhukhan, P. and Sil, P.C. (2014). Genistein: A phytoestrogen with multifaceted therapeutic properties. *Mini Rev. Med. Chem.* 14(11): 920–940.

110 Yang, Z., Kulkarni, K., Zhu, W. et al. (2012). Bioavailability and pharmacokinetics of genistein: Mechanistic studies on its ADME. *Anticancer Agents Med. Chem.* 12(10): 1264–1280.

111 Cimafranca, M.A., Davila, J., Ekman, G.C. et al. (2010). Acute and chronic effects of oral genistein administration in neonatal mice. *Biol. Reprod.* 83(1): 114–121.

112 Kwon, S.H., Kim, S.Y., Ha, K.W. et al. (2007). Pharmaceutical evaluation of genistein loaded pluronic micelles for oral delivery. *Arch. Pharm. Res.* 30(9): 1138–1143.

113 El-Marasy, S.A., Abdallah, H.M.I., El-Shenawy, S.M. et al. (2014). Anti-depressant effect of hesperidin indiabetic rats. *Can. J. Physiol. Pharm.* 92(11): 945–952.

114 Hernandez-Saavedra, D., Perez-Ramirez, I.F., Ramos-Gomez, M. et al. (2016). Phytochemical characterization and effect of Calendula officinalis, Hypericumperforatum, and Salvia officinalis infusions on obesity-associated cardiovascular risk. *Med. Chem. Res.* 25(1): 163–172.

115 Roohbakhsh, A., Parhiz, H., Soltani, F. et al. (2015). Molecular mechanisms behind the biological effects of hesperidin and hesperetin for the prevention of cancer and cardiovascular diseases. *Life Sci.* 124: 64–74.

116 Sharma, M., Akhtar, N., Sambhav, K. et al. (2015). Emerging potential of citrus flavanones as an antioxidant in diabetes and its complications. *Curr. Top. Med. Chem.* 15(2): 187–195.

117 Shi, X.P., Liao, S., Mi, H.J. et al. (2012). Hesperidin prevents retinal and plasma abnormalities in streptozotocin-induced diabetic rats. *Molecules.* 17(11): 12868–12881.

118 Actis-Goretta, L., Dew, T.P., Leveques, A. et al. (2015). Gastrointestinal absorption and metabolism of hesperetin-7-O-rutinoside and hesperetin-7-O-glucoside in healthy humans. *Mol. Nutr. Food Res.* 59(9): 1651–1662.

119 Escudero-Lopez, B., Calani, L., Fernandez-Pachon, M.S. et al. (2014). Absorption, metabolism, and excretion of fermented orange juice polyphenolsin rats. *Biofactors.* 40(3): 327–335.

120 Shete, G., Pawar, Y.B., Thanki, K. et al. (2015). Oral bioavailability and pharmacodynamic activity of hesperetin nanocrystals generated using a novel bottom-up technology. *Mol. Pharm.* 12(4): 1158–1170.

121 Salehi, B., Mishra, A.P., Nigam, M. et al. (2018). Resveratrol: A double-edged sword in health benefits. *Biomedicines.* 6: E91.

122 Gülçin, I. (2010). Antioxidant properties of resveratrol: A structure–activity insight. *Innov. Food Sci. Emerg. Technol.* 11: 210–218.

123 Wahl, D., Bernier, M., Simpson, S.J. et al. (2018). Future directions of resveratrol research. *Nutr. Healthy Aging.* 4: 287–290.

124 Berman, A.Y., Motechin, R.A., Wiesenfeld, M.Y. et al. (2017). The therapeutic potential of resveratrol: A review of clinical trials. *NPJ Precision Oncol.* 1: 35.

125 Thadhani, V.M. (2019). *Resveratrol in Management of Diabetes and Obesity: Clinical Applications, Bioavailability, and Nanotherapy. In Resveratrol-Adding Life to Years, Not Adding Years to Life.* 139–156. Rijeka, Croatia: Intech Open.

126 Chimento, A., DeAmicis, F., Sirianni, R. et al. (2019). Progress to improve oral bioavailability and beneficial effects of resveratrol. *Int. J. Mol. Sci.* 20: 1381.

127 Yücel, Ç., Karatoprak, G.S. and Akta, S.Y. (2018). Nanoliposomal resveratrol as a novel approach to treatment of diabetes mellitus. *J. Nanosci. Nanotechnol.* 18: 3856–3864.

128 Zhang, Z., Cui, C., Wei, F. et al. (2017). Improved solubility and oral bioavailability of apigenin via soluplus/pluronicF127 binary mixed micelles system. *Drug Dev. Ind. Pharm.* 43: 1276–1282.

129 Ding, S.M., Zhang, Z.H., Song, J. et al. (2014). Enhanced bioavailability of apigenin via preparation of a carbon nanopowder solid dispersion. *Int. J. Nanomed.* 9: 2327–2333.

130 Dang, H., Meng, M.H.W., Zhao, H. et al. (2014). Luteolin-loaded solid lipid nanoparticles synthesis, characterization, & improvement of bioavailability, pharmacokinetics in vitro and vivo studies. *J. Nanopart. Res.* 16: 2347.

131 Samadarsi, R. and Dutta, D. (2019). Design and characterization of mangiferin nanoparticles for oral delivery. *J. Food Eng.* 247: 80–94.

132 Ravichandran. R. (2014). Studies on gymnemic acids nanoparticulate formulations against diabetes mellitus. In: *Nanotechnology: Concepts, Methodologies, Tools, and Applications.* (ed. MP. Shah). 1276–1288. Hershey, PA, USA: IGI Global.

133 Madureira, A.R., Nunes, S., Campos, D.A. et al. (2016). Safety profile of solid lipid nanoparticles loaded with rosmarinic acid for oral use: In vitro and animal approaches. *Int. J. Nanomed.* 11: 3621–3640.

134 Grote, C.W. and Wright, D.E. (2016). A role for insulin in diabetic neuropathy. *Front. Neurosci.* 10: 581.

135 Liu, J., Lu, J.F., Kan, J. et al. (2014). Synthesis, characterization and in vitro anti-diabetic activity of catechin grafted inulin. *Int. J. Biol. Macromol.* 64: 76–83.

136 Zhu, W. and Zhang, Z. (2014). Preparation and characterization of catechin-grafted chitosan with antioxidant and anti-diabetic potential. *Int. J. Biol. Macromol.* 70: 150–155.

Part II

Bioactive Compounds Against Type 1 Diabetes Mellitus

6

Epidemiology and Genetics of Type 1 Diabetes Mellitus

The Effect of the Mediterranean Diet

*Bilge Nur Col and Muveddet Emel Alphan**

Department of Nutrition and Dietetics, Faculty of Health Science, Istanbul Atlas University, Istanbul, Turkey
* *Coresponding author*

6.1 Introduction

Diabetes mellitus (DM) is stated as one of the four noncommunicable diseases prioritized by the World Health Organization (WHO) [1]. It is one of the most important public health problems, which constitutes a serious global burden in terms of society's health and socio-economics, and whose prevalence is increasing in the world. The etiology of DM is associated with multifactorial complex interactions of genetic and environmental factors, behavioral, microbial, as well as immune systems varying between individuals [2]. The general classification of DM is gathered under four headings according to the basic criteria of the American Diabetes Association (ADA) updated in 2022: type 1 diabetes (T1DM; due to autoimmune β-cell destruction), type 2 diabetes (T2DM; caused by the progressive loss of insulin secretion produced by adequate β-cells against the background of insulin resistance), gestational diabetes mellitus (GDM; diabetes diagnosed in the second or third trimester of pregnancy that was not clearly overt diabetes prior to gestation), and some other specific types of diabetes due to other causes, e.g. monogenic diabetes syndromes (such as neonatal diabetes and maturity-onset diabetes of the young), diseases of the exocrine pancreas (such as cystic fibrosis and pancreatitis), and drug- or chemical-induced diabetes (such as with glucocorticoid use, in the treatment of HIV/AIDS, or after organ transplantation) [3].

Natural Products and their Bioactives in Antidiabetic Drug Discovery, First Edition.
Edited by Kanti Bhooshan Pandey, Maitree Suttajit and Pinar Atukeren.

6.2 T1DM

T1DM is a chronic disease caused by the destruction of the insulin-producing β-cells of the pancreas. The destruction of β-cells physiologically leads to absolute insulin deficiency. For this reason, exogenous insulin therapy is required for T1DM patients to continue their lives. In addition, monitoring the plasma glucose levels of patients is recommended. T1DM is usually diagnosed on the basis of clinical catabolic symptoms such as polyuria, polydipsia, weight loss, and marked hyperglycemia unresponsive to oral agents, suggestive of traditional insulin deficiency [4].

The most important task of insulin is to meet the energy needs of cells and to store energy sources in the cell. Its secretion occurs with the control of hormonal, neuronal, and substrate-related mechanisms following the ingestion of nutrients. In order to ensure normal metabolic control, insulin should show a normal release pattern in fasting and satiety states. Hyperglycemia in T1DM results from the inability of fat and muscle tissues to use or store glucose for their energy needs due to insulinopenia that develops due to the ongoing loss of insulin production from the β-cells of the pancreas. In cases with insulinopenia, glycogenolysis and gluconeogenesis increase, causing an increase in fasting blood sugars [5]. T1DM is etiologically classified under two headings [6].

1) Autoimmune (due to immune mechanisms) T1DM
2) Idiopathic T1DM

Differences in the initial onset of T1DM are observed in both youth and adults (including latent autoimmune diabetes in adulthood). Children are usually referred to health institutions acutely with symptoms of severe polyuria, polydipsia, and ketonemia [4]. In adults, T1DM appears with a more gradual onset with a clinical picture compatible with T2DM. Differentiating between T1DM and T2DM is very difficult at this stage. Traditionally, progressive β-cell destruction has been the hallmark of T1DM. However, regardless of whether the initial diagnosis was made in childhood or adulthood, it can be detected by residual C-peptide (an indicator required for insulin secretion) 40 years after the initial diagnosis. The definition of adult latent autoimmune diabetes (LADA), which is classified as a subgroup of T1DM in the current guidelines, is those who are positive for glutamic acid decarboxylase (GAD) or other islet autoantibodies (ICA, IA-2, IAA, ZnT8, etc.), although clinically classified as T2DM [7]. It was brought to the literature in 1995 to describe these cases. In 80% of these cases, insulin needs emerge within five years of diagnosis. These individuals are known to actually have the phenotype of T1DM. The main differences from classical T1DM are that insulin requirement occurs relatively late and diabetes starts at a slightly older age. Terms such as "type 1.5 diabetes", "autoimmune diabetes that does not require insulin", and "slow-progress T1DM", are also used to describe the forms of diabetes that fall between T1DM and T2DM. The definition of LADA includes a heterogeneous clinical picture [7, 8]. In its form, which is close to T1DM with its phenotypic features, there is a high titer of anti-GAD antibodies in the serum. In a study conducted in Australia, five characteristic features of LADA were determined. These features are: it is explained as the onset of diabetes before the age of 50, acute symptoms at diagnosis, BMI $<25\,kg/m^2$, and a personal and family history of autoimmune disease [9]. In this study, the authors demonstrated that the presence of at least two clinical features at diagnosis has 90% sensitivity and 71% specificity for detecting LADA, whereas a negative predictive value of 99% was detected for a LADA clinical risk score <2. Taken together, these data suggest that a systematic application of clinical risk score increases the efficiency of a screening programme for LADA [10]. No general recommendations on islet antibodies testing for adult-onset diabetes are available right now. Currently, these assays are performed by physicians only if a strong suspicion of LADA exists, usually on the basis of a normal or low BMI. From this perspective, normal-weight individuals are considered as potentially having LADA and may undergo immunological assays, whereas overweight and obese adults are supposed to have T2DM and are not investigated. However, this confuses the diagnosis of overweight/obese people with T1DM with diabetes that is not initially causing frailty (but whose course is T1DM or other autoimmune diseases), and leads to clinical prediction of weakly debilitating T1DM. A clinical study by Klingensmith et al. found evidence of islet autoimmunity in 10% of young patients aged 10–17 years with phenotypic T2DM. This suggests a possible diagnosis of T1DM [10, 11].

Thus, although leaner individuals are more likely to be diagnosed with T1DM, the potential for T1DM may be present in those who phenotypically appear to have T2DM. If hyperglycemia persists after treatment of newly diagnosed T2DM patients with non-insulin agents, this suggests T1DM. Islet antibody analysis (ICA) is recommended for definitive diagnosis and discrimination [12].

6.2.1 Autoimmune (Dependent on Immune Mechanisms) Diabetes

This type of diabetes, which was called "insulin-dependent diabetes" or "juvenile-onset diabetes" in previous years, constitutes 5–10% of all diabetics. Insulitis is observed in patients as a result of abnormal activation of the immune system mediated by T and B cells. The decrease in insulin secretion in T1DM of autoimmune origin occurs by two mechanisms. The first of these is the destruction of the β-cells of the pancreas, while the other mechanism is the reduction of insulin secretion from the β-cells of the pancreas by cytokines in the environment. Cell-mediated autoimmune destruction of pancreatic β-cells is mainly involved (Figure 6.1) [8].

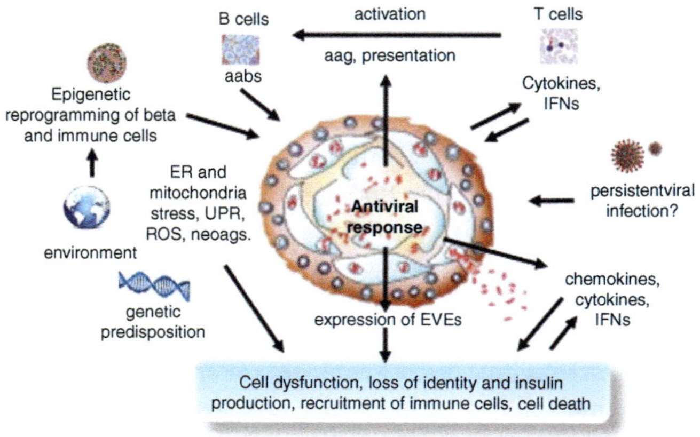

Figure 6.1 Simplified scheme of the pathogenic influences on β-cells in T1DM *Source:* [8] (Ana Zajec et al. 2022/MDPI/CC BY 4.0). aabs: autoantibodies; aag: autoantigen; ER: endoplasmic reticulum; EVEs: endogenous viral elements; IFNs: interferons; neoags: neoantigens; ROS: reactive oxygen species; UPR: unfolded protein response.

Islet cell antigen, GAD 65, insulin, and islet cell-associated phosphatase 2 (IA-2) are well-defined pancreatic antigens. Antibodies produced against some or all of these antigens are usually positive when fasting hyperglycemia is first detected in 85–90% of T1DM patients. In the diagnosis of immune-induced T1DM; detection of anti GAD antibody, islet cell antibody, anti insulin antibody (IAA), protein tyrosine phosphatase antibodies (ICA512 or IA2A), and antibodies against zinc carrier protein (ZnT8) is important [13].

It is also known that this disease is associated with strong human leukocyte antigens (human leukocyte antigen: HLA tissue group) and is associated with DQA and DQB genes. Some HLA-DR/DQ alleles increase the likelihood of the disease, while others may be protective. The rate of β-cell destruction is very variable, and sometimes it can develop very quickly (usually in infants and children) and sometimes slowly (mainly in adults). Clinical manifestations occur following a prodromal period lasting months to years after the onset of immunological disorders. Some patients show Diabetic Ketoacidosis (DKA) as the first symptom, others show moderate fasting hyperglycemia and progress rapidly to severe hyperglycemia or DKA in the presence of infection or other stresses. Sometimes, especially adults, people maintain β-cell function enough to prevent DKA for many years and eventually reach plasma C-peptide levels that are too low to survive without insulin. Rarely can this clinical picture be seen in the 1980s or 1990s [8, 14].

As a result of the interaction of various genetic predispositions with some environmental factors, autoimmune destruction of β-cells occurs. Genetic predisposition in these patients can also cause other autoimmune diseases such as Graves' disease, Hashimoto's thyroiditis, Myasthenia gravis, Addison's disease, pernicious anemia, and autoimmune hepatitis, celiac disease, and vitiligo. Although individuals are rarely obese at the onset of the disease, the presence of obesity does not exclude the diagnosis [8, 15].

6.2.2 Idiopathic Diabetes

Some forms of T1DM do not have a known etiology. It is known that this form of diabetes is genetically inherited. It is not associated with HLA. Some of these patients were found to be predisposed to DKA with insulinopenia. However, markers of β-cell autoimmunity are not seen. A small number of patients who fall into this category are of African or Asian descent and show episodic DKA. These patients may sometimes have absolute insulin requirements, and sometimes not. Pancreatic calcification can be seen in 5% of patients. More research is needed to determine the cause of β-cell destruction in this uncommon clinical picture [16].

6.3 Epidemiology of T1DM

T1DM accounts for approximately 5–10% of all diabetes patients. Although it can be seen at any age, it is usually detected in children or young adults. Ninety percent of diabetes seen in childhood and adolescence is T1DM. The disease is more common between the ages of 7 and 15. T1DM is more common in two periods, especially in 5–7 years (starting school and being exposed to infectious agents more intensely), and 10–14 years (high levels of sex hormones, growth hormone, and emotional stress in adolescence) [17]. The incidence of T1DM varies according to age, gender, family history, ethnicity, and geographical regions. It is known that various mechanisms mediate the global increase in the prevalence of T1DM, which has more than one etiology, and that there are significant differences in its incidence. Although it is known that autoimmune diseases are more common in girls, T1DM is seen with approximately equal frequency in children regardless of gender. In Europe, the incidence of T1DM is higher in boys, while in other countries it is more common in girls. Seasonal differences can also be seen in the emergence of new cases in T1DM. It is known that the incidence of T1DM peaks during the winter months. It is thought that this is because infections seen in the winter season are more common [18, 19].

The number of people with T1DM is increasing day by day in both rich and poor countries. The overall annual increase in the world is 3% on average [20]. The annual increase in the incidence of T1DM is reported between 2–5% in Europe, the Middle East, and Australia. At the same time, it is seen that the frequency of T1DM increases in most age and ethnic groups in the United States of America (USA) [21, 22]. The

Diabetes Mondiale (DIAMOND) Project, Epidemiology and Prevention of Diabetes (EURODIAB), and SEARCH for Diabetes in Youth (SEARCH) projects are among the most important studies carried out for childhood diabetes. The DIAMOND Project was initiated by the WHO in 1990 to identify the incidence of T1DM in children to address the public health implications of T1DM. The first report in 2000 reported that 4.5% of children aged 14 and under had diabetes between 1990 and 1994 in 50 countries (incidence: 19 164 cases per 75.1 million children). After examining 100 populations worldwide, China and South America had the lowest incidence of diabetes (<1/100 000 children per year). Sardinia, Finland, Sweden, Norway, Portugal, the United Kingdom, Canada, and New Zealand were reported as the countries with the highest incidence of diabetes (>20/per 100 000 children per year). According to the DIAMOND study in the USA, including Pennsylvania, Alabama, and Illinois, the incidence of diabetes was 10–20 people per 100 000 per year. An incidence of 5–10 people per 100 000 per year has been reported in half of the European population. Again, in the USA, in a multicenter incidence study called SEARCH, which investigated diabetes in youth (under 20 years of age), 1905 children and adolescents were diagnosed with T1DM in a population of more than 10 million individuals between 2002 and 2003 [23]. In the prevalence study, the prevalence of T1DM was found to be 2.28/per 1000 people under the age of 20. In Philadelphia, the rate of increase in children younger than 5 years between 1985 and 2004 was reported to be 70% [24].

In the EURODIAB study investigating the annual increase rates of T1DM in children aged 0–4, 5–9, and 10–14 in 17 countries in Europe, in these age groups between 1989 and 2003, respectively, increased rates of 5.4%, 4.3%, and 2.9% were determined. It has been reported that the average annual increase in the 15 years that followed was 3.9%. In the same study, it was reported that between 2005 and 2020, the highest increase occurred in children younger than 5 years of age, and the incidence would increase approximately twofold [25].

In general, it is predicted that there will be an increase of 60% in the 0–14 age group within 15 years. According to the 2017 data from the International Diabetes Federation (IDF), 132 600 new cases were diagnosed with T1DM every year in the world, and the number of T1DM cases between the ages of 0–19 was reported to be 1 106 500. According to the 2021 data of the IDF, it has been reported that 149 500 new cases are diagnosed with T1DM every year in the world, and the number of T1DM cases between the ages of 0–19 is 1 211 900. This proves that T1DM is increasing every year [18, 26].

The reason for the increase in T1DM is still unknown. Migrant populations are a good example of how important environmental factors are in the development of T1DM [27]. The incidence of T1DM in Japanese people living in Hawaii was five times higher than in those living in Japan. It was determined that the incidence of T1DM was two times higher in the people of Montreal whose ethnic origins are France and Italy. It has been determined that the incidence of T1DM in children of Israeli origin living in Canada is four times higher than that of children living in Israel [28]. It has been reported that the increasing incidence of T1DM cannot be explained by socioeconomic factors alone, showing that regional differences between and within the country, genetic, environmental factors, events in intrauterine life, diet in early life, and viral infections cause this increase [29]. Two different hypotheses have been proposed to explain the increase in T1DM cases since the second half of the twentieth century. Accelerator hypothesis: The fact that most people do not develop diabetes despite having a first-degree relative with T1DM has been explained by the protective role of environmental factors for many years. However, today, as a result of overnutrition, it is thought that autoimmunity is triggered by the increase in insulin resistance from childhood and this situation leads to the onset of the disease at a younger age. Hygiene hypothesis: The decrease in contact with antigenic stimuli as a result of the decrease in infections and parasitic diseases in developed societies causes delayed maturation of immune defense mechanisms in children. It has been suggested that, in this case, the susceptibility to allergic and autoimmune diseases may increase. Compulsory vaccination programs are applied from the first weeks of life since childhood diseases cannot be eradicated in developing societies. It is also suggested that vaccination strengthens the immune defense systems, so T1DM is less common. This idea indirectly supports the hygiene hypothesis [16]. An important point here is that the information on the incidence rates of T1DM is generally based on data from children under the age of 15. Archived data of people with newly diagnosed T1DM until their thirties and forties are almost non-existent. Therefore, the true lifetime incidence of T1DM is unknown [1]. In particular, data from Sweden and Belgium indicate that T1DM increases in children under the age of 15, whereas the frequency of T1DM decreases in the group over the age of 15. Indeed, considering that the incidence of T1DM peaks at puberty and the onset of puberty shifts to earlier ages in developed societies, these observations are considered to be correct [30, 31].

As a result, epidemiological information on T1DM indicates that the disease is more common in countries with developed health systems (Scandinavian countries), whereas it is relatively rare in third-world countries. The incidence of the disease is increasing in western countries and rapidly developing countries. The prevalence is parallel to the wealth of countries, and T1DM is more common in rich countries. It suggests that the decrease in some protective factors in the development and modernization process of public health may be responsible for the increase in the frequency of T1DM [16].

6.4 Genetics of T1DM

It is known that T1DM has a strong genetic background. It is defined as a complex and chronic disease in which genetic, epigenetic, and environmental factors are involved in its etiology. While the risk of developing T1DM in the general population is 0.4%, this risk may increase to 6–7% in the child of a patient with T1DM, 8% in dizygotic (fraternal) twins, and 67% in monozygotic (monozygotic) twins. These findings reflect the importance of genetic predisposition in the disease [32]. The genetic risk factors of the disease arise from the differences in the genetic material that affect the immune system. In 50% of patients with T1DM, changes in the "major histocompatibility complex (MHC)" locus on the short arm of chromosome 6, which encodes the antigen-presenting cell surface receptors, predisposes them to the development of the disease. This MHC gene locus contains closely sequenced genes involved in the immune response. The most common genetic abnormalities detected in patients with T1DM originate from the "human leukocyte antigen (HLA)" class II region. This region encodes the MHC regions, which are the surface receptors of antigen-presenting cells. It has been stated that amino acid variations (specific self-peptides) in each HLA class II allele are presented to T lymphocytes and thus may be associated with T lymphocyte-mediated pancreatic islet cell damage. In particular, variations in the DR and DQ regions facilitate the development of T1DM. DR3 and DR4 haplotypes are known as major risk factors for the development of T1DM. While 90% of patients with T1DM have one or both of these haplotypes, these haplotypes are found in 45–50% of the non-diabetic population. The most implicated alleles in this area are DQA1*0501, which is described as DQB1*0201 (when with DR3), and DQA1*0301 and DQB1*0302 (when with DR4) [33].

Similarly, some alleles in this region have been described to inhibit the development of T1DM. Individuals with the DRB1*1501, DQA1*0102, and DQB1*0602 alleles have a low risk of developing T1DM. Transmembrane T-cell receptors mediate immune stimulation and cell damage on contact with a specific antigen. During thymic (morphological change of the thymus) development, the body's own cells join these receptors and provide self-tolerance. Some polymorphisms in CTLA4, one of the gene regions involved in this process, have been associated with T1DM. Similarly, UBASH3A, another gene region responsible for tolerance, is also held responsible for the development of T1DM [34].

Although HLA class II alleles are responsible for 30–50% of the genetic risk of T1DM, multiple non-MHC loci contribute to disease risk with smaller effects. Among other non-HLA genetic causes associated with the development of T1DM, PTPN22, IL2RA, IFIH1, CAPSLIR7R, CLEC16A, STAT4, and PTPN2 mutations are shown. In conclusion, it is seen that genetic predisposition has an important weight in the development of T1DM [32, 35].

6.5 T1DM and Mediterranean Diet

Although genetic predisposition contributes to the pathogenesis of T1DM, many non-genetic factors such as environmental factors play a role in the development of this disease. Environmental triggers include dietary factors, toxins, and other factors such as the gut microbiota. Many nutritional factors such as cow's milk, animal protein, fruit, and fruit juices can contribute to the pathogenesis of T1DM. However, polyunsaturated omega-3 fatty acids play a protective role against disease by modulating the immune response. Recently, the gut microbiota, whose composition is largely influenced by diet, has been hypothesized to play an important role in the pathogenesis of T1DM by affecting gut permeability, molecular mimicry, and modulating the innate and adaptive immune systems [36].

Dietary components provide different substrates during fermentation processes that may result in several products. Changes in microbial makeup due to dietary modifications make some bacterial communities "genetically better equipped" to metabolize these substrates. Moreover, the same substrate can be used in different ways depending on the type or abundance and present frequency of bacteria colonizing the intestinal niche. An example of the previous statement is lactate. Depending on the type of microbiota, this substrate can be converted to butyrate or to short-chain fatty acids (SCFA) such as acetate, succinate, and propionate during anaerobic bacterial fermentation in the gut [36, 37].

The lactate model appears to be the strongest possible explanation for understanding the link between T1DM and dysbiosis. According to this model, the presence of lactic acid and butyrate-producing bacteria such as *Prevotella* and *Akkermansia* helps maintain a healthy epithelium. This is because butyrate contributes to mucin synthesis and the formation of tight junctions. It is known that these bacteria are common in the microbiota of healthy children around the world. In contrast, when microorganisms such as *Bacteroides* and *Veillonella* are abundant, this substrate follows the pathway to succinate, acetate, and propionate. These products disrupt mucin synthesis and increase extracellular permeability by changing tight junctions [38].

It is known that the nutritional style or the content of diets specific to the diseases affect the human gut microbiota both positively and negatively. The positive effects of the Mediterranean diet (MeD), which is known as the healthiest diet accepted in the world and provides a healthy life, have been proven on T1DM [39]. One of the most basic features of the MeD is its nutritional diversity. In this way, dietary intake of vegetables and fruits and various phytochemicals in its content has been increased, and the importance of its effectiveness in protecting health and treating diseases has been demonstrated by research. It is known that the MeD is characterized by high consumption of grains, legumes, vegetables, fruits, nuts, fish, and olive oil, moderate alcohol intake, and low consumption of red meat and saturated fat. The MeD diet consists of rich carbohydrates (55–60% of total energy), proteins (15% of total energy, mostly vegetable protein), low lipids (30% of total energy, of which about 19% monounsaturated fat, 5% comes from polyunsaturated fat and <7% saturated fat) and fiber (25 g/day). The main source of fat in the diet is extra virgin olive oil. It consists predominantly of monounsaturated fatty acids (MUFAs), polyunsaturated fatty acids (PUFAs) such as linoleic and α-linolenic acids, and saturated fatty acids (SFAs), mainly palmitic and stearic acids. It also contains vitamin E and polyphenols. The beneficial effects of olive oil can also be attributed to the presence of 3,3-dimethyl-1-butanol, a structural analog of choline. It plays a role in the inhibition of trimethylamine *N*-oxide (TMAO) formation. It is known as a phosphatidylcholine and L-carnitine metabolite. As reported in some studies, high TMAO levels contribute to increased cardiovascular risk. It is associated with mortality, coronary events, stroke, heart failure, and poor kidney function in those with T1DM [36, 40].

The MeD is characterized by high daily consumption of fibers, abundantly present in plant based foods that have different chemical structures and are resistant to enzymatic degradation in humans. In this way, it is important for heart health. It also contributes to the modulation of gut microbiota, increasing the number of *Firmicutes* relative to *Bacteroidetes*. This reduces the number of Bacteroides, which have been identified as the main genus causing dysbiosis associated with T1DM, and protects the microbiota [41]. De Filippis and co-workers have reported that high adherence to the MeD was associated with increased amounts of *Prevotellaceae* and reduced *Firmicutes-Bacteroidetes ratio*, higher SCFA levels, and lower TMAO concentrations [39].

Accordingly, in a study conducted in an elderly population, it was shown that the MeD alters the gut microbiota and promotes the growth of taxa such as *Faecalibacterium prausnitzii*, and *Bacteroides thetaiotaomicron* and *Prevotella copri*, which contribute to the formation of SCFAs. Furthermore, MeD-induced changes in the gut microbiota promote the production of more SCFAs, particularly butyrate, which have an immunomodulating effect and regulate oxidative stress and inflammation in the gut and other peripheral tissues [42, 43].

The gut microbiota is also modulated by probiotics, which contribute to the maintenance of physiological gut permeability. For this reason, in a study performed in a mouse model of T1DM, it was shown that *Lactobacillus johnsonii* administration delayed the development and progression of the disease and interferes with autoimmune pancreatic β-cell destruction [44]. T1DM is associated with high cardiovascular risk, which is greatly influenced by environmental factors such as diet. It is known that MeD, which is rich in monounsaturated fat and fibers, is protective against the risk of cardiovascular diseases (CVD) [45, 46]. A multicenter study conducted on a population of T1DM patients examined the effects of saturated fatty acid and dietary fiber intake on the onset of CVD, reported that increased fiber consumption, particularly soluble fiber consumption, can prevent the risk of both cardiovascular and all-cause mortality in these patients [47].

6.6 Conclusion

The prevalence of T1DM is increasing in all countries at an alarming rate, which not only compromises socioeconomic conditions but also the mental state of the population. Based on the tagline that "prevention is always better than cure", dependency on the MeD has shown a positive hope for T1DM patients. The interaction between the MeD and the gut microbiota and its effect on the development and progression of T1DM may lead to the hypothesis that the MeD may alter both the onset and progression of the disease by modulating the gut microflora and reducing the associated proinflammatory profile. High fiber intake may contribute to the environmental regulation of the microbiota, which increases the production of SCAFs, especially butyrate, leading to the maintenance of physiological intestinal permeability, thereby regulating the immune system response. In addition, high dietary fiber intake improves glycemic control and fat absorption and thus may delay the development of cardiovascular complications during T1DM.

References

1. Green, A., Hede, S.M., Patterson, C.C. et al. (2021). Type 1 diabetes in 2017: Global estimates of incident and prevalent cases in children and adults. *Diabetologia*. 64: 1–10.
2. Ilıksu Gozu, H. and Akcay, S. (2021). Diyabetin epidemiyoloji ve fizyopatolojisi. In: *Aile Hekimliginde Diyabet ve Komplikasyonları*. (ed.A. Dayan). 1st edn. 1–10. Ankara: Turk. Klinikleri.
3. American Diabetes Association Professional Practice Committee. (2022). 2. Classification and diagnosis of diabetes: Standards of medical care in diabetes. *Diabetes Care*. 45(1): 17–38.
4. Chiang, J.L., Kirkman, M.S., Laffel, L.M. et al. (2014). Type 1 diabetes through the life span: A position statement of the American Diabetes Association. *Diabetes Care*. 37: 2034–2054.
5. Sperling, M.A. (2002). Diabetes Mellitus. In: *Pediatric Endocrinology*. (ed. M.A. Sperling). 323–366. Pennsylvania (USA): Saunders Elsevier Science.
6. Patel, S.K., Ma, C.S., Fourlanos, S. et al. (2021). Autoantibody-negative type 1 diabetes: A neglected subtype. *Trends Endocrinol. Metab*. 32: 295–305.
7. Buzzetti, R., Zampetti, S. and Maddaloni, E. (2017). Adult-onset autoimmune diabetes: Current knowledge and implications for management. *Nat. Rev. Endocrinol*. 13: 674–686.
8. Zajec, A., Trebušak Podkrajšek, K., Tesovnik, T. et al. (2022). Pathogenesis of type 1 diabetes: Established facts and new insights. *Genes*. 13: 1–19.
9. Richard, I.G., Holt, J., DeVries, H. et al. (2021). The management of type 1 diabetes in adults. A consensus report by the American Diabetes Association (ADA) and the European Association for the Study of Diabetes (EASD). *Diabetes Care*. 44: 2589–2625.
10. Pieralice S, Pozzilli P. (2018). Latent autoimmune diabetes in adults: A review on clinical implications and management. *Diabetes Metab. J.* 42(6): 451–464.
11. Klingensmith, G.J., Pyle, L., Arslanian, S. et al. (2010). TODAY Study Group. The presence of GAD and IA-2 antibodies in youth with a type 2 diabetes phenotype: Results from the TODAY study. *Diabetes Care*. 33: 1970–1975.

12 Solis-Herrera, C., Triplitt, C., Reasner, C. et al. (2000). Classification of diabetes mellitus. In: *Endotext [Internet]*. (ed. K.R. Feingold, B. Anawalt, A. Boyce, et al). South Dartmouth (MA): MDText.com, Inc.

13 Belhiba, O., Aadam, Z., Jeddane, L. et al. (2020). Research of anti-GAD and anti-IA2 autoantibodies by ELISA test in a series of Moroccan pediatric patients with diabetes type 1. *Afr. Health Sci.* 20: 1337–1343.

14 Noble, J.A. and Valdes, A.M. (2011). Genetics of the HLA region in the prediction of type 1 diabetes. *Curr. Diab. Rep.* 11: 1–23.

15 Ramakrishna, B.S., Venugopal, G., Singh, A. et al. (2021). Human leukocyte antigen DQ (HLA-DQ) genotypes and haplotypes and their association with phenotype in patients with celiac disease in India. *JGH Open.* 5: 1190–1196.

16 Akalın, S. (2015). Diabetes Mellitus Sınıflaması ve Tanı. In: *Gecmisten Gelecege Diabetes Mellitus*. (ed. S. İmamoglu, I. Satman, S. Akalın, et al). 17–23. Ankara: Temd Dernegi.

17 Maahs, D.M., West, N.A., Lawrence, J.M. et al. (2010). Epidemiology of type 1 diabetes. *Endocrinol. Metab. Clin. North Am.* 39: 481–497.

18 10th Edition Committee. (2021). Estimating the incidence and prevalence of type 1 diabetes in children and adolescents. In: *IDF Diabetes Atlas*. (ed. E.J. Boyko, D.J. Magliano, S. Karuranga, et al). 25–26. International Diabetes Federation.

19 Sun, H., Saeedi, P., Karuranga, S. et al. (2022). IDF Diabetes Atlas: Global, regional and country-level diabetes prevalence estimates for 2021 and projections for 2045. *Diabetes Res. Clin. Pract.* 183: 109119.

20 Patterson, C.C., Karuranga, S., Salpea, P. et al. (2019). Worldwide estimates of incidence, prevalence and mortality of type 1 diabetes in children and adolescents: Results from the International diabetes federation diabetes atlas. *Diabetes Res. Clin. Pract.* 157: 1–9.

21 8th Edition Committee. (2017). Estimating type 1 diabetes in children and adolescents. In: *IDF Diabetes Atlas*. (ed. S. Karuranga, J.R. Fernandes, Y. Huang, et al). p. 38. International Diabetes Federation.

22 Mobasseri, M., Shirmohammadi, M., Amiri, T. et al. (2020). Prevalence and incidence of type 1 diabetes in the world: A systematic review and meta-analysis. *Health Promot. Perspect.* 10: 98–115.

23 Shojaeian, A. and Mehri-Ghahfarrokhi, A. (2018). An overview of the epidemiology of type 1 diabetes mellitus. *Int. J. Metab. Syndr.* 2: 1–5.

24 Lipman, T.H., Levitt Katz, L.E., Ratcliffe, S.J. et al. (2013). Increasing incidence of type 1 diabetes in youth: Twenty years of the Philadelphia pediatric diabetes registry. *Diabetes Care.* 36: 1597–1603.

25 Patterson, C.C., Dahlquist, G.G., Gyürüs, E. et al. (2009). Incidence trends for childhood type 1 diabetes in Europe during 1989–2003 and predicted new cases 2005–20: A multicentre prospective registration study. *Lancet* 373: 2027–2033.

26 Robert, A.A., Al-Dawish, A., Mujammami, M. et al. (2018). Type 1 diabetes mellitus in Saudi Arabia: A soaring epidemic. *Int. J. Pediatr.* 2018: 1–19.

27 Quinn, L.M., Wong, F.S. and Narendran, P. (2021). Environmental determinants of type 1 diabetes: From association to proving causality. *Front. Immunol.* 12: 1–15.

28 Abaci, A., Bober, E. and Buyukgebiz, A. (2007). Type 1 diabetes. *Current Pediatrics.* 5: 1–10.

29 Rewers, M. and Ludvigsson, J. (2016). Environmental risk factors for type 1 diabetes. *Lancet.* 387: 1–19.

30 Imperatore, G., Mayer-Davis, E.J., Orchard, T.J. et al. (2018). Prevalence and incidence of type 1 diabetes among children and adults in the United States and comparison with non-U.S. countries. In: *Diabetes in America.* (ed. C.C. Cowie, S.S. Casagrande, A. Menke, et al). 1–32. Bethesda (MD): National Institute of Diabetes and Digestive and Kidney Diseases (US).

31 Diaz-Valencia, P.A., Bougnères, P. and Valleron, A.J. (2015). Global epidemiology of type 1 diabetes in young adults and adults: A systematic review. *BMC Public Health.* 15: 1–19.

32 Steck, A.K. and Rewers, M.J. (2011). Genetics of type 1 diabetes. *Clin. Chem.* 57: 176–185.

33 Jerram, S.T. and Leslie, R.D. (2017). The genetic architecture of type 1 diabetes. *Genes.* 8: 1–22.

34 Pociot, F., Akolkar, N.B., Concannon, N.P. et al. (2010). Genetics of type 1 diabetes: What's next? *Diabetes.* 59: 1561–1571.

35 You, W.P. and Henneberg, M. (2016). Type 1 diabetes prevalence increasing globally and regionally: The role of natural selection and life expectancy at birth. *BMJ Open Diabetes Res. Care.* 4(1): e000161.

36 Calabrese, C.M., Valentini, A. and Calabrese, G. (2021). Gut microbiota and type 1 diabetes mellitus: The effect of Mediterranean diet. *Front. Nutr.* 7: 11–18.

37 Rowland, I., Gibson, G., Heinken, A. et al. (2018). Gut microbiota functions: Metabolism of nutrients and other food components. *Eur. J. Nutr.* 57: 1–24.

38 Mejía-León, M.E. and Barca, A.M. (2015). Diet, microbiota and immune system in type 1 diabetes development and evolution. *Nutrients.* 7: 9171–9184.

39 De Filippis, F., Pellegrini, N., Vannini, L. et al. (2016). High-level adherence to a Mediterranean diet beneficially impacts the gut microbiota and associated metabolome. *Gut.* 65(11): 1812–1821.

40 Koeth, R.A., Wang, Z., Levison, B.S. et al. (2013). Intestinal microbiota metabolism of L-carnitine, a nutrient in red meat, promotes atherosclerosis. *Nat. Med.* 19: 576–585.

41 Alkanani, A.K., Hara, N., Gottlieb, P.A. et al. (2015). Alterations in intestinal microbiota correlate with susceptibility to type 1 diabetes. *Diabetes.* 64: 3510–3520.

42 Ghosh, T.S., Rampelli, S., Jeffery, I.B. et al. (2020). Mediterranean diet intervention alters the gut microbiome in older people reducing frailty and improving health status: The NU-AGE 1-year dietary intervention across five European countries. *Gut.* 69: 1218–1228.

43 Bailey, M.A. and Holscher, H.D. (2018). Microbiome-mediated effects of the Mediterranean diet on inflammation. *Adv. Nutr.* 9: 193–206.

44 Gomes, A.C., Bueno, A.A., de Souza, R.G. et al. (2014). Gut microbiota, probiotics and diabetes. *Nutr. J.* 13: 1–24.

45 Leroux, C., Brazeau, A.S., Gingras, V. et al. (2014). Lifestyle and cardiometabolic risk in adults with type 1 diabetes: A review. *Can. J. Diabetes.* 38: 62–69.

46 Martínez-González, M.A., Gea, A. and Ruiz-Canela, M. (2019). The Mediterranean diet and cardiovascular health. *Circ. Res.* 124: 779–798.

47 Schoenaker, D.A., Toeller, M., Chaturvedi, N. et al. (2012). Dietary saturated fat and fibre and risk of cardiovascular disease and all-cause mortality among type 1 diabetic patients: The EURODIAB prospective complications study. *Diabetologia.* 55: 2132–2141.

7

The Emerging Role of Plant Polyphenols in the Management of Type 1 Diabetes Mellitus

S. Asha Devi[1,], Suparna Mandal[1], and S. Raja Sekhar[2]*

[1] *Laboratory of Gerontology, Department of Zoology, Bangalore University, Bangalore 560056, Karnataka, India*
[2] *Pathology Section, MPMMCC, Banaras Hindu University Campus, Varanasi 221005, U.P., India*
* *Corresponding author*

7.1 Introduction

The International Diabetic Federation has estimated that the incidence of diabetes mellitus (DM) might reach around 600 million cases by 2035; of these, approximately 5–10% are expected to be type 1 DM (T1DM) [1]. T1DM is a metabolic genetic disorder arising from the autoimmune destruction of the insulin (INS)-synthesizing pancreatic β-cells, auto-antibodies to INS, autoantibodies to major glutamate acid decarboxylase (GAD65), and tyrosine phosphatases, IA-2 and IA-2b [2].

In humans, T1DM is commonly diagnosed in children and adolescents with delayed symptoms and diagnosis. In such situations, insufficient INS synthesis and ketoacidosis are evident solely because of the immune destruction of β-cells in a progressive manner over months [3]. This situation causes heart diseases, kidney dysfunction, blurred vision, peripheral neuropathy, and fatigue, to name a few adverse effects. It is most often referred to as INS-dependent DM (IDDM). A typical therapy for T1DM subjects is INS injections, wherein tests exhibit auto antibody-positive and are INS resistant with age, and patients are categorized as double diabetic subjects [4].

Even though the exact etiology of T1DM is still unclear, there is unquestionably a genetic predisposition. It has a potential link with the polymorphic human leukocyte antigen (HLA), specifically with the HLA-DR (Human Leukocyte Antigen DR-Isotype), and HLA-DQ (Human Leukocyte Antigen DQ – Isotype) alleles, apart from many other genes that contribute to the genetic transmission of T1DM. An approximate estimate of the possibility of incidence of T1DM in the absence of family history is 0.4%. However, records indicate that the offspring of an affected mother has a 1–4% chance of developing T1DM; the incidence rate is 3–8% when

Natural Products and their Bioactives in Antidiabetic Drug Discovery, First Edition.
Edited by Kanti Bhooshan Pandey, Maitree Suttajit and Pinar Atukeren.

the father is affected. The incidence rate rises to 30% in offspring when both parents have T1DM. Among monozygotic twins, the risk of the first twin developing T1DM within the first ten years is 30%, and the approximate lifetime risk is 65% [5, 6].

Although genetics have a potential effect, for instance, in twins, environmental factors impact the progression of T1DM. Thus, the pathogenesis of the disease is most often unclear since both genetic and environmental factors interact in aggravating the inflammatory reactions and destruction of β-cells [7]. Humans may also be infected through drinking water contaminated with streptozotocin (STZ)-generating bacteria, *Streptomyces achromogenes*, which produce STZ and lead to pancreatic β-cell destruction. Infections through the soil bacteria from several sources of freshwater for drinking can cause transient T1DM due to low levels of STZ [8].

The possibility of the synthesis of autoantigens that can trigger autoimmune reactions in the regenerating β-cells that cause T1DM in humans cannot be overlooked [9]. The pancreas has exocrine and endocrine glands; the latter are referred to as the islets of Langerhans and lie between the exocrine acini. Humans have approximately 1–2 million islets with 50–300 endocrine cells per islet. Each has four cell types: α-cells, β-cells, δ-cells, and F-(PP) cells, which secrete glucagon, INS, somatostatin, and pancreatic polypeptide, respectively [10]. These hormones regulate each other's secretion via paracrine cell–cell interconnections. An islet has 60–75% β-cells that synthesize INS, while 20% are α-cells, which synthesize glucagon. δ-cells are less common than β- and α-cells. The F-cell is a rare cell type whose polypeptide inhibits the functions of the exocrine pancreas. The islets, with their rich blood capillaries, are innervated with both sympathetic and parasympathetic nerve fibers.

Although T1DM cannot be prevented, diabeticians worldwide have made great efforts to alleviate its severity. Though anti-diabetic drugs effectively alleviate T1DM symptoms, several side effects persist. Natural plant products are now being attempted as an alternative medicine to overcome these undesirable side effects. The focal points of this chapter are (i) to overview the scope of the use of animal models to understand INS-dependent T1DM in humans, (ii) to examine how natural polyphenols can slow the progression of pathogenic mechanisms and regulate glucose homeostasis in T1DM patients, and (iii) to explore the recent emerging role of nanoformulations as therapeutic natural anti-diabetic molecules in T1DM.

7.2 Pancreatic β-cell Destruction and T1DM

In T1DM, the islets of Langerhans are infiltrated with T-lymphocytes, causing inflammation and progressive damage to β-cells and leading to INS insufficiency and hyperglycemia. These changes are accompanied by high levels of ketone bodies in the liver resulting from increased fat breakdown and the beta-oxidation of free fatty acids (FFAs). Additionally, the breakdown of glycogen and proteins in the body is elevated, which further contributes to hyperglycemia (Figures 7.1 and 7.2).

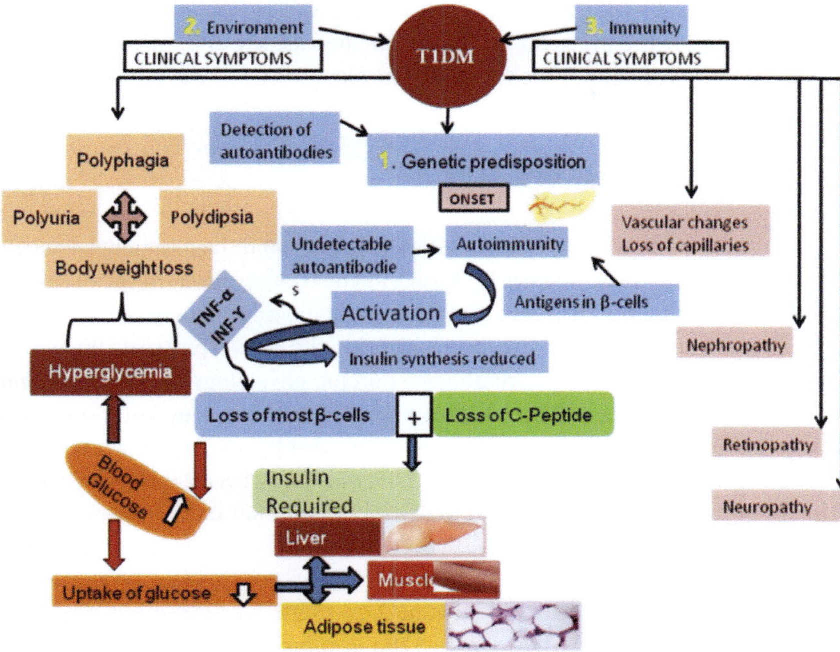

Figure 7.1 Factors involved in triggering type 1 diabetes (T1DM) and the impact of hyperglycemia on pivotal organs in humans.

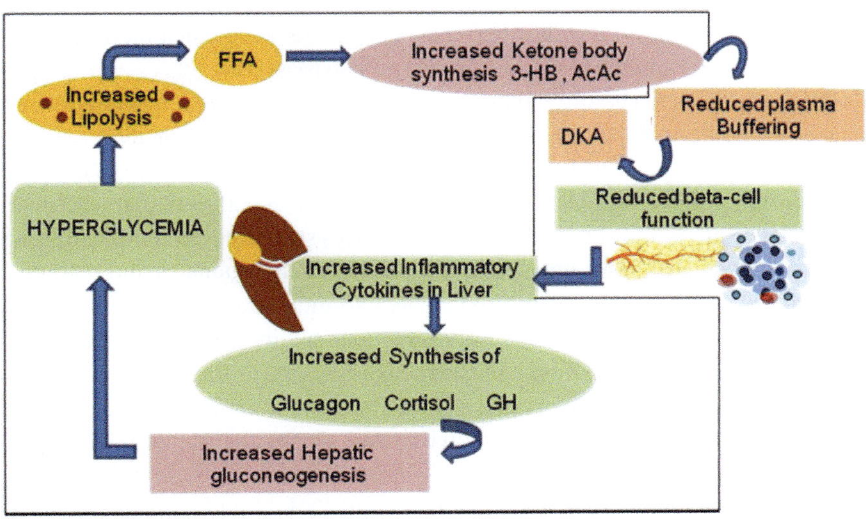

Figure 7.2 Major steps in the occurence of diabetic keto acidosis (DKA) from hepatic pathogenesis as a consequence of reduced β-cells in human T1DM.

The situation worsens due to elevated ketoacidosis levels (Figure 7.2), the rapid elevation in lipolysis, the transport of FFAs to the hepatic site and, in turn, a greater synthesis of ketone bodies, which acetoacetate along with β-hydroxybutyrate.

7.3 Animal Models of T1DM

Animal models of T1DM have been studied to understand the pathological mechanisms and therapeutic strategies for lessening the complications of various organ systems. Apart from the extensive use of small rat and mice models, large models, such as pigs and primates, have been used for the above purpose. Preclinical studies involving pigs are important since porcine physiology resembles human physiology. In addition to the similarities between morphology and position of the pig and human pancreas, porcine and INS differ by one amino acid. Rodent models of T1DM are not comparable to T1DM in humans because they are smaller animals with a shorter lifespan, and they have less similar clinical presentations than non-human primates (NHPs). Nevertheless, they are generally excellent models for mechanistic studies.

STZ and alloxan (5,5-dihydroxy pyrimidine-2,4,6-trione; ALX) are cytotoxic glucose analogs and are currently used to induce T1DM in animal models, such as rats, mice, rabbits, pigs, and dogs, although they differ in their mechanisms of induction. STZ and ALX models resemble human T1DM. However, effective T1DM induction is based not only on the doses of ALX and STZ but also on the route of administration in the animals as reviewed by Mostafavinia et al. [11]. Further, human β-cells are less responsive to STZ than rat and mice β-cells, as the latter have much more sensitive GLUT2 transporters in the islets compared to human islets [12].

STZ is an effective chemical approach to induce T1DM in rodents and destroy pancreatic islet β-cells [13, 14]. Alkylation damages the DNA and activates excess ribosylation of poly(ADP-ribose) polymerase-1 (PARP-1) polyADP [15]. NAD+ is an important molecule implicated in energy metabolism at the cellular level. The activation of Jun N-terminal kinase by STZ induces stress and precedes increased PARP-1 activity [16]. It inhibits DNA synthesis in microorganisms and mammalian cells by alkylation and cross-linking DNA strands, and it affects all stages of the mammalian cell cycle. STZ is mutagenic and carcinogenic in humans [17]. Further, the toxicity of STZ on the islet cells is related to its induction of high levels of nitric oxide and superoxide radicals [18, 19]. It has a short half-life and is rapidly metabolized in the hepatic tissue and eliminated by the renal tissue. The heart, muscles, and brain are largely affected by hyperglycemia but not by any further action of STZ on these pivotal organs [20]. In addition, the effects of STZ are gender-based as demonstrated in rodents. According to a previous study on C57BL/Ks mice, males were hyperglycemic at all time intervals following STZ injection, while females reached hyperglycemia only after 35 days owing to the protection of β-cell apoptosis from estradiol [21].

STZ is also effective in inducing T1DM in large animals such as pigs [22] and primates [23]. Based on their studies, Jensen-Waern et al. [24] reported that STZ-diabetic Swedish domestic pigs are similar to human T1DM patients considering the carbohydrate, amino acid, and fat metabolism in non-INS animals.

Interestingly, NHPs are genetically much closer to humans than pigs. Notably, reduced INS secretion and β-cell mass are two characteristics of T1DM humans. In NHPs, low doses of STZ and incomplete pancreatectomy induce diabetes, for instance, in rhesus macaques [25]. Further, STZ alone suffices in inducing diabetes in baboons [26], vervet monkeys [27], and macaques [28] with stringent ethical considerations [29], since primates could be a source of infection to humans. Undeniably, STZ-treated rhesus monkeys exhibiting hyperglycemia and autoantibodies have been verified as a preclinical diagnostic tool since they usually resemble the autoimmune conditions observed in human T1DM patients [30].

ALX is similar to glucose in terms of its structure and transport through the plasma membrane into the cytosol of β-islet cells by GLUT2. However, β-cells regenerate efficiently, and their response could be misinterpreted as anti-diabetic to certain compounds. In addition, unlike STZ-induced T1DM, ALX-induced T1DM is not confirmatory and does not resemble T1DM in humans [31].

7.4 Plant Polyphenols as Therapeutic Agents for T1DM

The increasing occurrence of T1DM has motivated scientists and clinicians to constantly study immunotherapy to prevent or minimize the complications associated with T1DM, such as nephropathy, neuropathy, atherogenesis, retinopathy, cognitive dysfunctions, and retinopathy.

Hence, while attempting to address the possibility of using polyphenols in a therapeutic approach for T1DM, preclinical studies use animal models. Such studies include tests to confirm the destruction of pancreatic β-cells, INS insufficiency, increased synthesis of reactive oxygen species (ROS), and inflammation. Findings imply the importance of plant polyphenols in maintaining blood glucose in T1DM patients because of their potential to protect against pancreatic β-cell damage by modulating INS-dependent and -independent signaling pathways [32, 33].

Large quantities of polyphenols are found in fruits and many plant-based beverages, vegetables, cereals, and legumes. Polyphenols have been frequently reported to reduce blood glucose levels in animal models and humans. In the former case, these effects are reported in chemically-induced diabetic models and in spontaneous diabetic rodent models as well. Further, animal T1DM models are often used to acquire insights into the screening of their therapeutic properties and possible treatments.

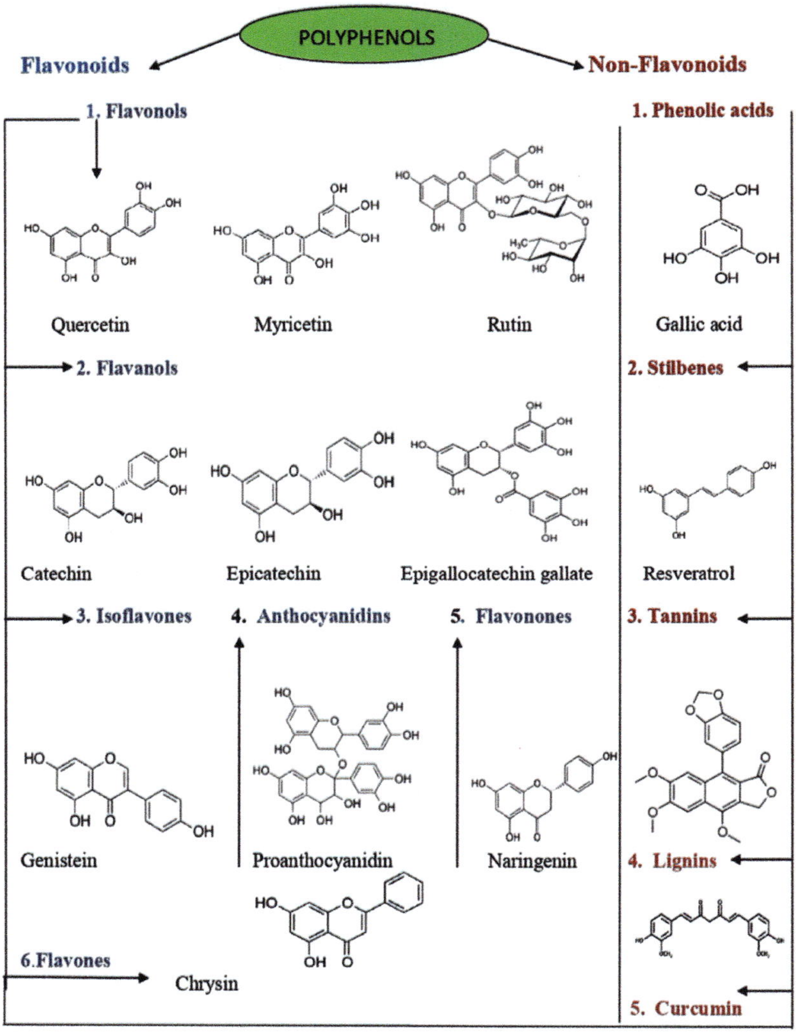

Figure 7.3 Classification of natural polyphenols obtained from fruits and vegetables.

Based on their structure, polyphenols are classified into two categories: flavonoids and non-flavonoids (Figure 7.3). B-type procyanidins are oligomers of monomeric flavanols, such as catechin and epicatechin [34] from grape seeds, resveratrol (RSV) from grape skin [35], and curcuminoids from turmeric [36]. A-type procyanidins are monomeric units that operate in conjugation through single linkages (either C4-C8 or C4-C6).

7.4.1 Flavonoids and T1DM

Humans consume a wide variety of flavonoid-containing foods and drinks, such as green tea, oranges, strawberries, mangoes, raspberries, red wine, and grapes, in large amounts. Flavonoids are highly protective against many metabolic disorders and needed for healthy aging. This is especially true of flavanols (catechin, epicatechin, and epigallocatechin), flavonols (quercetin (QUE), myricetin, and rutin), anthocyanidins (procyanidins), flavanones (naringenin), and isoflavones (genistein). Primarily, flavonoids are conjugates of glycosides and are converted as aglycone by beta-glucosidase. Glycosides cross the intestinal membrane and exist as aglycone. They are then further metabolized to their beta-D-glucuronide and sulfonate esters, which enter the blood and are seen in peripheral tissue cells [37].

Epigallocatechin gallate (EGCG) is most abundant in green tea and, unlike many other catechins, is found free (as opposed to in a conjugated form) in the plasma following its uptake in the intestine [38]. Human studies have demonstrated the non-toxicity of EGCG supplements [39]. The significance of T1DM in EGCG rodents is related to its inhibition of the intestinal absorption of glucose [40]. Studies on non-obese diabetic (NOD) mice have revealed that EGCG at 0.05% in drinking water (a dose corresponding to green tea consumption in humans) alleviates the initiation of T1DM by inhibiting the destruction of INS-synthesizing islet cells [41].

Further, Collins et al. [42] demonstrated the possibility of inhibiting gluconeogenesis by pure EGCG at concentrations of less than 1 μM in isolated primary cultures of hepatocytes derived from C57BL/6 mice. Gluconeogenesis was suppressed through the mediation of ROS, 5′AMP-activated protein kinase, and calcium/calmodulin-dependent protein kinase kinase. Also, improved glucose tolerance and reduced HbA1c were evident in green tea-supplemented STZ-induced diabetic rats [43].

Studies on STZ-induced T1DM in Wistar rats have revealed a minimization of oxidative stress (OS) and islet degeneration with GSPE and INS [44]. An elevation in Bcl-2 and the alleviation of Bax expressions in the pancreas and prefrontal cortex have been reported in these rats [45], along with an improved spatial working memory related to the synergetic influence of the combined treatment.

QUE is a well-documented bioflavonoid found in many foods; it is found in high concentrations in green tea, red wine, broccoli, apples, and onions. This flavonol exerts beneficial effects through its anti-inflammatory, antioxidant, and anti-diabetic [46] properties. *In vitro* and *in vivo* studies on STZ-induced diabetic rat models and mesenchymal stem cells (MSCs) and/or QUE and/or QUE/Zn, or their combination have revealed anti-inflammation, islet regeneration, and INS release [47]. These synergetic effects suggest that QUE/Zn is effective in MSC therapy to protect against OS and genotoxicity and in ameliorative therapy to protect against diabetes.

In rats, evidence against OS in the lenses of STZ-induced T1DM by naringenin further suggests the therapeutic potential of flavonoids in diabetic retinopathy [48]. Naringin has been reported to abrogate changes in the microstructure of bone trabecular number and separations while increasing the number of osteocalcin cells and adipocytes in STZ rats [49].

7.4.2 Non-flavonoids

Curcumin (CUR), an extract of the rhizome of the plant *Curcuma longa*, decelerated the progression of T1DM in a NOD*scid* female mice model compared to NOD mice through the autoimmunity against β-cell specific antigens observed in the CUR-supplemented mice. Further, in cyclophosphamide-induced T1DM NOD mice, CUR inhibited pancreatic leukocyte infiltration while preserving β-cells [50]. Oral RSV supplementation with nicotinamide in STZ-rats reduced blood glucose along with TNF-α, IL-1β, IL-6, NF-κB, and nitric oxide while increasing INS [51]. A comprehensive description of polyphenols in management of T1DM has been described in Table 7.1.

7.5 Nanotechnology for T1DM Treatment by Polyphenols

An important advantage of using polyphenols to treat T1DM as a therapeutic approach for hyperglycemia is the reduction of hepatic glucose production, cognitive impairment, and renal dysfunction, as well as efficient peripheral glucose disposal. Further, animal models highlight the importance of using specific models that are close to humans when evaluating the therapeutic significance of polyphenols in T1DM. However, the bioavailability of polyphenols as anti-diabetic compounds is more beneficial as nanoformulations owing to their prolonged circulation period and more specific target orientation [52, 53].

An important and easily biodegradable nanoparticle (NP) is non-toxic poly (lactic-*co*-glycolic acid) (PLGA) [54]. Further, nanoformulations have a greater surface area to volume ratio than conventional formulations, and they control the release of therapeutic molecules. The delivery of anti-diabetic polyphenols, naringenin, RSV, and QUE exert better anti-diabetic responses than non-formulations (Figure 7.4). A few inconsistencies have occurred in the outcome from *in vivo* and *in vitro* studies. Nevertheless, Torchilin [55] indicated that several pharmaceutical approaches, such as micronization, lipid emulsions, and hydrotrophy, have been adapted to improve their clinical applicability in diabetes.

The main concern with naringenin consumption is its poor solubility in water and low uptake in the intestine after oral intake, which is largely because of its rapid elimination by enzymes in the gut and liver [56]. However, NPs composed of alginate/chitosan (CEC) and naringenin are effective anti-diabetic

Table 7.1 Anti-diabetic role of flavonoids and non-flavonoids in Type 1 diabetes mellitus.

Polyphenol	Source	Dose	Cell culture/animal model /human	Findings	Reference
Flavanol					
Quercetin	Mangoes, raspberries, and apples	STZ, 70 mg/kg b.w. i.p. QUE (Q), 5 mg, 25 mg and 50 mg/kg b.w. i.p. for 40 days by gavage Control (C) QUE (Q) 5 mg/kg (CQ5); 25 mg/kg (CQ25); 50 mg/kg (CQ50). Diabetic (D), DQ5, DQ25, DQ50	Male Wistar rats	Decrease in serum TG and TBARS by QUE in diabetics Hepatic and Renal DQ50 group, it was observed a reduction of the enzyme activity	[38]
		STZ, 185 mg/kg b.w. i.p. Dietary 0.1% and 0.5% QUE	7-wk-old male BALB/c mice diet, AIN93G (CON) AIN93G +0.1%/0.5% QUE, 2 weeks AIN93G-STZ mice. Diabetic mice with AIN93G –0.1%/0.5% QUE	Increase in serum INS mice in 0.5% QUE diabetic mice. Hepatic gene expressions in 0.5% QUE reduced STZ-induced gene expression	[65]
		STZ, 70 mg/kg b.w. i.p. QUE (CQ), 150 μmole/kg b.w. i.p.	Male Wistar rats	Hepatic SOD and CAT increased in STZ rats but reduced with QUE	[66]

(Continued)

Table 7.1 (Continued)

Polyphenol	Source	Dose	Cell culture/animal model /human	Findings	Reference
	EGCG from Green Tea (GT)	STZ, 50 mg/kg b.w. ADSC injected at 1×10^6 cells/rat for 2 months through tail vein EGCG-pretreated ADSC injected at 1×10^6 cells/rat for 2 months	8 weeks of Male Wistar rats and stem cells from rat AD tissue. Grouped into- DM, single dose ADSC+DM E-ADSC+DM	Pre-incubation with ADSC with GT EGCG elevates Sirt1 expression, reduced ROS, less damage to pancreas Increased stem cell paracrine function and increased SOD gene expression Reduced neuropathic pain	[67]
	EGCG solution	2.5 mg/100 g b.w./day EGCG i.p. for 2 days	STZ, 60 mg/100 g b.w. i.p.	Reduced blood NOx and MDA, reduced TOS Plasma thiols, CAT and TAC improved in L-EGCG-pretreated diabetic rats	[68]
		STZ: 65 mg/kg b.w. EGCG at 100 mg/kg b.w. for 14 days prior to I/R injury	STZ-induced adult male Sprague-Dawley T1DM rats H9c2 cells exposed to 30 mM glucose followed by hypoxia/ reoxyg-enation injury	EGCG reduced infarct size, apoptosis, myocardial LDH, and MDA inT1DM rats EGCG decreased apoptosis, OS parameters in H9c2 cells.	[69]
Anthocyanin	GSPE	STZ: 55 mg/kg b.w.i.p. GSPE: 40 mg/kg/day	Male Wistar rats. 8 weeks of treadmill training with/ without GSPE	Myocardial FasL and Bcl2 increase in diabetics + GSPE with decreased sFas and sFas/FasL Exercise with grape seed extract further increases FasL and BcL-2 levels, and further decrease in sFas level and sFas/FasL Moderate intensity training with GSPE reduces apoptotic factors of myocytes in T1DM rats	[70]

		STZ, 70 mg/kg b.w. i.p. Animal, 250 mg/ kg b.w. PE, orally Cell lines, 50–300 μm PE)	Male Wistar rats Low INS: 1.26 nmol/animal, High INS: 9 nmol, i.p. PE+INS gp: 1.26 nmol PE+250mg/kg b.w. INS-sensitive cell lines, L6E9 myoblasts 3T3-L1 adipocytes	PE along with low dose of INS has antihyper-glycemic effect PE facilitated glucose uptake in dose-dependent manner	[71]
Flavanone	Naringin oranges, grapefruit, lime	STZ, 50 mg/kg b.w.i.p. Insulin, 8.0 I.U./kg b.w. twice daily Naringin, 50 mg/kg b.w. Ramipril, 3.0 mg/kg/b.w.	Male Sprague-Dawley rats Groups III, IV, V and VI- STZ + INS + Naringin + Ramipril	Untreated DM rats increase in OS, NADPH oxidase activity, LV cardiac fibrosis, MAPK, PKC-β and p38 expression Naringin decreased the above effects	[72]
		STZ, 60 mg/kg b.w. i.p. INS, 4.0 IU/kg b.w. Naringin 50 mg/kg b.w.	Sprague-Dawley rats Gp 1, D.W; Gp 2, Naringin; Gp, 3,4,5, STZ; Gp 3, INS; Gp 4, Naringin+STZ	STZ reduced hepatic glycogen, increased, blood pH and bicarbonate increased MDA Naringin improved plasma INS, hepatic glycogen Lowered MDA, blood pH and bicarbonate	[73]
		STZ at 50 mg/kg b.w. Naringenin, 5 mg/kg b.w. 10 mg/kg b.w. for 10 weeks	Male Wistar rats	Naringenin-treated diabetics reduced kidney apoptosis, cytokine-TGF-β1 expression, renal dysfunction reduced in diabetes	[74]

(Continued)

Table 7.1 (Continued)

Polyphenol	Source	Dose	Cell culture/animal model /human	Findings	Reference
Non-Flavonoids					
1. Stilbenes	RSV from grapes	STZ, 55 mg/kg i.p. for 5 days RSV, 50 mg/kg for 12 days, day 4 of STZ	Mice	NF-κB p65 in isolated islets CXCL16, NF-κB p65 and CD45 in islets of diabetics is evident and RSV inhibited these changes Normalized CXCL16 and CD45 in islets and spleen Macrophage infiltration in islets activates CXCL16/NF-κB p65 signaling pathway and RES inhibits this pathway in T1DM	[75]
		500 mg capsules, twice daily, 60 days	Male and female T1DM patients	Decreased FBS and HbA1c, decrease in MDA and increase in TAC Insignificant change in INS, HOMA-IR, HOMA-β RSV had strong anti-diabetic and antioxidant effects in patients with T1D	[76]
		STZ at 55 mg/kg b.w. RSV, 20 mg/kg i.p.	Male Wistar rats RSV, daily for 4 weeks	RSV reduced vascular MDA, IL-1β, and TNF-α in thoracic aorta IL-18 in cardiac muscle almost normal in RSV treated rats Cardiac muscle hypertrophy reduced	[77]

	STZ at 55 mg/kg b.w. RSV at 20 mg /kg/day i.p.	Male Wistar rats RSV, daily for 4 weeks	Elevated apoptotic markers in the hepatic Bcl-2, Bax and p53 and Inflammatory IL and TNF-α elevated in diabetic rats RSV suppressed the above markers and had hepatoprotective effects	[78]
	250 mg/kg RSV gavage, daily 25 mg/kg by injection on alternate days	Female NOD mice, 8 weeks of age Maintained up 32 weeks of age	RSV on alternate days was more effective in protecting from T1DM, ALT and AST and liver damage was not evident in treated mice RES reduced damage to pancreatic islets and free from insulitis Therapeutic potential lies in capacity of activating sirtuin 1 for treating T1DM in the NOD mouse Less absorption of RSV by oral route and therefore less effective against T1DM	[79]
Curcumin from turmeric	STZ, 65 mg/kg b.w. i.p. CUR, 100 mg/kg b.w.	Male Wistar rats, eight weeks of age male Wistar Maintained on CUR for 8 weeks	Increased TNF-α, IL1-β and IFN-γ Decreased Nrf-2, Glut-2, Caspase-12/9/8/3 CUR lowered inflammatory markers in β-cells and inhibited apoptosis	[80]

(Continued)

Table 7.1 (Continued)

Polyphenol	Source	Dose	Cell culture/animal model /human	Findings	Reference
		STZ, 60 mg/kg b.w. CUR, 200 mg/kg b.w.	Male mice Studies at 4, 8 and 12 weeks	Islet number reduced at 8 weeks in diabetics Increase in small islets increased in CUR-treated diabetics Lymphocytes visible in islets at 12 weeks in diabetics, and near the islets in CUR-diabetics without insulitis	[36]
Phenolic acid					
Gallic Acid	Grape seed Green White Tea Leaf Blueberry	STZ, 55 mg/kg b.w. i.p. 20 and 40 mg/kg b.w. STZ, 45 mg/kg b.w., i.v. GA, 25 mg/kg b.w.	Male Sprague Dawley rats 7–8 weeks of age Male Wistar rats 8 weeks treatment	GA reduced circulating TGF-β1 and TGF-β1 in kidneys of diabetics Renal pathology was alleviated by GA INS and HDL decreased MDA, H_2O_2, IL- MDA, H_2O_2, IL-1β TNF-α, Bax, caspase-9 and P-53 increased in diabetic hearts GA treatment ameliorated the above parameters in the STZ-induced diabetic rats	[81] [82]

	STZ, 50 mg/kg b.w. i.v. GA, 25 mg, 50 mg, and 100 mg/kg b.w./day	Male Wistar rats Treatment for 8 weeks	GA reduced polyphagia, polydipsia, hyperlipidemia, hypertension, bradycardia GA reduced increased force of contraction reduced in STZ rats Changes were dose-dependent	[83]

ADSC, Adipose-derived stem cells; ALT, alanine aminotransferase; AST, Aspartate aminotransferase; Bax, BCL2-Associated X Protein; Bcl-2, B-cell lymphoma 2; CXCL16, CXC chemokine ligand 16; EGCG, epigallocatechin gallate; FasL, Fas ligand; sFas, serum Fas; GA, Gallic acid; GSPE, Grape seed proanthocyanidin extract; GT, Green tea; HOMA-IR, homeostasis model assessment-estimated insulin resistance, HOMA-β, homeostasis model assessment; IL-18, Interleukin-18; Mir-26a, microRNA-26a; NF-κB, Nuclear factor kappa B; NOD mice; non-obese diabetic mice; RSV, Resveratrol; TAC, Total antioxidant capacity; TNF-α, Tumor necrosis factor; TGF-β1, Transforming growth factor beta 1; MAPK, Mitogen-activated protein kinase.

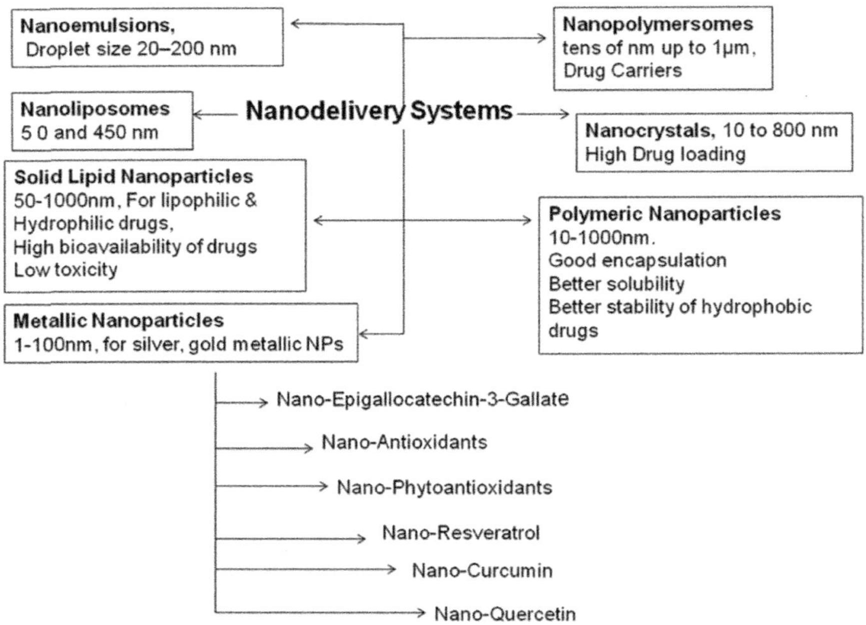

Figure 7.4 Polyphenol-loaded nanodelivery systems for therapeutic efficiency.

and non-toxic agents in STZ-induced diabetes [57]. Naringenin – for example, Eudragit® E (a copolymer) – is a hepatoprotective, anti-apoptotic, and antioxidant nanoformulation that supports medical therapeutic approaches for T1DM.

Nanoencapsulated RSV is more effective than free RSV in terms of stability and bioavailability in specific target organs [58]. PEGlycated nanoliposomes of RSV have been shown to improve glycemia and INS levels in β-cells for longer periods than purified RSV *in vitro* in β-cells [59]. Further, a study on diabetic retinopathy in rats fed gold nanoparticles (AuNPs) orally showed decreased activation of nuclear factor kappa B (NF-κB), tumor necrosis factor-alpha (TNF-α), interleukin 6 (IL-6), and IL-1β gene expressions, along with vascular endothelial growth factor, intracellular adhesion molecule 1, and extracellular signal-regulated factor. However, there is a lack of clinical evidence on RSV NPs for anti-diabetic therapy in human subjects with T1DM.

QUE has better potential than RSV to enhance therapeutic efficacy because of its improved bioavailability and target orientation when administered orally. Therefore, micro- and NP preparations are the most important approaches currently being investigated to improve the bioavailability of QUE [60]. Concerning the unique properties of NPs, a PLGA polymer for hydrophobic CUR has been proposed as a useful therapeutic approach [61]. In addition, PLGA is biodegradable and does not cause inflammation or immune reactions. Also, PLGA polymers have been approved by the USFDA for human therapeutic approaches [54].

Further, PLGA-nano-CUR (nCUR) has been identified as an anti-diabetic formulation in efforts to alleviate the loss of pancreatic β-cells. Studies on STZ-induced diabetic Sprague-Dawley rats showed reduced DNA damage from ROS in the pancreas and alleviated caspase-8 in insulinoma cells and RIN-m5f cells [62]. Furthermore, cytokines, IL-10, IL-17A, IL-1β, and TNF-α were suppressed by nCUR. Studies on STZ-induced T1DM C57BL/6 mice in which CEC protected against renal and cardiac inflammation also confirmed that the bioavailability of CUR improved by encapsulation with CEC NPs [63]. Further, oxidative stress-induced hepatic injury and inflammatory signals are effectively suppressed through CUR-PLA-PEG NPs compared to free CUR [64]. An optimistic approach employing polyphenol nanoformulations for alleviating T1DM symptoms may be strengthened further through efforts to alter the genes.

Based on much non-clinical evidence on polyphenols, sufficient clinical evidence supporting their therapeutic approaches is required from human T1DM trials. Incorporating complementary and alternative medicines inclusive of polyphenols with conventional medicine such as INS in T1DM therapy may also provide an alternative anti-diabetic therapy, especially in young children.

7.6 Conclusion

The advances made so far in experimental research on animal models of T1DM have provided further understanding of polyphenols and have revived scientists' and clinicians' interest in establishing effective T1DM therapies. A positive approach involves encapsulated nanoformulations of specific polyphenols to promote the efficient functioning of β-cells. Combinations of INS and polyphenol NPs might be more promising in the development of novel anti-diabetic therapeutics with the utmost consideration for patient compliance to the novel therapy. This concept is based on the fact that polyphenols can restore pancreatic function by slowing the progression of the disorder, especially in children before the pancreas has been severely damaged. More clinical trials are needed to protect and manage pancreatic islets and associated organs in T1DM in humans.

Acknowledgements

We acknowledge all lab members for their cooperation in reading the manuscript. The authors wish to extend their thanks to the UGC-CPEPA Scheme (F.No. 8-2/2008 (NS/PE) dt December 14, 2011), New Delhi for the financial support extended to our studies cited in this article and Bangalore University, Bangalore for supporting this work.

Conflicts of Interest

The authors declare that they have no conflict of interest.

References

1 Guariguata, L., Whiting, D.R., Hambleton, I. et al. (2014). Global estimates of diabetes prevalence for 2013 and projections for 2035. *Diabetes Res. Clin. Pract.* 103: 137–149.

2 American Diabetes Association (ADA). (2014). 2014 Standards of medical care in diabetes. *Diabetes Care.* 37(supplement 1): S14–S80.

3 Chiang, J.L., Maahs, D.M., Garvey, K.C. et al. (2018). Type 1 diabetes in children and adolescents: A position statement by the American Diabetes Association. *Diabetes Care.* 41: 2026–2044.

4 Cleland, S.J., Fisher, B.M., Colhoun, H.M. et al. (2013). Insulin resistance in type 1 diabetes: What is 'double diabetes' and what are the risks? *Diabetologia.* 56: 1462–1470.

5 Basina, M. and Maahs, D.M. (2018). Age at type 1 diabetes onset: A new risk factor and call for focused treatment. *Lancet.* 392: 453–454.

6 Redondo, M.J., Geyer, S., Steck, A.K. et al. (2018). Type 1 diabetes trialnet study group. A type 1 diabetes genetic risk score predicts progression of islet autoimmunity and development of type 1 diabetes in individuals at risk. *Diabetes Care.* 41: 1887–1894.

7 Katsarou, A., Gudbjornsdottir, S., Rawshani, A. et al. (2017). Type 1 diabetes mellitus. *Nat. Rev. Dis. Primers.* 3: 17016.

8 Roep, B.O. (2003). The role of T-cells in the pathogenesis of type-1 diabetes from cause to cure. *Diabetologia.* 46: 305–321.

9 Zhu, B.T. (2022). Pathogenic mechanism of autoimmune diabetes mellitus in humans: Potential role of streptozotocin-induced selective autoimmunity against human islet β-cells. *Cells.* 11: 492.

10 Xavier, G.D.S. (2018). The cells of the islets of Langerhans. *J. Clin. Med.* 12: 54.

11 Mostafavinia, A., Amini, A., Ghorishi, S.K. et al. (2016). The effects of dosage and the routes of administrations of streptozotocin and alloxan on induction rate of type1 diabetes mellitus and mortality rate in rats. *Lab. Anim. Res.* 32: 160–165.

12 McCulloch, L.J., Bunt, M., Braun, M. et al. (2011). GLUT2 (SLC2A2) is not the principal glucose transporter in human pancreatic beta cells: Implications for understanding genetic association signals at this locus. *Mol. Genet. Metab.* 104: 646–653.

13 Furman, B.L. (2015). Streptozotocin-induced diabetic models in mice and rats. *Curr. Protoc. Pharmacol.* 70: 47.1–5.47.20.

14 Goyal, S.N., Reddy, N.M., Patil, K.R. et al. (2016). Challenges and issues with streptozotocin-induced diabetes – A clinically relevant animal model to understand the diabetes pathogenesis and evaluate rherapeutics. *Chem. Biol. Interact.* 244: 49–63.

15 Kishore, L., Kajal, A. and Kaur, N. (2017). Role of nicotinamide in streptozotocin induced diabetes in animal models. *JETR.* 2(1): 555577.

16 Cheon, H., Cho, J.M., Kim, S. et al. (2010). Role of JNK activation in pancreatic beta-cell death by streptozotocin. *Mol. Cell. Endocrinol.* 321: 131–137.

17 Cepeda, V., Fuertes, M.A., Castilla, J. et al. (2006). Poly(ADP-ribose) polymerase-1 (PARP-1) inhibitors in cancer chemotherapy. *Recent Pat. Anticancer Drug Disco.* 1: 39–53.

18 Lenzen, S. (2008). The mechanisms of alloxan- and streptozotocin-induced diabetes. *Diabetologia.* 51: 216–226.

19 Lenzen, S. (2008). Oxidative stress: The vulnerable beta-cell. *Biochem. Soc. Trans.* 36: 343–347.

20 Eleazu, C.O., Eleazu, K.C., Chukwuma, S. et al. (2013). Review of the mechanism of cell death resulting from streptozotocin challenge in experimental animals, its practical use and potential risk to humans. *JDMDC.* 12: 60.

21 Le May, C., Chu, K., Hu, M. et al. (2006). Estrogens protect pancreatic beta-cells from apoptosis and prevent insulin-deficient diabetes mellitus in mice. *PNAS.* 103: 9232–9237.

22 Hara, H., Lin, J.Y., Zhu, X. et al. (2008). Safe induction of diabetes by high-dose streptozotocin in pigs. *Pancreas.* 36: 31–38.

23 Koulmanda, M., Qipo, A., Chebrolu, S. et al. (2003). The effect of low versus high dose of streptozotocin in cynomolgus monkeys (Macaca fascilularis). *Amer. J. Transplant.* 3: 267–272.

24 Jensen-Waern, M., Andersson, M. and Kruse, R. (2009). Effects of streptozotocin-induced diabetes in domestic pigs with focus on the amino acid metabolism. *Lab. Anim.* 43: 249–254.

25 Jin, S., Zeng, L., He, S. et al. (2010). Comparison of single high-dose streptozotocin with partial pancreatectomy combined with low-dose streptozotocin for diabetes induction in rhesus monkeys. *Exp. Biol. Med.* 235: 877–885.

26 Frost, F.A., Chen, S., Mezzies, M.J. et al. (2015). Successful pharmaceutical-grade streptozotocin (STZ)-induced hyperglycemia in a conscious tethered baboon (Papio hamadryas) model. *J. Med. Pharmacol.* 44: 202–217.

27 Zhu, H., Yu, L., He, Y. et al. (2014). Nonhuman primate models of type 1 diabetes mellitus for islet transplantation. *J. Diabetes Res.* 2014: 785948.

28 Graham, M.I., Mutch, L.A., Littredge, J.A. et al. (2012). Management of adverse side-effects after chemotherapy in macaques as exemplified by streptozotocin. Case studies and recommendations. *Lab. Anim.* 45: 178–192.

29 Prescott, M.J. (2020). Ethical and welfare implications of genetically altered non-human primates for biomedical research. *JAAE.* 2: 151–176.

30 Wei, L., Lu, Y., He, S. et al. Induction of diabetes with signs of autoimmunity in primates by the injection of multiple-low-dose streptozotocin. *Biochem. Biophys. Res. Commun.* 412: 373–378.

31 Misra, M. and Aiman, U. (2012). Alloxan: An unpredictable drug for diabetes induction? *Ind. J. Pharmacol.* 44: 538–539.

32 Bahadoran, Z., Mirmiran, P. and Azizi, F. (2013). Dietary polyphenols as potential nutraceuticals in management of diabetes: A review.. *J Diabetes Metab. Disord.* 12: 43.

33 Patel, D., Kumar, R., Laloo, D. et al. (2012). Diabetes mellitus: An overview on its pharmacological aspects and reported medicinal plants having antidiabetic activity. *Asian Pac. J. Trop. Biomed.* 2: 411–420.

34 González-Abuín, N., Pinent, M., Casanova-Mart, À. Et al. (2015). Procyanidins and their healthy protective effects against type 2 diabetes. *Curr. Med. Chem.* 22: 39–50.

35 Ku, C.R., Lee, H.J., Kim, S.K. et al. (2012). Resveratrol prevents streptozotocin-induced diabetes by inhibiting the apoptosis of pancreatic beta-cell and the cleavage of poly (ADP-ribose) polymerase. *Endocr. J.* 59: 103109.

36 Chanpoo, M., Petchpiboonthai, H., Panyarachun, B. et al. (2010). Effect of curcumin in the amelioration of pancreatic islets in streptozotocin-induced diabetic mice. *J. Med. Assoc. Thail.* 93(Suppl 6): S152–S159.

37 Barnes, S., Prasain, J., D'Alessandro, T. et al. (2011). The metabolism and analysis of isoflavones and other dietary polyphenols in foods and biological systems. *Food Funct.* 2: 235–244.

38 Maciel, R.M., Costa, M.M., Martins, D.B. et al. (2013). Antioxidant and anti-inflammatory effects of quercetin in functional and morphological alterations in streptozotocin-induced diabetic rats. *Res. Vet. Sci.* 95: 389–397.

39 Ishrucker, R.A., Bausch, J., Edwards, J.A. et al. (2006). Safety studies on epigallocatechin gallate (EGCG) preparations. Part 1: Genotoxicity. *Food Chem. Toxicol.* 44: 626–635.

40 Skopec, M.M., Green, A.K. and Karasov, W.H. (2010). Flavonoids have differential effects on glucose absorption in rats (Rattus norvegicus) and American robins (Turdis migratorius). *J. Chem. Ecol.* 36: 236–243.

41 Fu, Z., Zhen, W., Yuskavage, J. et al. (2011). Epigallocatechin gallate delayes the onset of type 1 diabetes in spontaneous non-obese diabetic mice. *Br. J. Nutr.* 105: 1218–1225.

42 Collins, Q.F., Liu, H., Pi, J. et al. (2007). Epigallocatechin-3-galllate (EGCG), a green tea polyphenol suppresses hepatic gluconeogenesis through 5-AMP=activated protein kinase. *J. Biol. Chem.* 282: 30143–30149.

43 Meng, J., Cao, S., Wei, X. et al. (2019). Effects and mechanisms of tea for the prevention and management of diabetes mellitus and diabetic complications: An updated review. *Antioxidants.* 8: 170.

44 Rajasekhar, S., Subramanyam, M.V. and Asha Devi, S. (2021). Grape seed proanthocyanidin extract suppresses oxidative stress in the rat pancreas of type-1diabetes. *Arch. Physio. Biochem*. 27: 1–13.

45 Rajasekhar, S., Subramanyam, M.V., Chandrasekhar Sagar, B.K. et al. (2019). Grape seed proanthocyanidin extract and insulin prevents cognitive decline in type 1 diabetic rat by impacting Bcl-2 and Bax in the prefrontal cortex. *Metab. Brain Dis*. 34: 103–117.

46 Vessal, M., Hemmati, M. and Vasei, M. (2003). Antidiabetic effects of quercetin in streptozocin-induced diabetic rats. *Comp. Biochem. Physiol. C Toxicol. Pharmacol*. 135C: 357–364.

47 Refat, M.S., Hamza, R.Z., Adam, A.M.A. et al. (2021). Quercetin/Zinc complex and stem cells: A new drug therapy to ameliorate glycometabolic control and pulmonary dysfunction in diabetes mellitus: Structural characterization and genetic studies. *PLoS ONE*. 16: e0246265.

48 Wojnar, W., Zych, M. and Kaczmarczyk-Sedlak, I. (2018). Antioxidative effect of flavonoid naringenin in the lenses of type 1 diabetic rats. *Biomed. Pharmacother*. 108: 974–984.

49 Rivoira, M., Rodríguez, V., Picotto, G. et al. (2018). Naringin prevents bone loss in a rat model of type 1 diabetes mellitus. *Arch. Biochem. Biophys*. 637: 56–63.

50 Castro, C.N., Barcala Tabarrozzi, A.E., Winnewisser, J. et al. (2014). Curcumin ameliorates autoimmune diabetes. Evidence in accelerated murine models of type 1 diabetes. *Clin. Exp. Immunol*. 177: 149–160.

51 Palsamy, P. and Subramaniam, S. (2010). Ameliorative potential of resveratrol on proinflammatory cytokines, hyperglycemia mediated oxidative stress, and pancreatic β-cell dysfunction in streptozotocin-nicotinamide-induced diabetic rats. *J. Cell. Physiol*. 224: 423–432.

52 Yang, D.K. and Kang, H.-S. (2018). Antidiabetic effect of cotreatment with quercetin and resveratrol in streptozotocin-induced diabetic rats. *Biomol. Ther*. 26: 130–138.

53 Jain, K.K. (2008). Nanomedicine: Application of nanobiotechnology in medical practice. *Med. Princ. Pract*. 17: 89–101.

54 Dechy-Cabaret, O., Martin-Vaca, B. and Bourissou, D. (2004). Controlled ring-opening polymerization of lactide and glycolide. *Chem. Rev*. 104: 6147–6176.

55 Paolino, D., Cosco, D., Cilurzo, F. et al. (2007). Innovative drug delivery systems for the administration of natural compounds. *Curr. Bioact. Compd*. 3: 262–277.

56 Siu, F.Y.K., Ye, H., Lin, H. et al. (2018). Galactosylated PLGA nanoparticles for the oral delivery of resveratrol enhanced bioavailability and in vitro anti-inflammatory activity. *Int. J. Nanomed*. 13: 4133–4144.

57 Torchilin, V.P. (2006). Multifunctional nanocarriers. *Adv. Drug Deliv. Rev*. 58: 1532–1555.

58 Manach, C., Scalbert, A., Morand, C. et al. (2004). Polyphenols: Food sources and bioavailability. *Am. J. Clin. Nutr*. 79: 727–747.

59 Maity, S., Mukhopadhyay, P., Kundu, P.P. et al. (2017). Alginate coated chitosan core-shell nanoparticles for efficient oral delivery of naringenin in diabetic animals—an in vitro and in vivo approach. *Carbohydr. Polym.* 170: 124–132.

60 Chimento, A., De Amicis, F., Sirianni, R. et al. (2019). Progress to improve oral bioavailability and beneficial effects of resveratrol. *Int. J. Mol. Sci.* 20: 1381.

61 Yücel, Ç., Karatoprak, G.Ş. and Aktaş, Y. (2018). Nanolipoosomal resveratrol as a novel approach to treatment of diabetes mellitus. *J. Nanosci. Nanotechnol.* 18: 3856–3864.

62 Ganagula, R., Arora, M., Jaisamut, P. et al. (2017). Nano-curcumin safely prevents streptozotocin-induced inflammation and apoptosis in pancreatic beta cells for effective management of Type 1 diabetes mellitus. *Brit. J. Pharmacol.* 124: 2074–2084.

63 Sudirman, S., Lai, C.S., Yan, Y.-L. et al. (2019). Histological evidence of chitosan-encapsulated curcumin suppresses heart and kidney damages on streptozotocin-induced type-1 diabetes in mice model. *Sci. Rep.* 9: 15233.

64 El-Naggar, M.E., Al-Joufi, F., Anwar, M. (2019). Curcumin-loaded PLA-PEG copolymer nanoparticles for treatment of liver inflammation in streptozotocin-induced diabetic rats. *Colloids Surf. B: Biointerfaces.* 177: 389–398.

65 Kobori, M., Masumoto, S., Akimoto, Y. et al. (2009). Dietary quercetin alleviates diabetic symptoms and reduces streptozotocin-induced disturbance of hepatic gene expression in mice. *Mol. Nutr. Food Res.* 53: 859–868.

66 Dias, A.S., Porawski, M., Alonso, A. et al. (2005). Quercetin decreases oxidative stress, NF-κB activation, and iNOS overexpression in liver of streptozotocin-induced diabetic rats. *J. Nutr.* 135: 2299–2304.

67 Chen, X., Le, Y., Tan,g S.-Q. et al. (2022). Painful diabetic neuropathy is associated with compromised microglial IGF-1 signaling which can be rescued by green tea polyphenol EGCG in mice. *Oxid. Med. Cell. Longev.* 2002: 6773662.

68 Bulboaca, A.L., Boarescu, P., Porfire, A.S. et al. (2020). The effect of nano-epigallocatechin-gallate on oxidative stress and matrix metalloproteinases in experimental diabetes mellitus. *Antioxidants.* 9: 172.

69 Wu, Y., Xia, Z.-Y., Zao, B. et al. (2017). (-)-Epigallocatechin-3-gallate attenuates myocardial injury induced by ischemia/reperfusion in diabetic rats and H9c2 cells under hyperglycemic conditions. *Int. J. Mol. Med.* 40: 389–399.

70 Abadi, A.G.N., Keshavarz, S., Kargarfard, M. et al. (2022). Effect of grape seed extract consumption along with moderate intensity aerobic training on some biomarkers of apoptosis of cardiomyocytes in Wistar rats with type 1 diabetes. *JBRMS.* 9: 16–24.

71 Pinent, M., Blay, M., Bladé, M.C. et al. (2004). Grape seed-derived procyanidins have an antihyperglycemic effect in streptozotocin-induced diabetic rats and insulinomimetic activity in insulin-sensitive cell lines. *Endocrinology.* 145: 4985–4990.

72 Adebiyi, O.A., Adebiy, O.O. and Owira, P.M. (2016). Naringin reduces hyperglycemia-induced cardiac fibrosis by relieving oxidative stress. *PLoS ONE*. 11: e0149890.

73 Murunga, A.N., Miruka, D.O., Driver, C. et al. (2016). Grapefruit derived flavonoid naringin improves ketoacidosis and lipid peroxidation in type 1 diabetes rat model. *PLoS ONE*. 11: e0153241.

74 Roy, S., Ahmed, F., Banerjee, S. et al. (2016). Naringenin ameliorates streptozotocin-induced diabetic rat renal impairment by downregulation of TGF-β1 and IL-1 via modulation of oxidative stress correlates with decreased apoptotic events. *Pharm. Biol.* 54: 1616–1627.

75 Darwish, M., Abo-youssef, A., Messiha, B.A.S. et al. (2021). Resveratrol inhibits macrophage infiltration of pancreatic islets in streptozotocin-induced type 1 diabetic mice via attenuation of the CXCL16/NF-κB p65 signaling pathway. *Life Sci.* 272: 119250.

76 Movahed, A., Raj, P., Nabipou, I. Et al. (2020). Efficacy and safety of resveratrol in type 1 diabetes patients: A two-month preliminary exploratory trial. *Nutrients*. 12: 161.

77 Pektaş, M.B., Sadi, G., Koca, M.B. et al. (2016). Resveratrol ameliorates the components of hepatic inflammation and apoptosis in a rat model of streptozotocin-induced diabetes. *Drug Dev. Res.* 77: 12–19.

78 Pektaş, A., Koc, H., Tosun, M. et al. (2017). Effects of resveratrol on diabetes-induced vascular tissue damage and inflammation in male rats. *Turkish J. Biochem*. 42: 451–458.

79 Lee, S.-M., Yang, H., Tarta, D.M. et al. (2011). Prevention and treatment of diabetes with resveratrol in a non-obese mouse model of type 1 diabetes. *Diabetologia*. 2011(54): 1136–1146.

80 Rashid, K. and Sil, P.C. (2015). Curcumin enhances recovery of pancreatic islets from cellular stress induced inflammation and apoptosis in diabetic rats. *Toxicol. Appl. Pharmacol*. 282: 297–310.

81 Garud, M.S. and Kulkarni, Y.A. (2018). Gallic acid attenuates type I diabetic nephropathy in rats. *Chem. Biol. Interact.* 282:69–76.

82 EL-Komy, M.M.A. (2016). Cardioameliorative effect of gallic acid on myocardial injury associated with diabetes induced by streptozotocin in male rats. *Egypt. J. Exp. Biol. (Zoo)*. 65: 141–158.

83 Patel, S.S. and Goyal, R.K. (2011). Cardioprotective effects of gallic acid in diabetes-induced myocardial dysfunction in rats. *Pharmacol. Res.* 3:239–245.

8

Bioactives as Modulators of β-cells and Immunity in Therapy of Type 1 Diabetes Mellitus

Shashank Kumar Ojha

Medicine Department, School of Pharmacy, Hebrew University of Jerusalem, Israel

8.1 Introduction

Type 1 diabetes mellitus [T1DM] is a lethal and multifactorial disease. It is characterized by autoimmune destruction of insulin-producing β-cells in the pancreatic islets of Langerhans that causes insulin deficiency/insufficiency and leads to diabetes [1]. More than 10 million people are affected worldwide with T1DM, which is responsible for more than 4% of yearly deaths [1, 2]. The complications involve heart diseases, chronic kidney disease, nerve damage, problems with feet, oral health, vision, hearing, and mental health. The etiology involves genetic, immunological, and environmental components. Autoimmunity is currently considered a significant factor in the pathophysiology of T1DM [1, 3].

Parts of the immune system, like cytotoxic and helper T-cells, B-cells, macrophages, and dendritic cells, have been implicated in autoimmunity and β-cell destruction in T1DM. Still, most involved processes depend upon the interaction between antigen-presenting cells and T-cells. A study has shown that both CD4+ and CD8+ T-cells are required, but activated CD4+ T-cells are both necessary and sufficient in causing disease [2]. Studies have shown that Th1-secreted cytokines initiate and propagate the inflammatory process in early diabetes and Th2-secreted cytokines like interleukins help inhibit inflammation in early diabetes [2]. Immune inactivation against specific antigens for treating autoimmune T1DM and targeting this would seem to be a reasonable approach to combat T1DM [4, 5].

Despite significant efforts in T1DM research, exogenous insulin injection is still one of the most efficient treatments for T1DM since the discovery of insulin in 1921 [4]. More than 1000 genes are known that are affected by drugs in the

Natural Products and their Bioactives in Antidiabetic Drug Discovery, First Edition.
Edited by Kanti Bhooshan Pandey, Maitree Suttajit and Pinar Atukeren.

treatment of T1DM through the involvement of the mechanism of immune cells and β-cells [5, 6]. The literature shows significant effects of plant compounds and extracts in treating T1DM by regulating different signaling pathways related to immunity and β-cells. Therefore, these bioactive affected signaling networks were analyzed and are discussed later in this chapter for their functional interplay in immune and β-cells.

Natural compounds have immense power and are a good source of drugs for treating different diseases. More than a thousand plants are known to cure diabetes with antihyperglycemic effects through other mechanisms [7]. Bioactives are considered safe because they are mostly delivered orally as a part of the diet [8]. Studies show that plant compounds like phenolics, alkaloids, flavonoids, terpenoids, glycosides, etc., serve as regulators and modulators of β-cell proliferation and survival. They also regulate insulin production and the overactivation of immune responses [9]. In the present chapter the role of bioactives in modulation of β-cells and immunity, thereby mediating the prevention of T1DM complications, are discussed.

8.2 Classes of Bioactives and Their Mechanisms of Action on Immune Cells and β-cells

Plant compounds or plant extracts affect β-cell homeostasis by reducing β-cell death and helping in β-cell regeneration [10]. β-cell failure results from excess infiltration of leukocytes in the pancreas. The infiltrated leukocytes lead to dysfunction and killing of β-cells with cytokines and released molecules through a different mechanism. Figure 8.1 shows different mechanisms of bioactive compounds that are involved in the modulation of β-cells and immunity. Select bioactives are discussed here in this section.

8.2.1 Alkaloids

Nicotinamide, a component of vitamin B3, isoform isonicotinamide, and trigonelline (obtained from fenugreek seeds), has been shown to protect β-cells from apoptosis by downregulating the caspase-3 and significantly reducing the level of serum tumor necrosis factor-α, interleukin-6, and interleukin-1β [11, 12]. The alkaloid conophylline (isolated from *Ervatamia microphylla*) and alkaloid harmine (isolated from *Peganum harmala)* have been shown to induce and promote β-cell differentiation [13–15].

An alkaloid affecting immune cells such as tetrandrine, isolated from *Stephania tetrandra*, is a bisbenzylisoquinoline alkaloid that was reported to be involved in inhibiting NFκB signaling in T-cells and calcium channel blockade, also improved NK cell count and, in turn, improved T1DM via pancreatic β-cell protection and regeneration [16].

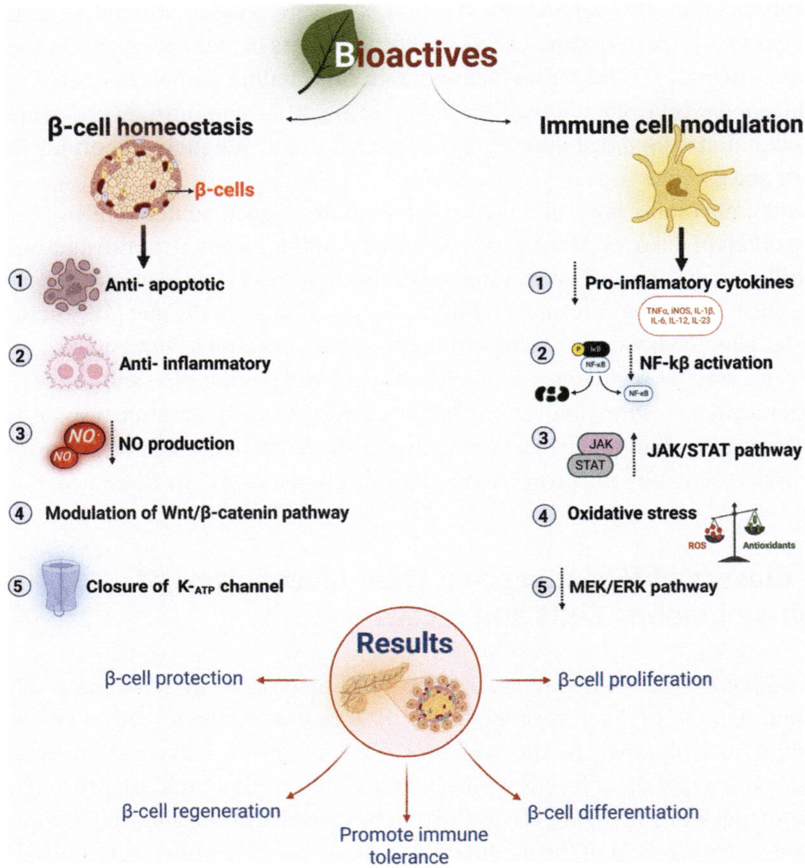

Figure 8.1 Schematic representation of bioactive modulating β-cells and immunity through different mechanisms.

8.2.2 Flavonoids

Cirsimaritin (a bioactive component of *Cirsium japonicum*), puerarin (a bioactive constituent of *Radix puerariae*), and kazinol U (bioactive component from *Broussonetia kazinoki*), and anthocyanins and anthocyanidins obtained from fruits and vegetables were known to protect β-cells from apoptosis by a different mechanism like PI3K/AKT signaling pathway and NFκB activation pathway [10, 17, 18]. Genistein is the most studied isoflavone concerning diabetes [19]. Genistein, found in several plants, including lupine, fava beans, soybeans, legumes, etc., was reported to increase β-cell proliferation, inhibit cytokine-induced β-cell apoptosis, and promote the oxidative capacity of β-cells [20, 21].

Daidzein is an isoflavonoid isolated from *Glycine tomentella*. It is reported to have the potential for upregulation of TG2 expression and Rac1 activity in infiltrating immune cells, and this helps in β-cell protection and regeneration [22]. Treatment with flavonoids like quercetin, luteolin, apigenin, kaempferol, and resveratrol lowered IL-1β- and IFN-γ-mediated cytotoxicity. Studies also showed that the flavonoids inhibit the nitrosative stress inhibiting inducible NO synthase (iNOS) gene expression via inhibition of NFκB activation and activation of the JAK/STAT pathway [23]. Brazillin, isolated from *Caesalpinia sappan*, augmented the immune cell environment in the pancreatic islets by increasing IL-2 production [24].

8.2.3 Quinones

Lupanine is isolated from the American *Lupinus* species. Lupanine and its derivatives act via blockage of the β-cell plasma membrane K_{ATP}-sensitive channels and enhanced insulin-releasing activity and were thus effective in β-cell protection in both autoimmune and metabolic diseases [25].

Emodin and physcion, isolated from *Fallopia Japonica*, have been shown to prevent islet destruction by suppressing inflammation and leukocyte chemotaxis by inactivating the MEK/ERK pathway [26]. Thymoquinone, a compound isolated from *Nigella sativa*, has been implicated in COX-2 suppression, thus oxidative stress suppression and β-cell preservation [26, 27].

8.2.4 Other Polyphenols

Gallic acid, a polyphenol commonly isolated from tea leaves, inhibits pancreatic β-cell apoptosis by stimulating GLUT4 translocation in adipocytes [28]. Silybin is an active constituent of silymarin. Silybin was shown to inhibit hyperglycemia and repair damaged pancreatic β-cells by promoting Sirt-1 expression in OVE26 transgenic mice, a T1DM and diabetic nephropathy model. Silymarin also induces the proliferation of insulin-producing β-cells by increasing the expression of pancreatic Nkx6.1 transcription factor and β-cell neogenesis [29, 30]. Cuminaldehyde and cuminol isolated from *Cuminum cyminum* were shown to exert their protective action in β-cells by maintaining the closure of the ATP-sensitive K (KATP) channel and increasing intracellular Ca^{2+} concentration in the islet T-cells, thus lowering glucose levels, without causing hypoglycemia [31].

Dihydroavenanthramide D isolated from oats and Butein isolated from anacardiaceae prevented cytokine-induced β-cell damage. They also help reduce iNOS expression and NO production, reducing β-cell destruction. They inhibit iNOS gene expression by inhibiting NFκB translocation. Thus, these polyphenols can be used to prevent functional β-cell damage and the progression of T1DM [18, 32].

8.2.5 Terpenoids

Ginsenosides, isolated from *Panax ginseng*, were demonstrated to effectively protect β-cells from cytokine-induced pancreatic islet damage via inhibiting NO production and regulating apoptosis-related genes, including Bax, Fas, and caspase-3 [33]. Withanolides, isolated from *Withania somnifera*, inhibit adipogenesis by modulating the Wnt/β-catenin pathway and β-cell protection and regeneration [34].

Curcumin, a bioactive compound from *Curcuma longa*, has acted as an antioxidant, anti-inflammatory, anti-viral, anti-bacterial, anti-fungal, and anticancer agent [35]. Curcumin also lowered pancreatic lipid peroxidation by cellular free radical scavenging enzymes and suppressed the level of serum TNF-α and IL-1 [36]. Nymphayol, isolated from the extracts of *Nymphaea stellata* flowers, commonly called Egyptian lotus, was known to stimulate insulin secretion in β-cells via enhancing the antioxidant activity against reactive oxygen species (ROS) produced under hyperglycemic conditions [37]. Guggulsterone, derived from the gum resin *Commiphora Mukul* (Guggulu), and crude extracts from the gum resin of *Boswellia serrata*, suppressed inflammation via inhibition of NFκB translocation. It increases pancreatic cell survival by lowering IL-1β- and IFN-γ-stimulated NFκB and JAK-STAT cascades [38, 39].

8.2.6 Cannabinoids

Cannabidiol, a compound isolated from *Cannabis sativa*, acts as a cannabinoid (CB) receptor agonist and can increase IL-4 and IL-10 serum levels. Blockade of CB1 receptors inhibited apoptosis in β-cells under stress and is a valuable strategy in preventing β-cell apoptosis in T1DM. The cannabinoids D9-THC and cannabidiol are unique terpenophenols isolated from *Cannabis* that cause a significant reduction in plasma levels of the pro-inflammatory cytokines IFN-γ and TNF-α. Moreover, both cannabinoids reduced cytokines of helper T-cells (Th1) but increased Th2 cytokines, specifically IL-4 and IL-10. The differentiation of Th0 cells to Th2 cells rather than Th1 helps the pancreatic cell from insulitis. Cannabinoids help preserve, protect, and regenerate β-cells [40, 41].

8.2.7 Vitamins

Vitamins are bioactive compounds that cannot be synthesized by an organism and so must be obtained in the diet and are known to play a role in β-cell function, growth, and development. Vitamin D is a group of essential steroid hormones synthesized from sunlight exposure and absorbed from foods like fatty fish and their liver oils, eggs, and fortified foods. Several studies have revealed the role of vitamin D in pancreatic β-cell function [42]. Vitamin A or retinol is an essential dietary nutrient for average growth, reproduction, and vision. Root vegetables and greens, such as squash, carrots, pumpkins, collards, and beet greens, are good sources of vitamin A. The effects of vitamin A on insulin secretion are dependent on its metabolites, which showed that all-trans retinoic acid (1000 nM) increased insulin secretion and insulin content [43]. Vitamin C (ascorbic acid), an antioxidant vitamin, is widely distributed in fresh fruits and vegetables and plays a vital role in protecting against free radical-induced damage. Glucose and ascorbic acid are structurally similar, so the transport and accumulation of ascorbic acid in pancreatic β-cells affect insulin secretion [10]. A comprehensive description of bioactive compounds of different classes, their sources, and biological roles are shown in Table 8.1.

Table 8.1 Different classes of bioactive compounds and their biological actions relevant in type 1 diabetes therapy.

Classes	Bioactive components	Sources	Mechanism of action/ Effect observed	References
Alkaloids	Nicotinamide Trigonelline Conophylline Harmine Tetrandrin	Peanuts, avocado, chicken liver Fenugreek seeds *Ervatamia microphylla* *Peganum harmala* *Stephania tetrandra*	Reduces the level of caspase-3, tumor necrosis factor-α, interleukin-6, and interleukin-1β Promote β-cell differentiation Inhibits NFκB signaling	[11, 12] [13–15] [16]
Flavonoids	Cirsimaritin Puerarin Kazinol U Genistein Daidzein Brazilin	*Cirsium japonicum* *Radix puerariae* *Broussonetia* kazinoki lupine, fava beans, soybeans, legumes *Glycine tomentella* *Caesalpinia sappan*	It affects the PI3K/AKT signaling pathway and NFκB activation pathway Increase β-cell proliferation, inhibit cytokine-induced β-cell apoptosis β-cell protection and regeneration Increase IL-2 production	[17] [19, 21, 22] [22] [24]

(Continued)

Table 8.1 (Continued)

Classes	Bioactive components	Sources	Mechanism of action/ Effect observed	References
Quinones	Lupanine Emodin and physcion Thymoquinone	Lupinus *Fallopia Japonica* *Nigella* sativa	Blockage of the β-cell plasma membrane and effective in β-cell protection Inactivating MEK/ ERK pathway COX-2 suppression	[25] [26] [27]
Polyphenols	Gallic acid Silybin Cuminaldehyde and cuminol Dihydroavenan-thramide D Butein	tea leaves silymarin *Cuminum cyminum* Oats Anacardiaceae	Inhibit pancreatic β-cell apoptosis Induce proliferation of insulin-producing β-cells Protective action in β-cells Reducing iNOS expression and NO production so reduced β-cell destruction	[28] [29, 30] [31] [32]
Terpenoids	Ginsenosides Withanolides Curcumin Nymphayol Guggulsterone	*Panax ginseng* *Withania somnifera* *Curcuma longa* *Nymphaea stellata* *Commiphora mukul*	Inhibiting NO production β- cell protection and regeneration Suppressed the level of serum TNF-α and IL-1 Stimulated secretion of insulin in β-cells Inhibition of NFκB translocation	[33] [34] [35, 36] [37] [38, 39]
Cannabinoids	Cannabidiol	*Cannabis sativa*	Preventing β-cell apoptosis	[40, 41]
Vitamins	Vitamin D Vitamin A (all-trans-retinoic acid) Vitamin C (ascorbic acid)	fatty fish and their liver oils, and eggs Root vegetables and carrots, pumpkins, fresh fruits and vegetables	Pancreatic β-cell function Increase insulin secretion Glucose-induced insulin secretion	

Abbreviations NF-κB: Nuclear factor-κB; PI3K/Akt: the phosphatidylinositol-3-kinase; IL-2: Interleukin-2; MEK/ERK: mitogen-activated protein kinases/extracellular signal-regulated kinases; iNOS: Inducible nitric oxide synthase.

8.3 Conclusion and Future Direction

Due to advancements in technology and innovative approaches, pharmacological knowledge of plant extracts/bioactives is increasing. At present, many effective plant-based medicines are known to cure diseases. In the future, bioactives can significantly contribute to the creation of novel drugs and the resolution of global health issues. T1DM is a severe public health issue, and the rising prevalence of diabetes worldwide is putting a financial strain on each country's economy. Traditional plant-derived bioactive chemical treatments for diabetes are currently being promoted due to their safety and lack of adverse effects compared to synthetic drugs. The mechanisms of action of the reported phytochemicals and plant extracts attributed to antidiabetic activity also need to be explored at depth in order to develop bioactives as potential therapeutic agents against T1DM.

References

1 Atkinson, M.A., Eisenbarth, G.S. and Michels, A.W. (2014). Type 1 diabetes. *Lancet*. 383(9911): 69–82.

2 Hassan, G.A., Sliem, H.A., Ellethy, A.T. et al. (2012). Role of immune system modulation in prevention of type 1 diabetes mellitus. *Indian J. Endocrinol. Metab.* 16(6): 904.

3 Bluestone, J.A., Herold, K. and Eisenbarth, G. (2010). Genetics, pathogenesis and clinical interventions in type 1 diabetes. *Nature*. 464(7293): 1293–1300.

4 Brooks-Worrell, B. and Palmer, J.P. (2013). Prevention versus intervention of type 1 diabetes. *Clin. Immunol.* 149(3): 332–338.

5 Ortutay, C., Siermala, M. and Vihinen, M. (2007). Molecular characterization of the immune system: Emergence of proteins, processes, and domains. *Immunogenetics*. 59(5): 333–348.

6 Bojadzic, D. and Buchwald P. (2018). Toward small-molecule inhibition of protein–protein interactions: General aspects and recent progress in targeting costimulatory and coinhibitory (immune checkpoint) interactions. *Curr. Top. Med. Chem.* 18(8): 674–699.

7 Bartolome, A.P., Villaseñor, I.M. and Yang, W.C. (2013). Bidens pilosa L. (Asteraceae): Botanical properties, traditional uses, phytochemistry, and pharmacology. *Evid. Based Complement. Alternat. Med.* 2013: 340215.

8 Hung, A.M., Roumie, C.L., Greevy, R.A. et al. (2012). Comparative effectiveness of incident oral antidiabetic drugs on kidney function. *Kidney Int.* 81(7): 698–706.

9 Chang, C.L., Chen, Y.C., Chen, H.M. et al. (2013). Natural cures for type 1 diabetes: A review of phytochemicals, biological actions, and clinical potential. *Curr. Med. Chem.* 20(7): 899–907.

10 Oh, Y.S. and Jun, H.S. (2014). Role of bioactive food components in diabetes prevention: Effects on beta-cell function and preservation. *Nutr. Metab. Insights.* 7: 51–59.

11 Fukaya, M., Tamura, Y., Chiba, Y. et al. (2013). Protective effects of a nicotinamide derivative, isonicotinamide, against streptozotocin-induced β-cell damage and diabetes in mice. *Biochem. Biophys. Res. Commun.* 442(1–2): 92–98.

12 Liu, L., Du, X., Zhang, Z. et al. (2018). Trigonelline inhibits caspase 3 to protect β cells apoptosis in streptozotocin-induced type 1 diabetic mice. *Eur. J. Pharmacol.* 836: 115–121.

13 Moloudizargari, M., Mikaili, P., Aghajanshakeri, S. et al. (2013). Pharmacological and therapeutic effects of Peganum harmala and its main alkaloids. *Pharmacogn. Rev.* 7(14): 199–212.

14 Wang, P., Alvarez-Perez, J.C., Felsenfeld, D.P. et al. (2015). A high-throughput chemical screen reveals that harmine-mediated inhibition of DYRK1A increases human pancreatic beta cell replication. *Nat. Med.* 21(4): 383–388.

15 Umezawa, K., Hiroki, A., Kawakami, M. et al. (2003). Induction of insulin production in rat pancreatic acinar carcinoma cells by conophylline. *Biomed. Pharmacother.* 57(8): 341–350.

16 Lai, J.H. (2002). Immunomodulatory effects and mechanisms of plant alkaloid tetrandrine in autoimmune diseases. *Acta Pharmacol. Sin.* 23(12): 1093–1101.

17 Lee, J., Rodriguez, J.P., Lee, K.H. et al. (2017). Determination of flavonoids from Cirsium japonicum var. maackii and their inhibitory activities against aldose reductase. *Appl. Biol. Chem.* 60(5): 487–496.

18 Apaya, M.K., Kuo, T.F., Yang, M.T. et al. (2020). Phytochemicals as modulators of β-cells and immunity for the therapy of type 1 diabetes: Recent discoveries in pharmacological mechanisms and clinical potential. *Pharmacol. Res.* 156: 104754.

19 Gilbert, E.R. and Liu, D. (2013). Anti-diabetic functions of soy isoflavone genistein: Mechanisms underlying its effects on pancreatic β-cell function. *Food Funct.* 4(2): 200–212.

20 Sharp, A. and Burkhardt, B.R. (2018). Cornus officinalis increases proliferation, inhibits cytokine-induced apoptosis, and promotes the oxidative capacity of the human 1.1 B4 pancreatic cell line. *Diabetes.* 67(Supplement_1): 2467–PUB.

21 Fu, Z., Zhang, W., Zhen, W. et al. (2010). Genistein induces pancreatic β-cell proliferation through activation of multiple signaling pathways and prevents insulin-deficient diabetes in mice. *Endocrinology.* 151(7): 3026–3037.

22 Choi, M.S., Jung, U.J., Yeo, J.K. et al. (2008). Genistein and daidzein prevent diabetes onset by elevating insulin level and altering hepatic gluconeogenic and lipogenic enzyme activities in non-obese diabetic (NOD) mice. *Diabetes Metab. Res. Rev.* 24(1): 74–81.

23 Tai, Z.F., Zhang, G.L. and Wang, F. (2012). Identification of small molecule activators of the janus kinase/signal transducer and activator of transcription pathway using a cell-based screen. *Biol. Pharm. Bull.* 35(1): 65–71.

24 Yang, K.M., Jeon, S.D., So, D.S. et al. (2000). Brazilin augments cellular immunity in multiple low dose streptozotocin (MLD-STZ) induced type I diabetic mice. *Arch. Pharm. Res.* 23(6): 626–632.

25 López, P.M., de la Mora, P.G., Wysocka, W. et al. (2004). Quinolizidine alkaloids isolated from Lupinus species enhance insulin secretion. *Eur. J. Pharmacol.* 504(1–2): 139–142.

26 Cha, T.L., Qiu, L., Chen, C.T. et al. (2005). Emodin down-regulates androgen receptor and inhibits prostate cancer cell growth. *Cancer Res.* 65(6): 2287–2295.

27 Shen, M.Y., Liu, Y.J., Don, M.J. et al. (2011). Combined phytochemistry and chemotaxis assays for identification and mechanistic analysis of anti-inflammatory phytochemicals in Fallopia japonica. *PLoS One.* 6(11): e27480.

28 Prasad, C.V., Anjana, T., Banerji, A. et al. (2010). Gallic acid induces GLUT4 translocation and glucose uptake activity in 3T3-L1 cells. *FEBS Lett.* 584(3): 531–536.

29 Khazim, K., Gorin, Y., Cavaglieri, R.C. et al. (2013). The antioxidant silybin prevents high glucose-induced oxidative stress and podocyte injury in vitro and in vivo. *Am. J. Physiol. Renal Physiol.* 305(5): F691–700.

30 Soto, C., Raya, L., Juárez, J. et al. (2014). Effect of Silymarin in Pdx-1 expression and the proliferation of pancreatic β-cells in a pancreatectomy model. *Phytomed.* 21(3): 233–239.

31 Patil, S.B., Takalikar, S.S., Joglekar, M.M. et al. (2013). Insulinotropic and β-cell protective action of cuminaldehyde, cuminol and an inhibitor isolated from cuminum cyminum in streptozotocin-induced diabetic rats. *Br. J. Nutr.* 110(8): 1434–1443.

32 Lv, N., Song, M.Y., Lee, Y.R. et al. (2009). Dihydroavenanthramide D protects pancreatic β-cells from cytokine and streptozotocin toxicity. *Biochem. Biophys. Res. Commun.* 387(1): 97–102.

33 Chen, F., Chen, Y., Kang, X. et al. (2012). Anti-apoptotic function and mechanism of ginseng saponins in Rattus pancreatic β-cells. *Biol. Pharm. Bull.* 35(9): 1568–1573.

34 Beg, M., Chauhan, P., Varshney, S. et al. (2014). A withanolide coagulin-L inhibits adipogenesis modulating Wnt/β-catenin pathway and cell cycle in mitotic clonal expansion. *Phytomed.* 21(4): 406–414.

35 Aggarwal, B.B. and Harikumar, K.B. (2009). Potential therapeutic effects of curcumin, the anti-inflammatory agent, against neurodegenerative, cardiovascular, pulmonary, metabolic, autoimmune and neoplastic diseases. *Int. J. Biochem. Cell Biol.* 41(1): 40–59.

36 Aggarwal, S., Ichikawa, H., Takada, Y. et al. (2006). Curcumin (diferuloylmethane) down-regulates expression of cell proliferation and antiapoptotic and metastatic gene products through suppression of IκBα kinase and Akt activation. *Mol. Pharmacol.* 69(1): 195–206.

37 Subash-Babu, P., Ignacimuthu, S., Agastian, P. et al. (2009). Partial regeneration of β-cells in the islets of Langerhans by Nymphayol a sterol isolated from Nymphaea stellata (Willd.). *Flowers Bioorg. Med. Chem.* 17(7): 2864–2870.

38 Sinal, C.J. and Gonzalez, F.J. (2002). Guggulsterone: An old approach to a new problem. *Trends Endocrinol. Metab.* 13(7): 275–276.

39 Ammon, H.P. (2016). Boswellic acids and their role in chronic inflammatory diseases. *Adv. Exp. Med. Biol.* 928: 291–327.

40 Murillo-Rodríguez, E., Millán-Aldaco, D., Palomero-Rivero, M. et al. (2006). Cannabidiol, a constituent of Cannabis sativa, modulates sleep in rats. *FEBS Lett.* 580(18): 4337–4345.

41 Di Marzo, V., Piscitelli, F. and Mechoulam, R. (2011). Cannabinoids and endocannabinoids in metabolic disorders with focus on diabetes. *Handb. Exp. Pharmacol.* 2011(203):75–104.

42 Sandler, S., Buschard, K. and Bendtzen, K. (1994). Effects of 1, 25-dihydroxyvitamin D3 and the analogues MC903 and KH1060 on interleukin-1β-induced inhibition of rat pancreatic islet β-cell function in vitro. *Immunol. Lett.* 41(1): 73–77.

43 Chertow, B.S., Goking, N.Q., Driscoll, H.K. et al. (1997). Effects of all-trans-retinoic acid (ATRA) and retinoic acid receptor (RAR) expression on secretion, growth, and apoptosis of insulin-secreting RINm5F cells. *Pancreas.* 15(2): 122–131.

9

Obesity in Type 1 Diabetes Mellitus

Clinical Impact and Nutritional Therapy

Gülen Ecem Kalkan[1] and Burcu Ateş Özcan[2,]*

[1] *Department of Nutrition and Dietetics, Faculty of Health Science, Istanbul Atlas University, Istanbul, Turkey*
[2] *Department of Nutrition and Dietetics, Faculty of Health Science, Marmara University, Istanbul, Turkey*
* *Corresponding author*

9.1 Introduction

Diabetes Mellitus (DM) is a chronic endocrine disease characterized by hyperglycemia leading to abnormalities in the secretion or sensitivity of insulin produced in the islets of Langerhans of the pancreas, resulting in abnormalities in carbohydrate, protein, and fat metabolism [1]. Type 1 diabetes mellitus (T1DM) is a chronic autoimmune disease characterized by the autoimmune destruction of pancreatic β-cells and hyperglycemia resulting from insulin deficiency. Although T-cell-mediated destruction of β-cells appears to be the cause of the disease, it has been reported that the disease results from complex interactions between environmental factors, microbiomes, genomes, metabolism, and the immune system [2]. T1DM patients account for 5–10% of all diabetic patients, and the prevalence is gradually increasing [3]. Worldwide, the incidence and prevalence of T1DM are increasing by 2–3% annually [4].

Obesity is a chronic disease caused by increased adipose tissue in the body. Obesity is a global public health problem that affects populations of all ages and has increased dramatically in all socioeconomic groups over the past 30 years and continues to increase [5]. In the last 20 years, the prevalence of obesity has nearly tripled worldwide and is now considered an epidemic [6]. Obesity, defined as a body mass index (BMI) of ≥30 kg/m^2, affects approximately 35% of men and 40% of women in the United States of America (USA) [7]. While there are many studies today on the physiology and consequences of obesity associated with type 2 diabetes

Natural Products and their Bioactives in Antidiabetic Drug Discovery, First Edition.
Edited by Kanti Bhooshan Pandey, Maitree Suttajit and Pinar Atukeren.

(T2DM), the frequency of studies on obesity in T1DM is not the same as in T2DM. According to data from the "Diabetes Search Study in Young People" as part of the National Health and Nutrition Examination, 22.1% of young people with T1DM and 16.1% without T1DM were slightly overweight [8]. There are data that continuous subcutaneous insulin infusion (CSII), intensive insulin therapy, and multiple insulin injections used to treat T1DM lead to weight gain. Increased body weight in T1DM may lead to insulin resistance (IR) and affect metabolic control [9]. Increased insulin dose due to IR may lead to weight gain and promote obesity [10].

Double diabetes (DD) is a new term for individuals with T1DM who have clinical signs of T2DM, such as obesity and IR. Nowadays, the prevalence of people with DD symptoms is increasing significantly because of the increase in mild obesity/obesity, decreased physical activity, and unhealthy lifestyle changes. However, epidemiologic data are limited [11]. Recent studies show that DD increases the risk of developing all macrovascular comorbidities and microvascular complications in individuals with T1DM and IR, and is associated with poor glycemic control and mortality [12]. Therefore, it is important to accurately determine the potential risk factors and mechanisms to define the clinical and biochemical findings of DD. In addition, it is recommended that more studies be conducted to determine whether DD factors impact the increased risk of developing diabetes complications [13]. Despite all the advances in insulin therapy, nutrition therapy is one of the most important steps in treating T1DM and obesity.

9.2 T1DM Definition and Prevalence

T1DM is a chronic autoimmune disease that begins with pancreatic β-cell damage, causes defects in insulin secretion and/or action, is characterized by hyperglycemia, and affects carbohydrate, protein, and lipid metabolism [14]. In the vast majority (70–90%) of diabetics, loss of β-cells is the result of T1DM-associated autoimmunity (with the formation of T1DM-associated autoantibodies) (type 1A). In a smaller proportion, no immune response or autoantibodies are detected, and the cause of β-cell destruction is not fully known; this version has a strong genetic component (type 1 B) [14]. Immune-mediated T1DM (type 1A) was previously referred to as "insulin-dependent diabetes" or "young-onset diabetes" and accounts for 5–10% of all diabetic patients [15].

The etiology of T1DM is multifactorial, but the genetic susceptibility, environmental factors, immune system, and pathogenic processes underlying T1DM remain unclear [16]. The number of adults with diabetes worldwide has increased from 108 million in 1980 to 422 million in 2014. According to the International Diabetes Federation (IDF), this number is expected to reach 693 million by 2045, and the rate will increase from 8.8% to 9.9% [17]. T1DM, the most common endocrine disorder in childhood and adolescence, accounts for 5–10% of all diabetes cases. It is estimated that approximately 96 000 children under 15 years of age develop T1DM worldwide each year [18]. Europe was the region with the highest number of children and adolescents with T1DM at 28.4%, followed by North America and the Caribbean at 21.5%. The prevalence of T1DM, the number of deaths attributable to diabetes, and the health care expenditures due to T1DM continue to increase worldwide, with significant social, financial, and public health consequences [17].

9.3 Prevalence of Obesity in Patients with T1DM

Obesity is a positive energy balance in which energy intake is greater than energy expenditure. It interacts with many physiological, psychological, social, metabolic, behavioral, and genetic factors. The prevalence of childhood obesity has increased dramatically worldwide over the past three decades [19]. In the USA, obesity increased from 7% to 18% in the 6- to 11-year-old age group, and from 5% to 18% in the 12- to 19-year-old age group between 1980 and 2010. Today, approximately one-third of children and adolescents in the USA are classified as overweight or obese [20]. A simulation study conducted in the USA showed that by 2030, 55–60% of today's children will be obese [21].

The temporal evolution of overweight and obese people in the general population is well documented, but obesity rates in populations with preexisting conditions such as T1DM have not been fully studied. Traditionally, the phenotype of patients with T1DM has been considered normal weight or lean, but recent studies have shown otherwise [22]. The trend of increasing obesity prevalence has been faster in T1DM patients compared with the general population. The Pediatric Diabetes Consortium (PDC) report published in 2010 reported that 21% of adolescents with T1DM are mildly overweight and obese. Currently, about 50% of T1DM patients are either overweight or obese. In the Pittsburgh Diabetes Complications Epidemiology (EDC) study, in which adult patients with T1DM were followed for an average of 18 years, the prevalence of those overweight increased sevenfold from 29% to 42% and obese from 3% to 23%. Weight gain was unrelated to age and associated with clinical factors such as insulin therapy [23]. In a cohort study of 507 adolescents (8–16 years) with T1DM divided into four groups for ten years, 33% were overweight or obese. This suggests that the problem of obesity is increasing in T1DM patients [24].

Gilbertson et al. reported a prevalence of mild obesity and obesity in children with T1DM of 28.8% and 5.8%, respectively, in the 8- to 11-year-old age group, and 30.6% and 5.0%, respectively, in the 12- to 15-year-old age group [25]. In Canadian and Polish studies, 35% and 30% of children and adolescents with T1DM were classified as mildly overweight and obese, respectively. In addition, children and adolescents with T1DM are more overweight and obese than their healthy peers [26]. Studies of children with T1DM have shown that after 18 years of follow-up, the prevalence of obesity increases sevenfold in patients with T1DM, and insulin levels are positively related to weight gain [27]. It has been reported that a person with T1DM may gain body weight due to intensive insulin therapy throughout their life, excessive insulin consumption, recurrent hypoglycemia, and associated snacking [28].

9.4 Pathophysiology of Obesity in T1DM

Until 1993, when the Diabetes Control and Complications Trial (DCCT) results were published, individuals with T1DM were described as having normal or low body weight, a low tendency to develop metabolic syndrome and IR, and a higher risk of microangiopathic complications than cardiovascular disease. Since the publication of this study, intensive insulin therapy has been used to significantly reduce the incidence of microvascular and macrovascular complications and positively affect metabolic control. However, one of this treatment's most important side effects was increased body weight [29]. In treating T1DM, frequent blood glucose checks along with multiple daily injections or intensive insulin use, e.g. with an insulin pump, are part of the standard treatment protocol. However, intensive insulin therapy has been found to lead to excessive weight gain, abdominal obesity, IR, dyslipidemia, elevated blood pressure, and adverse cardiovascular outcomes [25].

In addition, T1DM patients are also affected by high-energy diets, eating larger portions, and sedentary lifestyles, which are the main causes of obesity worldwide [30]. Therefore, one of the triggers for the increasing prevalence of obesity in T1DM is the current obesogenic environment. In addition, recent scientific evidence has shown that other endogenous biological factors that cannot be altered, such as gut hormone profile, genetic predisposition, epigenetics, or gut microbiota, may play a role in the development of obesity [31].

Regarding the effect of insulin on obesity, it has been found that with the initiation of insulin therapy in individuals with T1DM, decreased urinary glucose loss and metabolic switch from catabolism to anabolism may lead to an increase in body weight [32]. Insulin is an anabolic hormone that inhibits protein catabolism, stimulates lipogenesis, and slows basal metabolic rate, resulting in increased fat accumulation [23]. Insulin increases lipogenesis and protein production by decreasing lipolysis and protein catabolism, and basal insulin is known to cause an increase in body weight by increasing anabolic activity [33]. Endogenous insulin enters the liver via the portal vein to suppress gluconeogenesis, whereas exogenous insulin initially circulates systemically and disproportionately affects muscle and adipose tissue compared with the liver [10].

Another situation related to biological factors involves the hormones that regulate the hunger-satiety mechanism. Gastrointestinal hormones such as ghrelin, cholecystokinin, gastric inhibitory peptide (GIP), glucagon-like peptide-1 (GLP-1), oxyntomodulin, amylin, insulin, leptin, pancreatic peptide, and pancreatic polypeptide Y (PPY) are important regulators of food intake. However, the secretion of insulin, amylin (which delays gastric emptying and inhibits food intake), and glucagon are markedly unregulated in T1DM [34]. In addition to the loss of amylin and insulin secretion induced by β-cells, α-cells are also dysfunctional, such that inadequate glucagon secretion during hypoglycemia and glucagon suppression during satiety have been observed [35]. These natural hormonal imbalances can potentially contribute to altered food intake and thus promote weight gain [36, 37].

In addition, decreased physical activity levels in patients with T1DM increases adipose tissue mass and negatively affect muscle mass and quality. This further compromises metabolic control, cardiovascular health, and cardiorespiratory fitness [38]. Patients who adhere to physical activity guidelines have better body composition, lower HbA1c levels, and a better lipid profile [39].

Obesity risk and current difficulties with weight control similarly promote disordered eating behaviors in T1DM. Numerous studies have described disordered eating patterns and eating disorders in individuals with T1DM [27]. For example, they may use insulin restriction/manipulation, which is an extremely dangerous method of weight control. Strategic insulin restriction as a “calorie-reducing” means to prevent weight gain leads to chronic hyperglycemia with symptoms including dehydration, loss of lean body tissue, and glycosuria. Glucose loss inhibits glucose utilization by muscle and fat, leading to further deterioration of glycemic control, breakdown of muscle and fat for energy, and ketogenesis [40]. It also promotes adverse medical consequences such as high rates of diabetic ketoacidosis, chronic daily hyperglycemia, suboptimal glycemic control, nephropathy, and foot problems [41].

9.5 Impact of Obesity on Complications of T1DM

The effects of obesity on macrovascular and microvascular complications in T1DM have been studied by many groups [42]. The Swedish National Diabetes Registry of nearly 21 000 adults with T1DM claims that obesity is significantly associated with an increased risk of hospitalization for heart failure [42]. Studies show that mildly overweight or obese individuals with T1DM have higher cardiometabolic risks such as hypertension, abnormal lipid profile, IR, and vascular dysfunction [43]. In another observational study in adults with T1DM, obesity was associated with the presence and progression of coronary artery calcification, a marker of subclinical atherosclerosis [44]. Similarly, data from the DCCT study have shown an increase in cardiometabolic risk factors (lipids, blood pressure) and more diffuse atherosclerosis in

T1DM patients with the greatest weight gain. Moreover, after 14 years, cardiovascular events were significantly higher in the group with intensive insulin therapy and the greatest weight gain than in the group with minimal weight gain [45].

It has also been reported that weight gain can lead to nonalcoholic fatty liver disease and changes in sex hormones, increase the risk of psychosocial problems, and thus reduce the quality of life of T1DM patients [46]. Metabolic abnormalities such as IR and metabolic syndrome frequently co-occur with obesity in the general population. These abnormalities are now considered common phenotypes in obesity and T1DM patients [47].

In addition, individuals with T1DM who are mildly overweight or obese are at higher risk for IR. Decreased insulin sensitivity contributes to both the development and progression of macrovascular and microvascular complications in T1DM [48]. It has been reported that components of metabolic syndromes, such as arterial hypertension and dyslipidemia, are more common in mildly overweight and obese adolescents with T1DM [12, 13]. In a study of the Pittsburgh EDC cohort (mean age 30.8 years), the prevalence of metabolic syndrome in T1DM ranged from 8% to 21%, depending on the definition [47]. Assessment of the components of the metabolic syndrome in 326 individuals with T1DM aged 5 to 34 years found that the prevalence of low high-density lipoprotein cholesterol was three times higher in obese individuals and of hypertension four times higher than in normal-weight individuals.

9.6 Nutritional Treatment of Obesity in T1DM

Nutrition therapy is one of the basic components of diabetes and obesity treatment. It is used for glycemic control, lipid profile, weight control, and mass muscle maintenance. Several scientific societies have published dietary recommendations for nutrition management in T1DM [2]. Lifestyle changes are essential for managing obesity in T1DM, and these changes must begin with a dietary shift to a healthy diet. Various diets can lead to weight loss in overweight or obese adults with and without T2DM (Mediterranean diet, low-carbohydrate, vegetarian, or plant-based diet) [49]; however, very little research supports any of these approaches in T1DM.

International guidelines emphasize that nutritional management should be planned to ensure optimal glycemic control and growth development, maintain ideal body weight, and prevent diabetes-related complications in people with T1DM [50]. To maintain nutritional management, dietary planning should be individualized and consider the individual's dietary habits, sociocultural structure, food preferences, family structure, and physical activity status [51].

The American Diabetes Association (ADA) reports that the key to individualized nutrition therapy for patients with T1DM is to estimate the carbohydrate content of a meal based on experience or to monitor carbohydrate intake by carbohydrate counting. This method is recommended for individuals with T1DM who receive multiple doses of insulin therapy and continuous subcutaneous insulin infusion therapy [1]. Studies have demonstrated that various eating plans, varying in macronutrient composition, can be used effectively and safely in the short term (1–2 years) to achieve weight loss in people with T1DM. These plans include structured low-calorie meal plans with meal replacements, a Mediterranean-style eating pattern, and low-carbohydrate meal plans with additional support [52].

Australia and Canada recommend a carbohydrate intake with an energy content of at least 45% for diabetics. The clinical consensus is that carbohydrate intake in older, overweight, or obese adolescents can be a lower carbohydrate intake (40% energy) with a higher protein intake (25% energy) [53]. The ISPAD-recommended macronutrient distribution for pediatric and adolescent type 1 DM is shown in Table 9.1 [53].

9.6.1 Carbohydrates

Macronutrients should be studied because they require insulin for metabolism and influence healthy eating [54]. The glycemic response to food depends on the amount of macronutrients in the meal and, in particular, the amount, type, and quantity of carbohydrates. The amount of fiber in a carbohydrate-containing food, the type of monosaccharides it contains, the structure of the starch, and the method of preparation are the most important factors affecting blood glucose levels [55]. Powers et al. reported that the proportion of energy derived from carbohydrates in individuals with T1DM compared with the general population in the same age group is 44.9% and 50.2%, respectively, and they have lower carbohydrate intakes. Although individuals with T1DM pay close attention to the carbohydrate content of their meals as part of diabetes management, the low carbohydrate intake is noteworthy. Specifically, individuals who count carbohydrates may turn to foods high in protein and fat to reduce the bolus insulin dose

Table 9.1 ISPAD clinical practice consensus guidelines 2018: nutritional management in children and adolescents with diabetes.

A guide to the distribution of macronutrients	
Carbohydrate	45% to 55% energy Moderate sucrose intake (up to 10% total energy)
Fat	30% to 35% energy <10% saturated fat + trans fatty acids
Protein	15% to 20% energy

they need during the meal. In addition, the Western diet, which is now widespread due to industrialization, contains large amounts of saturated fat, protein, omega-6 fatty acids, sodium, refined sugars, and low amounts of fiber and omega-3 fatty acids [56]. Nutritional studies of children with diabetes in many industrialized countries have shown that children with reduced carbohydrate intake tend to eat a low-quality diet [57]. A reduced-carbohydrate diet may increase the risk of hypoglycemia or decrease the effect of glucagon in treating hypoglycemia [58]. The patient's dietary habits, body weight, glycemic control, and cholesterol and triglyceride levels should be considered when determining the appropriate amount of carbohydrate to administer as part of nutrition therapy [59].

9.6.2 Dietary Fibers

Dietary fibers are an important component of dietary interventions promoting glycemic control. It is recommended that diabetic patients consume more dietary fiber because it can increase the feeling of satiety, prolong gastrointestinal transit time, and have beneficial effects on blood glucose levels [60]. It is known that a high-carbohydrate diet without increasing dietary fibers increases the risk of cardiovascular diseases by increasing the concentrations of low-density lipoproteins (LDL) and triglycerides. Therefore, the consumption of dietary fiber should be promoted in people with diabetes [59]. In 2019, a meta-analysis of 28 studies showed that soluble fiber intake affects HbA1c levels and HOMA-IR, and improves glycemic control. In particular, soluble fiber intake has a significant effect on blood lipids and the postprandial glucose curve, and compared to insoluble fiber intake, it is also significant in weight control as it increases satiety in gastric volume [1].

Dietary fiber may influence feelings of satiety by altering processes in the head, stomach, and intestinal phases during swallowing, digestion, and absorption [61]. In addition, soluble dietary fiber affects metabolism. Dietary fiber supplementation positively affects the health of overweight and obese adults by improving anthropometric and metabolic outcomes by regulating body composition and blood glucose levels. However, because of the heterogeneity of the studies, these results should be interpreted with caution [62]. Increased fiber consumption with dietary sources such as vegetables, fruits, grains, and legumes is critical in reducing or eliminating the obesity epidemic [63].

9.6.3 Protein

Eating protein has minimal acute effects on glycemic response, lipids, and hormones, generally consumed by people with diabetes and has no long-term effects on insulin requirements. According to evidence-based dietary recommendations for preventing and managing diabetes, a protein intake of 0.8–1 g/kg/day is recommended for adult individuals with diabetes. In the presence of microvascular complications, when the glomerular filtration rate begins to decline – it is 0.6–42 0.8 g/kg/day for individuals not receiving medical renal replacement therapy, 1.0–1.2 g/kg/day for individuals undergoing hemodialysis, and 1.2–1.4 g/kg/day for individuals undergoing peritoneal dialysis – protein supplementation is recommended [64]. Excessive protein consumption plays a role in the development of diabetic nephropathy. Therefore, it is recommended that daily protein consumption is kept within normal limits to delay the development of nephropathy [65].

Dietary protein may affect metabolic control in diabetes by altering the presence of gluconeogenic substrates and insulin and contraindication hormone secretion [66]. Protein breakdown and conversion to glucose in T1DM patients depend on glycemic control and the amount of insulin administered. If less insulin is administered than the optimal insulin requirement, glycemia cannot be controlled because the conversion of proteins to glucose is accelerated [59]. This suggests that the effect of dietary proteins on the delayed glycemic fluctuation after meals and the continuous glycemic fluctuation in T1DM is due to the alteration of hormones that affect glucose homeostasis and the conversion of amino acids to glucose through glycogenic pathways.

According to epidemiological and clinical studies, increased protein intake in the diet indirectly leads to decreased carbohydrate consumption and a diet rich in saturated fat. In this dietary model, there is a possibility of an increase in IR [67]. In addition to the amount of protein consumed, the quality of the protein is also important. The consumption of protein of animal origin, which is high in essential amino acids, should be adjusted to meet at least 50% of the daily protein requirement, especially in children of growing and developing age [68]. Lean meat, chicken, fish, eggs, low-fat dairy products, legumes, and oilseeds can be used as protein sources [69].

9.6.4 Lipids

The ideal amount of dietary fat for people with diabetes is controversial. The IOM (The Institute of Medicine) states that a total fat intake of 20–35% of daily energy intake for the general population is acceptable [70]. However, high total fat intake has been shown to increase the risk of being overweight and obese, and high intakes of saturated fat and trans fat have been associated with an increased risk of cardiovascular disease [71]. In terms of metabolic targets and cardiovascular disease risk in diabetic patients, the type of fat consumed is more important than the total amount of fat [72].

The main purpose of dietary fat in clinical practice is to ensure that saturated fat, trans fat, and total fat intakes do not exceed guideline recommendations [73]. To reduce the risk of cardiovascular diseases in persons with T1DM, it is recommended that 10% of energy be derived from polyunsaturated fatty acids and 20% from monounsaturated fatty acids, that saturated fat is limited to less than 7%, and that consumption of trans fats be avoided [72]. A diet similar to the Mediterranean diet (with monounsaturated fats, whole grain carbohydrates, plant foods, and less red and processed meats) is likely to result in long-term health and lower risk of cardiovascular disease [74].

Studies in diabetic patients have found that omega-3 fatty acids, i.e. long-chain polyunsaturated fatty acids from fish oils containing eicosapentaenoic acid (EPA) and docosahexaenoic acid (DHA), have no effect on glycemic control but improve the lipid profile of diabetic patients and alter platelet aggregation and reduce cardiovascular mortality [72]. Results from the DCCT showed that high fish consumption and consumption of omega-3 fatty acids, one of the long-chain polyunsaturated fatty acids, reduced albuminuria in people with T1DM [75].

9.6.5 Micronutrients and Dietary Supplements

There is no clear evidence of vitamin or mineral supplement benefits in persons with diabetes who are not substantially deficient (compared with the general population) [54]. For this reason, it is concluded that there is no need for vitamin and mineral supplementation in persons following medical nutrition therapy unless a deficiency is present [69, 76]. Currently, there is insufficient evidence to support the routine use of cinnamon, other herbs/supplements, and micronutrients such as chromium, magnesium, and vitamin D to treat diabetes. In addition, herbal products are not standardized and contain different active ingredients that may interact with other drugs and enhance their effectiveness. For this reason, diabetic patients must report their dietary supplements and herbal products to medical personnel/RDN [54].

Because the risk of developing hypertension is higher in patients with T1DM than in the general population, it is recommended that sodium intake be limited. The recommended amount for sodium intake is 1 gram per 1000 calories. It is recommended that daily sodium intake not exceed 3 grams, individuals with mild and moderate hypertension should consume less than 2.4 grams per day, and patients with hypertension and nephropathy should consume less than 2 grams per day [77]. A multivitamin may be necessary for certain populations, such as pregnant or lactating women, older adults, vegetarians, and people on very low-calorie or low-carbohydrate diets [52].

9.6.6 Natural Food Bioactives in the Prevention of Obesity in T1DM

Functional foods, when consumed as part of a balanced diet, have been proposed as a possible alternative method of weight management, obesity prevention, and improving the metabolic consequences of obesity [78]. In recent years, many animal studies, as well as human clinical trials and epidemiological studies, performed in order to investigate the possible effect of specific functional foods and their bioactive compounds on weight control and effects of blood sugar, have proposed protecting effects that were exercised by different mechanisms [79].

Tea (*Camellia sinensis*) is a widely consumed beverage around the world. Green tea is unfermented and is rich in polyphenolic compounds. Black tea is fermented and contains mostly theaflavinsand polyphenols [80]. In a meta-analysis study conducted with a large number of participants, it was observed that when the catechins in green tea were administered together with caffeine, statistically significant reductions were observed in the participants' BMI, body weight, and waist circumference, while the application of green tea alone did not positively change the anthropometric measurements [81]. Foods rich in polyphenols have been reported safe in diabetes-related obesity prevention and treatment [82]. *In vivo* and *in vitro* studies have shown that berry consumption regulates glucose, lipid, and protein metabolism by providing restorative effects on insulin signaling, secretion, and sensitivity; it has also shown that it reduces oxidative stress and inflammatory cytokine levels [83]. In some human studies, specific kinds of berries, as well as pomegranate, led to increased fat oxidation, decreased body weight, and waist circumference, while preventing obesity's metabolic consequences [79]. However, more studies are needed for further investigation of the effects of natural food bioactives on weight management in T1DM patients.

9.7 Conclusion

The prevalence of obesity has increased more rapidly in patients with T1DM than in the general population. While intensive insulin therapy, lack of physical activity, and the development of dual diabetes may explain some mechanisms of weight gain in patients with T1DM, little research has been conducted on effective weight control interventions in this long-established population.

Nutritional therapy and physical activity are important supportive treatments for diabetes because they can improve glycemic control and control body weight. However, specific nutritional interventions for T1DM are not yet found in the literature. No randomized, controlled clinical trials have examined dietary approaches to optimize weight management and glycemic control in T1DM. There is a need for studies specifically for this group and nutritional therapy protocols for patients with T1DM who have obesity.

References

1 American Diabetes Association. (2021). Classification and diagnosis of diabetes: Standards of medical care in diabetes. *Diabetes Care*. 44(Suppl. 1): S15–S33.

2 DiMeglio, L.A., Acerini, C.L., Codner, E. et al. (2018). ISPAD clinical practice consensus guidelines 2018: Glycemic control targets and glucose monitoring for children, adolescents, and young adults with diabetes. *Pediatr. Diabetes*. 19(Suppl. 27): 105–114.

3 Fullerton, B., Jeitler, K., Seitz, M. et al. (2014). Intensive glucose control versus conventional glucose control for type 1 diabetes mellitus. *Cochrane Database Syst. Rev.* 2014(2): CD009122.

4 Snouffer, E. (2018). An inexplicable upsurge: The rise in type 1 diabetes. *Diabetes Res. Clin. Pract.* 137: 242–244.

5 Greydanus, D.E., Agana, M., Kamboj, M.K. et al. (2018). Pediatric obesity: Current concepts. *Dis. Mon.* 64(4): 98–156.

6 Kjaer, I.G., Kolle, E., Hansen, B.H. et al. (2015). Obesity prevalence in Norwegian adults assessed by body mass index, waist circumference and fat mass percentage. *Clin. Obes.* 5(4): 211–218.

7 Flegal, K.M., Kruszon-Moran, D., Carroll, M.D. et al. (2016). Trends in obesity among adults in the United States, 2005 to 2014. *JAMA.* 315(21): 2284–2291.

8 DuBose, S.N., Hermann, J.M., Tamborlane, W.V. et al. (2015). Type 1 diabetes exchange clinic network and diabetes prospective follow-up registry. Obesity in youth with type 1 diabetes in Germany, Austria, and the United States. *J. Pediatr.* 167(3): 627–632.

9 Prinz, N., Schwandt, A., Becker, M. et al. (2018). Trajectories of body mass index from childhood to young adulthood among patients with type 1 diabetes – a longitudinal group-based modeling approach based on the DPV registry. *J. Pediatr.* 201: 78–85.

10 Mottalib, A., Kasetty, M., Mar, J.Y. et al. (2017). Weight management in patients with type 1 diabetes and obesity. *Curr. Diab. Rep.* 17(10): 92.

11 Merger, S.R., Kerner, W., Stadler, M. et al. (2016). Prevalence and comorbidities of double diabetes. *Diabetes Res. Clin. Pract.* 119: 48–56.

12 Pinhas-Hamiel, O., Levek-Motola, N., Kaidar, K. et al. (2015). Prevalence of overweight, obesity and metabolic syndrome components in children, adolescents and young adults with type 1 diabetes mellitus. *Diabetes Metab. Res. Rev.* 31(1): 76–84.

13 Kietsiriroje, N., Pearson, S., Campbell, M. et al. (2019). Double diabetes: A distinct high-risk group? *Diabetes Obes. Metab.* 21(12): 2609–2618.

14 American Association of Diabetes. (2021). Classification and diagnosis of diabetes: Standards of medical care in diabetes-2021. *Diabetes Care.* 44(Suppl. 1): S21–226.

15 American Diabetes Association. (2020). Classification and diagnosis of diabetes: Standards of medical care in diabetes-2020. *Diabetes Care.* 43(Suppl. 1): S14–S31.

16 Lernmark, Å. (2021). Etiology of autoimmune islet disease: Timing is everything. *Diabetes.* 70(7): 1431–1439.

17 IDF Diabetes Atlas. (2021). Brussels, Belgium: International Diabetes Federation. https://diabetesatlas.org (accesed 30 November 2022).

18 Mayer-Davis, E.J., Kahkoska, A.R., Jefferies, C. et al. (2018). ISPAD clinical practice consensus guidelines 2018: Definition, epidemiology, and classification of diabetes in children and adolescents. *Pediatr. Diabetes.* 19(Suppl. 27): 7–19.

19 Word Health Organization. (2020). Obesity and overweight. https://www.who.int/news-room/fact-sheets/detail/obesity-and-overweight (accessed 13 March 2023).

20 Güngör, N.K. (2014). Overweight and obesity in children and adolescents. *J. Clin. Res. Pediatr. Endocrinol.* 6(3): 129–143.

21 Kumar, S. and Kelly, A. (2017). Review of childhood obesity: From epidemiology, etiology, and comorbidities to clinical assessment and treatment. *Mayo Clin. Proc.* 92(2): 251–265.

22 Hemmingsson, E. (2018). Early childhood obesity risk factors: Socioeconomic adversity, family dysfunction, offspring distress, and junk food selfmedication. *Curr. Obes. Rep.* 7(2): 204–209.

23 Szadkowska, A., Madej, A., Ziółkowska, K. et al. (2015). Gender and age – Dependent effect of type 1 diabetes on obesity and altered body composition in young adults. *Ann. Agric. Environ. Med.* 22(1): 124–128.

24 Conway, B., Miller, R.G., Costacou, T. et al. (2010). Temporal patterns in overweight and obesity in type 1 diabetes. *Diabet Med.* 27(4): 398–404.

25 Baskaran C, Volkening LK, Diaz M, Laffel LM. A decade of temporal trends in overweight/obesity in youth with type 1 diabetes after the diabetes control and complications trial. *Pediatr. Diabetes.* 16(4): 263–270.

26 Gilbertson, H.R., Reed, K., Clark, S. et al. (2018). An audit of the dietary intake of Australian children with type 1 diabetes. *Nutr. Diabetes.* 8(1):10.

27 Driscoll, K.A., Corbin, K.D., Maahs, D.M. et al. (2017). Advancing care for type 1 diabetes and obesity network (ACT1ON). Biopsychosocial aspects of weight management in type 1 diabetes: A review and next steps. *Curr. Diab. Rep.* 17(8):58.

28 Bae, J.P., Lage, M.J., Mo, D. et al. (2016). Obesity and glycemic control in patients with diabetes mellitus: Analysis of physician electronic health records in the US from 2009–2011. *J. Diabetes Comp.* 30(2): 212–220.

29 Priya, G. and Kalra, S. (2018). A review of insulin resistance in type 1 diabetes: Is there a place for adjunctive metformin? *Diabetes Ther.* 9(1): 349–361.

30 Solmi, M., Ioannidis, J.P.A. and Carvalho, A.F. (2019). Environmental risk factors and interventions for obesity. *Eur. J. Clin. Invest.* 49(5): e13080.

31 Durrer Schutz, D., Busetto, L., Dicker, D. et al. (2019). European practical and patient-centered guidelines for adult obesity management in primary care. *Obes. Facts.* 12(1): 40–66.

32 Jacob, A.N., Salinas, K., Adams-Huet, B. et al. (2006). Potential causes of weight gain in type 1 diabetes mellitus. *Diabetes Obes. Metab.* 8(4): 404–411.

33 Russell-Jones, D. and Khan, R. (2007). Insulin-associated weight gain in diabetes– causes, effects, and coping strategies. *Diabetes Obes. Metab.* 9(6): 799–812.

34 Woods, S.C., Lutz, T.A., Geary, N. et al. (2006). Pancreatic signals controlling food intake; insulin, glucagon, and amylin. *Philos. Trans. R Soc. Lond. B Biol. Sci.* 361(1471): 1219–1235.

35 McCrimmon, R.J. and Sherwin, R.S. (2010). Hypoglycemia in type 1 diabetes. *Diabetes.* 59(10): 2333–2339.

36 Dinneen, S., Alzaid, A., Turk, D. et al. (1995). Failure of glucagon suppression contributes to postprandial hyperglycemia in IDDM. *Diabetologia.* 38: 337–343.

37 Robertson, K., Riddell, M.C., Guinhouya, B.C. et al. (2014). International society for pediatric and adolescent diabetes. ISPAD clinical practice consensus guidelines 2014. Exercise in children and adolescents with diabetes. *Pediatr. Diabetes.* 15(Suppl. 20): 203–223.

38 Krause, M.P., Riddell, M.C. and Hawke, T.J. (2011). Effects of type 1 diabetes mellitus on skeletal muscle: Clinical observations and physiological mechanisms. *Pediatr. Diabetes.* 12(4 Pt 1): 345–364.

39 Bohn, B., Herbst, A., Pfeifer, M. et al. (2015). DPV Initiative. Impact of physical activity on glycemic control and prevalence of cardiovascular risk factors in adults with type 1 diabetes: A cross-sectional multicenter study of 18,028 patients. *Diabetes Care.* 38(8): 1536–1543.

40 Goebel-Fabbri, A.E., Fikkan, J., Franko, D.L. et al. (2008). Insulin restriction and associated morbidity and mortality in women with type 1 diabetes. *Diabetes Care.* 31(3): 415–419.

41 Neumark-Sztainer, D., Patterson, J., Mellin, A. et al. (2002). Weight control practices and disordered eating behaviors among adolescent females and males with type 1 diabetes: Associations with sociodemographics, weight concerns, familial factors, and metabolic outcomes. *Diabetes Care.* 25(8): 1289–1296.

42 Vestberg, D., Rosengren, A. and Olsson, M. (2013). Relationship between overweight and obesity with hospitalization for heart failure in 20 985 patients with type 1 diabetes: A population-based study from the Swedish National diabetes registry. *Diabetes Care.* 36: 2857–2861.

43 Calella, P., Gallè, F., Fornelli, G. et al. (2020). Type 1 diabetes and body composition in youth: A systematic review. *Diabetes Metab. Res. Rev.* 36: e3211.

44 Rodrigues, T.C., Veyna, A.M., Haarhues, M.D. et al. (2011). Obesity and coronary artery calcium in diabetes: The Coronary Artery Calcification in Type 1 Diabetes (CACTI) study. *Diabetes Technol. Ther.* 13: 991–996.

45 Purnell, J.Q., Hokanson, J.E., Marcovina, S.M. et al. (1998). Effect of excessive weight gain with intensive therapy of type 1 diabetes on lipid levels and blood pressure: Results from the DCCT. Diabetes Control and Complications Trial. *JAMA.* 280(2): 140–146.

46 Birkebaek, N.H., Kahlert, J., Bjarnason, R. et al. (2018). Body mass index standard deviation score and obesity in children with type 1 diabetes in the Nordic Countries. HbA 1c and other predictors of increasing BMI SDS. *Pediatr. Diabetes.* 19(7): 1198–1205.

47 Corbin, K.D., Driscoll, K.A., Pratley, R.E. et al. (2018). Advancing Care for Type 1 Diabetes and Obesity Network (ACT1ON), obesity in type 1 diabetes: Pathophysiology, clinical impact, and mechanisms. *Endocr. Rev.* 39(5): 629–663.

48 Bjornstad, P., Snell-Bergeon, J.K., Nadeau, K.J. et al. (2015). Insulin sensitivity and complications in type 1 diabetes: New insights. *World J. Diabetes*. 6(1): 8–16.

49 American Diabetes Association. (2021). Facilitating behavior change and well-being to improve health outcomes: Standards of medical care in diabetes-2021. *Diabetes Care*. 44(Suppl.1): S53–S72.

50 Tascini, G., Berioli, M.G., Cerquiglini, L. et al. (2018). Carbohydrate counting in children and adolescents with type 1 diabetes. *Nutrients*. 10(1): 109.

51 Ewers, B., Vilsboll, T., Andersen, H.U. et al. (2019). The dietary education trial in carbohydrate counting. *BMJ Open*. 9(9): e029859.

52 American Diabetes Association Professional Practice Committee. (2022). Facilitating behavior change and well-being to improve health outcomes: Standards of medical care in diabetes—2022. *Diabetes Care*. 45(Suppl.1): S60–S82.

53 Smart, C.E., Annan, F., Higgins, L.A. et al. (2018). ISPAD clinical practice consensus guidelines 2018: Nutritional management in children and adolescents with diabetes. *Pediatr. Diabetes*. 19: 136–154.

54 Evert, A.B., Dennison, M., Gardner, C.D. et al. (2019). Nutrition therapy for adults with diabetes or prediabetes: A consensus report. *Diabetes Care*. 42: 731–754.

55 Ayyıldız, F. and Akbulut, G. (2019). Medical Nutrition Therapy in Endocrine and Cardiometabolic Diseases. 1. (ed. G. Akbulut). 180–274. Ankara: Ankara Nobel Tıp Kitapevi.

56 Powers, M.A., Gal, R.L., Connor, C.G et al. (2018). Eating patterns and food intake of persons with type 1 diabetes within the T1D exchange. *Diabetes Res. Clin. Pract*. 141: 217–228.

57 Øverby, N.C., Flaaten, V., Veierød, M.B. et al. (2007). Children and adolescents with type 1 diabetes eat a more atherosclerosis-prone diet than healthy control subjects. *Diabetologia*. 50(2): 307–316.

58 Ranjan, A., Schmidt, S., Damm-Frydenberg, C. et al. (2017). Low-carbohydrate diet impairs the effect of glucagon in the treatment of insulin-induced mild hypoglycemia: A randomized crossover study. *Diabetes Care*. 40(1): 132–135.

59 Tüfekçi Alphan, E. (2013). Diabetes mellitus and nutritional treatment. In: *Nutritional Therapy in Diseases* (ed. E. Alphan Tüfekçi), 415–509. 1st edn. Ankara: Hatipoğlu Publishing House.

60 Nansel, T.R., Lipsky, L.M. and Liu, A. (2016). Greater diet quality is associated with more optimal glycemic control in a longitudinal study of youth with type 1 diabetes. *Am. J. Clin. Nutr*. 104(1): 81–87.

61 Burton-Freeman, B. (2000). Dietary fiber and energy regulation. *J. Nutr*. 13: 272–275.

62 Thompson, S.V., Hannon, B.B., An, R. et al. (2017). Effects of isolated soluble fiber supplementation on body weight, glycemia, and insulinemia in adults with overweight and obesity: A systematic review and meta-analysis of randomized controlled trials. *Am. J. Clin. Nutr*. 106(6): 1514–1528.

63 Slavin, J.L. (2005). Dietary fiber and body weight. *Nutrition*. 21(3): 411–418.

64 Evert, A.B., Dennison, M., Gardner, C.D. et al. (2019). Nutrition therapy for adults with diabetes or prediabetes: A consensus report. *Diabetes Care*. 42(5): 731–754.

65 Jovanovic, S.M. (2013). Nutritional management in type 1 diabetes mellitus. In: *Type 1 Diabetes [Internet]* (ed. A.P. Escher and A Li). London: IntechOpen [cited 2022 Dec 02].

66 Paterson, M., Bell, K.J., O'Connell, S.M. et al. (2015). The role of dietary protein and fat in glycaemic control in type 1 diabetes: Implications for intensive diabetes management. *Curr. Diab. Rep*. 15(9): 1–9.

67 Franz, M.J., Boucher, J.L. and Evert, A.B. (2014). Evidence-based diabetes nutrition therapy recommendations are effective: The key is individualization. *Diabetes Metab. Syndr. Obes. Targets Ther*. 7: 65–72.

68 Bilen, H. (2012). T1DM medical nutrition therapy. *Turkey Clin. J. Endocrin-Special Top*. 5(3): 47–52.

69 Abdelghaffar, S. (2015). Nutritional management of type 1 diabetes. In: *Major Topics in Type 1 Diabetes* (ed. K.P Nunes) 121–135.Frankfurt: InTech.

70 American Diabetes Association. (2017). Standards of medical care in diabetes-2017. *Diabetes Care*. 40(1): 1–132.

71 Sacks, F.M., Lichtenstein, A.H., Wu, J.H.Y. et al. (2017). Dietary fats and cardiovascular disease: A presidential advisory from the American Heart Association. *Circulation*. 136(3): e1–23.

72 Dworatzek, P.D., Arcudi, K., Gougeon, R. et al. (2013). Nutrition therapy. *Can. J. Diabetes*. 37: 45–55.

73 Becker, W., Lyhne, N., Pedersen, A.N. et al. (2004). Nordic nutrition recommendations 2004 - integrating nutrition and physical activity. *Scand. J. Nutr*. 48: 178–187.

74 Cadario, F., Prodam, F., Pasqualicchio, S. et al. (2012). Lipid profile and nutritional intake in children and adolescents with type 1 diabetes improve after a structured dietician training to a Mediterranean-style diet. *J. Endocrinol. Invest*. 35(2): 160–168.

75 Lee, C.C., Sharp, S., Wexler, D. et al. (2010). Dietary intake of eicosapentaenoic and docosahexaenoic acid and diabetic nephropathy: Cohort analysis of the diabetes control and complications trial. *Diabetes Care*. 33: 1454–1456.

76 Turkish Endocrinology and Metabolism Society. (2016). Diabetes Mellitus and its Complications Diagnosis, Treatment and Follow-up Guide 2016, 15–93. Ankara: Miki Printing.

77 Nutrition Recommendations and Interventions for Diabetes. (2008). A position statement of the American Diabetes Association. *Diabetes Care*. 31(Suppl. 1): S61.

78 Ozen, A.E., Pons, A. and Tur, J.A. (2012). Worldwide consumption of functional foods: A systematic review. *Nutr. Rev*. 70: 472–481.

79 Konstantinidi, M. and Koutelidakis, A.E. (2019). Functional foods and bioactive compounds: A review of its possible role on weight management and obesity's metabolic consequences. *Medicines (Basel).* 6(3): 94.

80 Yuda, N., Tanaka, M., Suzuki, M. et al. (2012). Polyphenols extracted from black tea (Camellia sinensis) residue by hot-compressed water and their inhibitory effect on pancreatic lipase in vitro. *J. Food Sci.* 77(12): H254–261.

81 Phung, O.J., Baker, W.L., Matthews, L.J. et al. (2010). Effect of green tea catechins with or without caffeine on anthropometric measures: A systematic review and meta-analysis. *Am. J. Clin. Nutr.* 91(1): 73–81.

82 Ley, S.H., Hamdy, O., Mohan, V. et al. (2012). Prevention and management of type 2 diabetes: Dietary components and nutritional strategies. *Lancet.* 383(9933): 1999–2007.

83 Hameed, A., Galli, M., Adamska-Patruno, E. et al. (2020). Select polyphenol-rich berry consumption to defer or deter diabetes and diabetes-related complications. *Nutrients.* 12(9): 2538.

10

Protective Effects of Natural Non-insulin Drugs against Type 1 Diabetes Mellitus

Waseem Hassan[1,*], *Faraza Javaid*[2], *Andleeb Shahzadi*[3], *and Sahar Bakht*[4]

[1] *Department of Pharmacy, COMSATS University Islamabad, Lahore Campus, Lahore 54000, Pakistan*
[2] *Quaid-e-Azam College of Pharmacy, Sahiwal, Pakistan*
[3] *Department of Medical Pharmacology, Cerrahpaşa Medical Faculty, Istanbul University-Cerrahpaşa, Istanbul, Turkey.*
[4] *Department of Pharmaceutical Chemistry, Government College University Faisalabad, Faisalabad, Pakistan*
[*] *Corresponding author*

10.1 Introduction

Autoimmune illnesses are both debilitating and costly to treat. The National Institutes of Health in the United States of America estimated in 2005 that 8% of the population, or 23.5 million individuals, suffer from autoimmune illnesses, with direct health-care expenses surpassing $100 billion each year. A total of over 100 autoimmune illnesses have been discovered [1]. Type-1 diabetes mellitus (T1DM), commonly known as autoimmune diabetes, affects an estimated 10 million people around the world. T1DM is triggered and initiated by environmental, genetic, and immunological variables, just like many other autoimmune illnesses [2]. Infectious agents, commensal enterobacteria, sex hormones, and dietary variables all play a role in the immune system's formation and (re)shaping [3]. Several genes, including major histocompatibility complex class II (MHC II), cytotoxic T-lymphocyte antigen 4 (CLTA4), insulin, among others have been linked to the immunological response that controls T1DM outcome. The immunological response in T1DM patients is also influenced by immune system dysregulation [4, 5]. T1DM is caused by invading immune cells killing insulin-producing pancreatic cells, resulting in hypoinsulinemia, hyperglycemia, and life-threatening consequences including blindness, chronic ulceration, kidney disease, and heart disease. Slow progress has been made in T1DM therapeutic research and development due to the intricacy of the immune and endocrine systems. Because insulin is now the sole medicine authorized to treat

Natural Products and their Bioactives in Antidiabetic Drug Discovery, First Edition.
Edited by Kanti Bhooshan Pandey, Maitree Suttajit and Pinar Atukeren.

T1DM, new therapeutic agents for T1DM treatment are urgently needed. Plants have long been regarded as a helpful "green pharmacy" that provides natural treatments for T1DM and a variety of other disorders. Phytochemicals are prospective pharmacological leads for T1DM treatment because they have the ability to control an overactive immune response by modulating immune cells and/or enhancing the (re)generation and/or survival of cells [6]. The bioactivities and clinical potential of plant drugs and extracts for the treatment of T1DM are discussed in this chapter.

10.2 Pathogenesis of T1DM

T1DM pathogenesis is complex, involving several pathways and coordinated by the immunological and endocrine systems. The main variables that contribute to T1DM development include genetic, immunological, and/or environmental events. The outcome of T1DM and associated complications is influenced by dysregulation of immunity and cell homeostasis [7]. Antigen-presenting cells (APC), dendritic cells (DC), B cells, and macrophages capture and deliver auto-antigens (such as insulin, glutamic acid decarboxylase (GAD65), and other proteins) to T-cells during T1DM development. T-cells develop into multiple T-cell subsets after activation [8]. Insulitis occurs when these cells invade the pancreas, causing the release of inflammatory cytokines such as interferon (IFN), tumor necrosis factor (TNF), and perforin. Other immune cells, such as regulatory T (Treg) cells, myeloid-derived suppressor cells (MDSC), natural killer (NK) cells, and others, can contribute to misdirected autoimmunity. As a result, the cells suffer apoptosis, which causes them to malfunction [9]. The pathogenic condition of diabetes is linked to the loss of cell bulk and function. Restoration of antigen-specific immunological tolerance appears to be a viable strategy for combating T1DM from the standpoint of immunity. Inactivation of the misdirected immune response and improvements of immunological tolerance in distinct immune cells have been shown to be efficacious in animal models of T1DM. However, in human clinical trials, such tactics have so far failed. The clinic's limited time window during "the honeymoon stage", when leftover cells are sufficient for insulin production, is one reason for this failure. A new advancement has enhanced early T1DM diagnosis and aids in accurately identifying people before the honeymoon period and preventing the condition [10]. Immune inactivation/tolerance has been linked to self-antigens, major histocompatibility complexes (MHCs), antigen receptors, co-receptors, cytokines, chemokines, reactive oxygen species (ROS), death ligands/receptors, and their downstream mediators (caspases, nuclear factors, kinases, and so on). For the prevention and treatment of T1DM, maintaining cell bulk and functionality is just as crucial as restoring immunological tolerance. Targeting and modifying cell growth, proliferation, survival, and death of cells, differentiation from cell progenitors, and control of cell homeostasis would appear to be reasonable anti-T1DM treatments from this perspective (the endocrine perspective) [11]. In the embryonic and neonatal phases, pancreatic cells replicate at an exponential pace. Unless certain criteria are met, replication becomes exceedingly low after these phases. Interactions between metabolic signals, growth factors, cytokines, and hormones have been shown to promote proliferation, survival, and function in pancreatic ß-cells while maintaining homeostasis. Downstream factors hypothesized to contribute to cell homeostasis include nuclear factors, kinases, cell cycle regulators, ER stress regulators, death ligands/receptors/adaptors, and so on [9]. Immunotherapy and insulin/cell replacement therapy are now being investigated as T1DM treatment alternatives. More and more evidence suggests that early-stage T1DM combination therapy that (re)establishes both immunological tolerance and cell homeostasis may be an effective T1DM prevention and treatment method. This method may be able to treat the disease at its earliest stages. Despite minor efforts in T1DM research, there is currently no T1DM prevention or cure. For T1DM patients, insulin injection is currently the only alternative. Researchers in academia and the pharmaceutical sector have put in a lot of time and effort to find potential therapeutic targets for immunological disorders like T1DM. As a result of this research, many druggable genes linked to immunity have been proposed [10]. Surprisingly, some signaling molecules and/or pathways connected to immune cells and/or pancreatic ß-cells have been observed to be influenced by these substances and extracts. As a result, the functional interplay of these compounds and their signaling networks in immune cells and cells was investigated, as detailed in [12]. The goal of this chapter is to provide new findings in T1DM phytotherapy, with an emphasis on unique ethnopharmacological techniques, contemporary research on specific plants and isolated chemicals, and, most importantly, recent contributions from clinical trials.

10.3 Medicinal Plants in the Treatment of T1DM

The use of medicinal plants in the treatment of diabetes mellitus is once again being investigated. Many pharmaceuticals are created from prototypic compounds found in medicinal plants [13]. Over 400 traditional plant remedies for diabetes have been reported to date, but only a tiny percentage of these have been scientifically and medically evaluated to determine their efficacy. In human and animal models of T1DM, the hypoglycemic impact of several plant extracts has been proven. Traditional medicinal plants should be researched further to find some remedy against T1DM, according to the World Health Organization's Expert

Committee on Diabetes. The absence of scientific and clinical data showing herbal medicine's efficacy and safety is a major impediment to its incorporation into modern medical procedures [6]. Clinical research on herbal medications is required, as is the development of simple bioassays for biological standardization, pharmacological and toxicological evaluation, and the development of numerous animal models for toxicity and safety testing. The majority of investigations on herbal and natural products have found that they have greater anti-diabetic effects than synthetic medications at a lower cost and with fewer adverse effects. The majority of anti-diabetic herbs can enhance ß-cell function and boost the islet of Langerhans' capacity to secrete insulin, which is highly antioxidant. Some of the traditional plants with significant findings in the management of T1DM are summarized in Table 10.1.

Table 10.1 Traditional plants used in the management of diabetes mellitus type-1.

Plant Name	Common Name	Parts Used
Panax ginseng	Korean Ginseng	Whole plant [14]
Anoectochilus roxburghii	Copper Glint Orchid	Whole plant [15]
Camellia sinensis	Chinese Tea	Leaves [16]
Tabernaemontana divaricata	Pinwheel Flower	Leaves, flowers [17, 18]
Carcia papaya	Papaya, Ketalah	Leaves [19]
Momordica charantia	Karela	Fruit [20]
Pandanus amaryllifolius	Pandan	Leaves [21]
Fallopia japonica	Japanese Knotweed	Whole plant [1]
Tripterygium wilfordii	Thunder Duke Vine	Whole plant [22]
Helleborus purpurascens	Lenten Rose	Whole plant [23]
Olea europaea	Common Olive	Leaves, fruit [24]
Allium cepa	Pyaz	Fruit [20]
Allium sativum	Lehsan	Fruit [20]
Psacalium peltatum	Indian Bush	Whole plant [1]
Bidens pilosa	Black Jack	Whole plant [25]
Annona squamosa	Sugar Apple	Leaves [26]
Beta vulgaris	Chukkander	Leaves [27]
Trigonella foenum graecum	Methi	Seeds [28]
Cinnamon zylanicum	Darcheeni	Bark [20]
Boerhavia diffusa	Punarnava	Roots [29]
Eugenia Jambo	Jamun	Fruit [30]

(Continued)

Table 10.1 (Continued)

Plant Name	Common Name	Parts Used
Mangifera indica	Mango	Leaves [20]
Nigella sativa	Kalonji	Seeds [31]
Hibiscus rosa-sinensis	Gudhal or Jasson	Flowers [32]
Ipomoea batatas	Sakkargand	Fruit [33]
Swertia chirayita	Chirata	Leaves [34]
Tinospora crispa	Makabuhai	Whole plant [35]

10.4 Herbal Drug Formulations

Herbal formulations have traditionally played a pivotal role in the long-term and short-term management of T1DM. These products are marketed to improve glucose consumption in the peripheral circulation, boost hepatic and muscle glucagon levels, enhance β-cell repair and regeneration, and raise c peptide levels. Most of the established herbal products claim to protect β-cells from oxidative stress and possess antioxidant effects. Moreover, these products often reduce glycated hemoglobin levels, normalize microalbuminuria, and modulate the lipid profile, acting similarly to insulin, subsequently reducing the risk of long-term diabetes problems [20]. Admark Herbals Ltd. claims that Dia-Care is beneficial for Type 1 and Type 2 diabetes within 90 days of treatment and cures within 18 months. The treatment is divided into six phases, each lasting 90 days. Five grams (1 teaspoon) of powder is combined with half a glass of water, thoroughly mixed, and stored overnight. Only the water, not the sediment, should be consumed first thing in the morning on an empty stomach. Fresh water is poured into the remaining medicine, which is preserved for the entire day and swallowed half an hour before dinner [36].

Plethico Laboratory's Syndrex comprises fenugreek seed extracts that have been germinated. Fenugreek has been utilized in traditional formulas for over 1000 years. We are currently investigating the mechanism of action of this anti-diabetic medication utilizing both animal models and cultured islet cells [37].

Antihyperlipidemic and antioxidant activities are also present in the polyherbal antihyperglycemic formulation known as Glyoherb. It is advertised to avoid, postpone, or treat side effects like retinopathy, neuropathy, and retinopathy. Results from studies on animals suggested that it improves glucose tolerance and reduces blood glucose levels in T1DM rats induced by Streptozotocin (STZ). Additionally, compared to control groups, Glyoherb therapy has been proven to lower blood cholesterol, triglycerides (TGs), very-low-density lipoprotein (VLDL), and low-density lipoprotein (LDL) levels; however, it is not as effective as glibenclamide. Additionally, it has been demonstrated to raise high-density lipoprotein (HDL) levels. In STZ-induced impaired kidney and liver functions, glyoherb did not exhibit any harmful effects; instead, it enhanced kidney and liver functions.

Additionally, Diabeta Plus is a strong immunomodulator, an antihyperlipidemic, an antidepressant, and a hepatoprotective formulation with anti-diabetic properties. Utilizing Diabeta alone can avoid or delay difficulties in pre-diabetics by combating several variables that can lead to the development of diabetes. When combined with oral hypoglycemic medications, it aids in reducing resistance and purports to relieve complaints of pruritus, discomfort, and polyuria brought on by diabetes mellitus [20].

Many different plants have been utilized to treat diabetes and its consequences, either individually or in formulations. The active ingredients in this herbal mixture are not well described, which is one of the primary flaws. It's critical to understand the active ingredient and the molecular interactions, as this will aid in determining the product's therapeutic efficacy and standardization. Model systems are currently being used to examine the mechanisms of action of some of these plants.

10.5 Phytochemicals Constituents in the Treatment of T1DM

Phytochemicals are chemical molecules derived from live organisms. Plants are the principal suppliers of natural chemicals. Secondary metabolites are natural compounds synthesized by plants, and their function in plants is gaining attention due to their use as dyes, glues, oils, waxes, flavoring agents, drugs, and perfumes, as well as potential sources of natural drugs, antibiotics, insecticides, and herbicides [20]. Medicinal plants have emerged as promising solutions for both preventing and treating a variety of disorders. About 150 phytochemicals have undergone extensive research, and more than 4000 phytochemicals have been compiled in a database [38]. Alkaloids, terpenes, flavonoids, lignans, steroids, curcumin, saponin, flavonoids, glucosides, and phenolics are some of the main fundamental components. Polyphenols make up the most numerous and widely dispersed class of bioactive chemicals. A complex class of plant compounds known as polyphenols includes one or more benzene rings as well as various numbers of hydroxyl (OH), carbonyl (CO), and carboxylic acid (COOH) groups. These often have one or more sugar residues connected to them (i.e. they are conjugated). The most prevalent class of polyphenols is flavonoids. These substances can reduce the negative effects of ROS and free radicals, which can contribute to the premature death of immune cells. The formation of numerous types of immune cells and the phagocytic activity of macrophages are both promoted by bioactive food ingredients, enhancing disease resistance [12]. Some of the isolated phytocompounds used in the management of T1DM are summarized in Table 10.2.

Table 10.2 Phytochemicals used in the management of T1DM.

Name	Classification
Ginsenosides	Saponin [39]
Kinsenoside	Glucoside [1]
Berberine	Alkaloid [40]
Epigallocatechin 3-gallate (EGCG)	Flavanol [41]
Conophylline	Alkaloid [42]
Nicotinamide	Pyridine [43]
Nymphayol	Steroids [44]
Silymarin	Flavonoid [45]
Diacerhein	Anthraquinone [46]
Cannabidiol	Cannabinoids [47]
Resveratrol	Polyphenol [1]
Capsaicin	Alkaloid [48]
Harmine	Alkaloid [49]
Trans-*N*-Caffeoyltyramine	Polyphenol [50]
Cyanidin-3-*O*-glucoside	Polyphenol [51]
Kazinol C	Polyphenol [52]
Silybin	Flavonoid [53]
Steviol glycoside	Terpenes [54]
Gallic acid	Phenolic acid [55]
Cuminol	Aromatic terpenoids [56]
Lupanine	Alkaloid [57]
Shikonin	Quinone [58]
Withanolide	Steroids [23]
Alfacalcidol	Vitamin [59]
Salvianolic acid B	Flavonoid [60]
Thymoquinone	Quinone [61]
Naringin	Flavonoid [62]
Curcumin	Polyphenol [63]
Palbinone	Terpenes [64]
Borapetoside A	Terpenes [23]
Guggulsterone	Steroids [23]
Methyl caffeate	Polyphenol [65]

Table 10.2 (Continued)

Name	Classification
Catechin	Flavonoid [23]
Quercetin	Flavonoid [66]
Kaempferol	Flavonoid [67]
Pycnanthuquinone	Quinone [23]
Nymphayol	Steroids [1]

10.6 Conclusion

Natural products offer a great deal of potential for clinical application in the future as alternative therapies for the management of T1DM and its associated problems due to their antioxidant and even blood glucose-lowering properties. Although natural products have a promising future in the treatment of diabetes, there are a few obstacles that should be carefully evaluated before using them in clinical studies. Natural products can be used on several targets. This fact could result in off-target consequences that are both advantageous and destructive. As a result, more focused compounds should be created. It is necessary to understand the intricate mechanisms, the outcome of which can serve as a strong foundation for the creation of innovative, highly selective molecules. Additionally, some natural products' ability to reduce blood sugar levels is advantageous. However, it is a concern to know whether or not natural ingredients cause diabetic patients' blood glucose levels to fluctuate. Therefore, when treating T1DM or its complications, special attention should be paid to the precise regulation of blood glucose levels using a combination of conventional medications and natural products as alternative therapies.

References

1 Chang, C.L., Chen, Y.C., Chen, H.M. et al. (2013). Natural cures for type 1 diabetes: A review of phytochemicals, biological actions, and clinical potential. *Curr. Med. Chem.* 20(7): 899–907.

2 Norris, J.M., Johnson, R.K. and Stene, L.C. (2020). Type 1 diabetes-early life origins and changing epidemiology. *Lancet Diabetes Endocrinol.* 8(3): 226–238.

3 Li, L., Liu, S. and Yu, J. (2020). Autoimmune thyroid disease and type 1 diabetes mellitus: Same pathogenesis; new perspective? *Ther. Adv. Endocrinol. Metab.* 11: 2042018820958329.

4 Nekoua, M.P., Dechaumes, A., Sane, F. et al. (2020). Enteroviral pathogenesis of type 1 diabetes: The role of natural killer cells. *Microorganisms*. 2020; 8:(7) 989.
5 Warshauer, J.T., Bluestone, J.A. and Anderson, M.S. (2020). New frontiers in the treatment of type 1 diabetes. *Cell Metab.* 31(1): 46–61.
6 Shanmugam, K.R., Shanmugam, B., Subbaiah, G.V. et al. (2021). Medicinal plants and bioactive compounds for diabetes management: Important advances in drug discovery. *Curr. Pharm. Des.* 27(6): 763–774.
7 Lim, S., Bae, J.H., Kwon, H.S. et al. (2021). COVID-19 and diabetes mellitus: From pathophysiology to clinical management. *Nat. Rev. Endocrinol.* 17(1): 11–30.
8 Roep, B.O. (2003). The role of T-cells in the pathogenesis of type 1 diabetes: From cause to cure. *Diabetologia*. 46(3): 305–321.
9 Darenskaya, M.A., Kolesnikova, L.I. and Kolesnikov, S.I. (2021). Oxidative stress: Pathogenetic role in diabetes mellitus and its complications and therapeutic approaches to correction. *Bull. Exp. Biol. Med.* 171(2): 179–189.
10 Li, M., Song, L.J. and Qin, X.Y. (2014). Advances in the cellular immunological pathogenesis of type 1 diabetes. *J. Cell. Mol. Med.* 18(5): 749–758.
11 Clark, M., Kroger, C.J. and Tisch, R.M. (2017). Type 1 diabetes: A chronic anti-self-inflammatory response. *Front. Immunol.* 8: 1898.
12 Samtiya, M., Aluko, R.E., Dhewa, T. et al. (2021). Potential health benefits of plant food-derived bioactive components: An overview. *Foods (Basel, Switzerland).* 10(4): 839.
13 Hachkova, H., Nagalievska, M., Soliljak, Z. et al. (2021). Medicinal plants galega officinalis L. and Yacon leaves as potential sources of antidiabetic drugs. *Antioxidants (Basel, Switzerland).* 10(9): 1362.
14 Hong, Y.J., Kim, N., Lee, K. et al. (2012). Korean red ginseng (Panax ginseng) ameliorates type 1 diabetes and restores immune cell compartments. *J. Ethnopharmacol.* 144(2): 225–233.
15 Cui, S.C., Yu, J., Zhang, X.H. et al. (2013). Antihyperglycemic and antioxidant activity of water extract from anoectochilus roxburghii in experimental diabetes. *Exp. Toxicol. Pathol.* 65(5): 485–488.
16 Sharangi, A. (2009). Medicinal and therapeutic potentialities of tea (Camellia sinensis L.)–A review. *Food Res. Int.* 42(5–6): 529–535.
17 Rahman, M., Sayeed, M.A., Biplab, K.P. et al. (2012) Antidiabetic and cytotoxic activities of methanolic extract of Tabernaemontana divaricata (L.) leaves in alloxan induced mice. *Asian J. Pharm. Clin. Res.* 5: 49–52.
18 Muzammil, S., Andleeb, R., Hayat, S. et al. (2022). Integrating in silico and in vitro approaches to screen the antidiabetic properties from Tabernaemontana divaricata (Jasmine) flowers. *Evid. Based Complement. Alternat. Med: eCAM.* 2022: 4616815.
19 Juárez-Rojop, I.E., Díaz-Zagoya, J.C., Ble-Castillo, J.L. et al. (2012). Hypoglycemic effect of carica papaya leaves in streptozotocin-induced diabetic rats. *BMC Complement. Altern. Med.* 12: 236.

20 Adedapo, A. and Ogunmiluyi, I. (2020). The use of natural products in the management of diabetes: The current trends. *J. Drug Deliv. Ther.* 10(1): 153–162.

21 Suryani, C., Tamaroh, S. and Budipitojo, T. (2018). Increased of hypoglycemic effect and pancreatic regeneration of pandanus amaryfollius leaves ethyl acetate extract in streptozotocin-induced diabetic rats. *Int. Food Res. J.* 25(5): 1792.

22 Shi, N.W., Wu, Y.J. and Wang, Y.F. (2006). Study on the mechanism of tripterygium wilfordii ployglycosidium in preventing type 1 nonobese diabetic mice. Zhongguo Zhong Xi Yi Jie He Za Zhi. 26(Suppl): 79–83.

23 Apaya, M.K., Kuo, T.F., Yang, M.T. et al. (2020). Phytochemicals as modulators of β-cells and immunity for the therapy of type 1 diabetes: Recent discoveries in pharmacological mechanisms and clinical potential. *Pharmacol. Res.* 156: 104754.

24 Alesci, A., Miller, A., Tardugno, R. et al. (2022). Chemical analysis, biological and therapeutic activities of Olea europaea L. *Extracts Nat. Prod. Res.* 36(11): 2932–2945.

25 Yang, W.-C. (2014). Botanical, pharmacological, phytochemical, and toxicological aspects of the antidiabetic plant *Bidens pilosa* L. *Evid -based Complement. Altern. Med: eCAM.* 2014: 698617.

26 Gupta, R.K., Kesari, A.N., Murthy, P. et al. (2005). Hypoglycemic and antidiabetic effect of ethanolic extract of leaves of annona squamosa L. in experimental animals. *J. Ethnopharmacol.* 99(1): 75–81.

27 Abd El-Ghffar, E.A., Hegazi, N.M., Saad, H.H. et al. (2019). HPLC-ESI- MS/MS analysis of beet (Beta vulgaris) leaves and its beneficial properties in type 1 diabetic rats. *Biomed. Pharmacother.* 120: 109541.

28 Kumkum, M., Rokeya, C.A., Tahsin, M.R. et al. (2021). An evaluation of dose and source-dependent activity of ethanolic extract of trigonellafoenum-graecum against diabetes using alloxan-induced rat model. *Health Sci. J.* 15(1): 805.

29 Akhter, F., Alvi, S.S., Ahmad, P. et al. (2019). Therapeutic efficacy of Boerhaavia diffusa (Linn.) root methanolic extract in attenuating streptozotocin-induced diabetes, diabetes-linked hyperlipidemia and oxidative-stress in rats. *Biomed. Res. Ther.* 6(7): 3293–3306.

30 Sharma, B., Balomajumder, C. and Roy, P. (2008). Hypoglycemic and hypolipidemic effects of flavonoid rich extract from Eugenia jambolana seeds on streptozotocin induced diabetic rats. *Food Chem. Toxicol.* 46(7): 2376–2383.

31 Kaur, G., Invally, M., Khan, M.K. et al. (2018). A nutraceutical combination of Cinnamomum cassia & Nigella sativa for type 1 diabetes mellitus. *J. Ayurveda Integr. Med.* 9(1): 27–37.

32 Pethe, M., Yelwatkar, S., Manchalwar, S. et al. (2017). Evaluation of biological effects of hydroalcoholic extract of Hibiscus Rosa Sinensis flowers on alloxan induced diabetes in rats. *Drug Res.* 67(8): 485–492.

33 Akhtar, N., Akram, M., Daniyal, M. et al. (2018). Evaluation of antidiabetic activity of Ipomoea batatas L. extract in alloxan-induced diabetic rats. *Int. J. Immunopathol. Pharmacol.* 32: 2058738418814678.

34 Dey, P., Singh, J., Suluvoy, J.K. et al. (2020). Utilization of Swertia chirayita plant extracts for management of diabetes and associated disorders: Present status, future prospects and limitations. *Nat. Prod. Bioprospect.* 10(6): 431–443.

35 Thomas, A., Rajesh, E.K. and Kumar, D.S. (2016). The significance of Tinospora crispa in treatment of diabetes mellitus. *Phytother. Res.* 30(3): 357–366.

36 Reddy, K.S., Reddy, Y.P. and Devanna, N. (2012). Antidiabetic activity of a polyherbal formulation (DIA-CARE). *Adv. Pharmacol. Toxicol.* 13(3).

37 Dixit, P.P., Misar, A., Mujumdar, A.M. et al. (2010). Pre-treatment of Syndrex protects mice from becoming diabetic after streptozotocin injection. *Fitoterapia.* 81(5): 403–412.

38 Sarkar, I.N., Law, W. and Balick, M.J. (2019). Identifying phytochemicals from biomedical literature utilizing semantic knowledge sources. *Stud. Health Technol. Inform.* 264: 278–282.

39 Kim, H.Y. and Kim, K. (2007). Protective effect of ginseng on cytokine-induced apoptosis in pancreatic beta-cells. *J. Agric. Food Chem.* 55(8): 2816–2823.

40 Baska, A., Leis, K. and Gałązka, P. (2021). Berberine in the treatment of diabetes mellitus: A review. *Endocr. Metab. Immune Disord. Drug Targets.* 21(8): 1379–1386.

41 Fu, Z., Zhen, W., Yuskavage, J. et al. (2011). Epigallocatechin gallate delays the onset of type 1 diabetes in spontaneous non-obese diabetic mice. *Br. J. Nutr.* 105(8): 1218–1225.

42 Han-ya, Y., Tokuyama, H. and Fukuyama, T. (2011). Total synthesis of (-)-conophylline and (-)-conophyllidine. *Angew. Chem. Int. Ed.* 50(21): 4884–4887.

43 Cruz, P.L., Moraes-Silva, I.C., Ribeiro, A.A. et al. Nicotinamide attenuates streptozotocin-induced diabetes complications and increases survival rate in rats: Role of autonomic nervous system. *BMC Endocr. Disord.* 21(1): 133.

44 Smoak, P., Burke, S.J. and Collier, J.J. (2021). Botanical interventions to improve glucose control and options for diabetes therapy. *SN Compr. Clin. Med.* 3(12): 2465–2491.

45 Derosa, G., D'Angelo, A. and Maffioli, P. (2016). The role of a fixed Berberis aristata/Silybum marianum combination in the treatment of type 1 diabetes mellitus. *Clin. Nutr.* 35(5): 1091–1095.

46 Martorell, M., Castro, N., Victoriano, M. et al. (2021). An update of anthraquinone derivatives emodin, diacerein, and catenarin in diabetes. *Evid-based Complement. Altern. Med: eCAM.* 2021: 3313419.

47 Weiss, L., Zeira, M., Reich, S. et al. (2008). Cannabidiol arrests onset of autoimmune diabetes in NOD mice. *Neuropharmacol.* 54(1): 244–249.

48 Foshati, S., Moradi, S., Tavassoly, M. et al. (2021). Short- and long-term effects of capsaicin supplementation on glycemic control: A systematic review and meta-analysis of controlled trials. *Food Funct.* 12(12): 5236–5246.

49 Krentz, N.A.J., Shea, L.D., Huising, M.O. et al. (2021). Restoring normal islet mass and function in type 1 diabetes through regenerative medicine and tissue engineering. *Lancet Diabetes Endocrinol.* 9(10): 708–724.

50 Yalamanchili, C., Chittiboyina, A.G., Haider, S. et al. (2020). In search for potential antidiabetic compounds from natural sources: Docking, synthesis and biological screening of small molecules from Lycium spp. (Goji). *Heliyon*. 6(1): e02782.

51 Croden, J., Silva, J.R., Huang, W. et al. (2021). Cyanidin-3-O-Glucoside improves the viability of human islet cells treated with amylin or Aβ1-42 in vitro. *PloS one*. 16(10): e0258208.

52 El-Nashar, H.A.S., Mostafa, N.M., El-Shazly, M. et al. (2021). The role of plant-derived compounds in managing diabetes mellitus: A review of literature from 2014 to 2019. *Curr. Med. Chem*. 28(23): 4694–4730.

53 Nikkhah, S., Hafshejan, R.J., Hajivar, F.G. et al. (2021). Effects of Silymarin on blood glucose concentration, hepatic histopathological changes and FOXA2 and FOXA3 gene expression in Streptozotocin-induced diabetic male Wistar rats. *Genet. Appl*. 5(2): 17–23.

54 Li, Y., Zhu, W., Cai, J. et al. (2021). The role of metabolites of steviol glycosides and their glucosylated derivatives against diabetes-related metabolic disorders. *Food Funct*. 12(18): 8248–8259.

55 Xu, Y., Tang, G., Zhang, C. et al. (2021). Gallic acid and diabetes mellitus: Its association with oxidative stress. *Molecules*. 26(23): 7115.

56 Majeed, Y., Shaukat, M.B., Abbasi, K.Y. et al. (2021). Indigenous plants of Pakistan for the treatment of diabetes: A review. *Agrobiol. Rec*. 4: 44–63.

57 Bobkiewicz-Kozłowska, T., Dworacka, M., Kuczyński, S. et al. (2007). Hypoglycaemic effect of quinolizidine alkaloids–lupanine and 2-thionosparteine on non-diabetic and streptozotocin-induced diabetic rats. *EurJ. Pharmacol*. 565(1–3): 240–244.

58 Saeed, M., Shoaib, A., Tasleem, M. et al. (2021). Assessment of antidiabetic activity of the Shikonin by allosteric inhibition of protein-tyrosine phosphatase 1B (PTP1B) using state of art: An in silico and in vitro tactics. *Molecules*. 26(13): 3996.

59 Hu, X.B., Duan, T.T., Liu, J. et al. (2020). Effect of vitamin D supplementation on pancreatic β-cell destruction and type 1 diabetes. *Chin. Med. J*. 134(1): 41–43.

60 Qiang, G., Yang, X., Shi, L. et al. (2015). Antidiabetic effect of Salvianolic acid A on diabetic animal models via AMPK activation and mitochondrial regulation. *Cell Physiol. Biochem*. 36(1): 395–408.

61 Faisal Lutfi, M., Abdel-Moneim, A.H., Alsharidah, A.S. et al. (2021). Thymoquinone lowers blood glucose and reduces oxidative stress in a rat model of diabetes. *Molecules*. 26(8): 2348.

62 Kulkarni, Y.A. and Suryavanshi, S.V. (2021). Combination of Naringenin and Lisinopril ameliorates nephropathy in type-1 diabetic rats. *Endocr. Metab. Immune Disord. Drug Targets*. 21(1): 173–182.

63 Marton, L.T., Pescinini, E.S.L.M., Camargo, M.E.C. et al. (2021). The effects of curcumin on diabetes mellitus: A systematic review. *Front. Endocrinol*. 12: 669448.

64 Tuan, D.T., Thu, N.B., Nhiem, N.X. et al. (2009). Palbinone and triterpenes from Moutan Cortex (Paeonia suffruticosa, Paeoniaceae) stimulate glucose uptake and glycogen synthesis via activation of AMPK in insulin-resistant human HepG2 cells. *Bioorganic Med. Chem. Lett.* 19(19): 5556–5559.

65 Camaya, I., Donnelly, S. and O'Brien, B. (2022). Targeting the PI3K/Akt signaling pathway in pancreatic β-cells to enhance their survival and function: An emerging therapeutic strategy for type 1 diabetes. *J. Diabetes.* 14(4): 247–260.

66 Dong, B., Shi, Z., Dong, Y. et al. (2022). Quercetin ameliorates oxidative stress-induced cell apoptosis of seminal vesicles via activating Nrf2 in type 1 diabetic rats. *Biomed. Pharmacother.* 151: 113108.

67 Luo, W., Chen, X., Ye, L. et al. (2021). Kaempferol attenuates streptozotocin-induced diabetic nephropathy by downregulating TRAF6 expression: The role of TRAF6 in diabetic nephropathy. *J. Ethnopharmacol.* 268: 113553.

Part III

Bioactive Compounds Against Type 2 Diabetes Mellitus

11

Age-induced Biomarkers of Oxidative Stress in Type 2 Diabetes Mellitus

Role of Plant Polyphenols

Brahm Kumar Tiwari[1] *and Kanti Bhooshan Pandey*[2,3,*]

[1] *Department of Paramedical Sciences, Faculty of Allied Health Sciences, SGT University, Gurugram-122505, Haryana, India*
[2] *CSIR-Central Salt & Marine Chemicals Research Institute, Bhavnagar, Gujarat-364002, India*
[3] *Academy of Scientific and Innovative Research (AcSIR), Ghaziabad-201002, India*
[*] *Corresponding author*

11.1 Introduction

Type 2 diabetes mellitus (T2DM) is drastically increasing worldwide and now has become a major health concern. According to the International Diabetes Federation (IDF) report 2021, 541 million people live with Impaired Glucose Tolerance (IGT), which will put them at high risk of T2DM [1]. The risk factors associated with T2DM are non-modifiable (age, genetics, ethnicity, and metabolic risk factors) and modifiable (sedentary lifestyle/physical inactivity, inadequate sleep, obesity/increasing weight, unhealthy diet, and exposure to environmental pollutants) [2].

Aging is one of the strongest factors that exists with T2DM [3]. Aging associated with oxidative stress leads to progression of T2DM and affects older individuals. Among the many described factors associated with the causes of aging, oxidative stress, due to an imbalance in the generation of reactive oxygen and nitrogen species (ROS and RNS) and the antioxidant defense system, which causes accumulation of oxidized products of macromolecules like lipids, proteins and DNA, is considered to be the major one. The oxidation of cellular biomolecules might be responsible for the macro and microvascular complications that cause tissue damage and organ dysfunction and ultimately results in aging and other associated complications including T2DM [4–6].

Natural Products and their Bioactives in Antidiabetic Drug Discovery, First Edition.
Edited by Kanti Bhooshan Pandey, Maitree Suttajit and Pinar Atukeren.

Many conventional synthetic drugs are available to manage T2DM such as pharmacological drugs (3-(diaminomethylidene)-1,1-dimethylguanidine, 1-butyl-3-(4-methylphenyl)sulfonylurea, (2S)-3-phenyl-2-[(4-propan-2-ylcyclohexane carbonyl)amino]propanoic acid, *N*-[2-[4-(cyclohexyl carbonyl sulfamoyl)phenyl] ethyl]-5-methylpyrazine-2-carboxamide, etc.); however, the relief provided by these medicines is temporary and their usage is associated with side effects. In addition, repeated doses, cost, and restricted availability are the other issues of concern that have raised the demand of natural therapies to manage this metabolic disorder [7, 8].

Evidence from in vitro, in vivo, and clinical studies has revealed the effective role of plant polyphenols in the prevention of many human diseases and maintainence of overall human health by the positive modulation of energy metabolism, increasing plasma antioxidant potential, and enhancing immunity [9, 10]. The role of polyphenols on different biomarkers of T2DM has been documented in many studies [11, 12]; however, the effects of polyphenols on age-induced biomarkers of oxidative stress during T2DM are either less, or controversial. This chapter explores the latest understanding of the role of polyphenols against the age-induced oxidative stress biomarkers in T2DM.

11.2 Age is an Important Risk Factor for T2DM

According to the World Health Organization (WHO), aging occurs due to the long-time accumulation of the products of cellular and molecular damage, which leads to the gradual progression of physical and mental inactivity and makes the body prone to diseases and finally death [13]. Some studies tend to associate various factors such as generation of ROS and RNS, cumulative destruction of mitochondrial DNA (mtDNA) and mitochondria by ROS/RNS, as factors responsible for aging [14–16]. These factors are also responsible for the prevalence of T2DM that instigates insulin resistance (IR), increase in the sensitivity of apoptosis, and decrease in the β-cell proliferation capacity [17, 18]. Annually, about a 0.7% reduction of the first and second phase of secretion of insulin was reported during aging [19]. The prevalence of T2DM in elderly people is lead through impaired fasting glycemia (IFG) along with an elevation of glucose intolerance [20]. Oxidative stress during aging induces a decrease in insulin susceptibility and β-cell functioning and proliferation with an increase in IR. These consequences induce a persistent hyperglycemia, which later on turns into T2DM. Some of the results showed other factors contributing to the prevalence of age-associated diabetic complications such as a sedentary lifestyle, increased obesity, and detrimental distribution of fat. All of these modifiable factors also contribute to IR and hyperinsulinemia.

11.3 Oxidative Stress Mediated Pathogenesis of DM

Gradual decline in the plasma antioxidant potential during aging invites the prevalence of ROS and RNS in the cells, due to which the condition of oxidative stress originates that stimulates the oxidative modification of vital molecules of the cells including proteins, lipids, and nucleic acids [21, 22]. Specific biomarkers of oxidative damage to biomolecules of the cells have been reported in different biological specimens (plasma, urine, and tissues) of T2DM patients, which correlates with aging and increased probability of development of diabetes [6, 23].

Several studies revealed that the synthesis and release of cytokines such as tumor necrosis factor-a (TNF-a), interleukin-6 (IL-6), and free radicals increases during aging and T2DM, which enhances the risk of other related complications [24, 25]. Persistent hyperglycemia during diabetes is responsible for the activation of several metabolic pathways such as glucose autoxidation, nonenzymatic protein glycosylation, modification of protein kinase C activity, sorbitol pathway, pseudohypoxia, lipoprotein-altered metabolism, and cytokine-associated alteration, which significantly participate in the production of ROS and as a result cause the prevalence of oxidative stress [26, 27]. There are many molecular mechanisms that have been proposed to explain diabetes-induced complications [28].

Polyol pathways play a major role in T2DM by counterbalancing the redox state and leading to a 30% glucose influx in the body [29]. There are two key enzymes: the first is aldose reductase (AR), which converts glucose to sorbitol and NADPH to $NADP^+$, and the second is sorbitol dehydrogenase (SD), responsible for the conversion of sorbitol to fructose and makes NADH from NAD^+. The sorbitol pathway plays a key role in glucose metabolism and is involved in maintaining the redox homeostasis by utilizing substrates such as NADPH and NAD^+ [30]. Activated polyol pathway mediated depletion in NADPH content and thus surplus NADH impairs biological reactions such as metabolism of lipids, growth factor formation, and activity of protein kinase C, and generates complications in T2DM.

Aging also induces the elevation of advanced glycation end products (AGEs), which also mediate the pathophysiology of diabetes. The non-enzymatic interaction between protein and reducing sugar or lipids leads to the generation of AGEs [31]. There are external and internal sources that lead to the AGEs' accumulation in the body and this generates free radicals, which are responsible for oxidative stress associated with cell damage. The external sources of AGEs are heat-treated food, food rich in protein and fat, cooked oil, etc. It has been observed that an AGE-rich diet is responsible for elevation in serum AGEs' content and increased crosslinking of AGEs in diabetic individuals. It has been observed that during DM, an AGE-rich diet increased the activities and expression of NF-κB, MAPK, and VCAM-1 in comparison to an AGE-low diet [32, 33]. According to the literature, the intracellular AGEs' formation increases in endothelial cells during hyperglycemia and aging [34].

Researchers revealed the correlation of the polyol pathway with protein kinase C- (PKC-) diacylglycerol (DAG) pathway during hyperglycemia. According to the studies, the activated PKC-DAG pathway enhances the polyol pathway that increases the oxidative stress and AGE content, and all these are responsible for organ specific diabetic complications such as renal, retinal, and cardiovascular problems [35]. Excess fatty acid oxidation, induced by hyperglycemia and IR, increases the flux of fructose 6-phosphate into the hexosamine pathway, which contributes in the pathogenesis of DM-associated complications. In this pathway, the first and rate-limiting enzyme, glutamine: fructose-6-phosphate amidotransferase (GFAT), converts fructose-6-phosphate and glutamine into glucosamine-6-phosphate and glutamate, which is then converted into a major end-product, i.e. UDP-N-acetylglucosamine (UDP-GlcNAc) [36, 37].

11.4 Biomarkers of Oxidative Stress in T2DM during Aging

Oxidative stress leads to the pathogenesis of T2DM with increasing vascular complications. The extent of oxidative stress can be measured by enzymatic and non-enzymatic biomarkers in different biological specimens of diabetic patients that are involved in damage [6, 38]. Cu-Zn superoxide dismutase (Cu-Zn-SOD) is a well-described antioxidant enzyme, found in all aerobic cells. It catalyzes the dismutation of superoxide ($O_2^{\bullet -}$) to hydrogen peroxide (H_2O_2) and oxygen in the space of the cytoplasm and may be responsible for the protection of intracellular organelles and DNA from the destructive effect of ROS. Three forms of SOD are present in mammalian cells: Cu-Zn-SOD (present in cytosol), Mn-SOD (localized in mitochondria), and extracellular EC-SOD. $O_2^{\bullet -}$ is capable of rapidly reacting with nitric oxide (NO), reducing NO bioactivity and producing the oxidative peroxynitrite radical. SOD is a potent biomarker of oxidative injury. Studies have reported that the low level of H_2O_2 works as a messenger for insulin signaling whereas its increased level explores toxic effects on pancreatic cells and also damages proteins, DNA, RNA, and lipids, expanding diabetic complications [6, 39].

Catalase is another key antioxidant enzyme that exists in all living organisms. Catalase enzymatically neutralizes the effect of H_2O_2 through the conversion of H_2O_2 into oxygen and water molecules. Catalase deficiency or malfunctioning is associated with T2DM. It has been reported that H_2O_2 is responsible for β-cell damage and interrupting the insulin production through the signaling pathway [27]. Along with SOD, catalase is an indices of oxidative stress.

Reduced glutathione (GSH; γ-glutamyl cysteinyl glycine) is a major intracellular antioxidant present in mammalian tissues. A large body of research on different model systems, including humans, has shown that reduced content of GSH is direct evidence of oxidative load in the system. In almost all diseases including aging and diabetes, compromised GSH content has been documented [40]. GSH has the ability to maintain the protein -SH group in reduced state, participates in the transportation of amino acids, neutralizes free radicals, and works as coenzymes in several reactions and prevents tissue damage. In the state of oxidative stress, GSH becomes oxidized to oxidized glutathione (GSSG), and thus provides reducing equivalents to the stressors and neutralizes them. Glutathione reductase (GR) leads to the reduction of GSSG to GSH by oxidation of NADPH to NAD^+. The antioxidant function is not performed by the GSSG; however, GR facilitates the conversion of GSH with the help of NADPH as a cofactor. The decreased GSH content due to the uncontrolled management of oxidative injury in T2DM during aging leads to microvascular complications.

Malondialdehyde (MDA) and 4-Hydroxynonenal (HNE) are considered as major lipid peroxidation markers due to their nature to react with macro biomolecules such as nucleic acids and proteins [41]. The oxidative lipid by-products are generated through the ROS-mediated oxidative degradation. Elevated levels of HNE and MDA demonstrate that peroxidative injury may induce apoptosis, necrosis in β pancreatic cells, and may contribute to the progression of pancreatic dysfunction and IR [42]. Researchers have proposed that increased MDA, HNE, thiobarbituric acid reactive substances (TBARS), lipid hydroperoxides, and lipoperoxides in diabetic subjects, are all well-established biomarkers of oxidative stress [43, 44].

The poor management of oxidative stress in T2DM in elderly people may contribute to the increased level of protein oxidation products such as advanced oxidation protein products (AOPP), protein carbonyls (PCO), and decreased total thiols (TSH) [45]. AOPP and PCO are the early markers of oxidative stress mediated protein damage; carbonyl moieties are generated through the ROS mediated modification of amino acids such as histidine, proline, lysine, and arginine, and AOPP is dityrosine containing a cross-linked protein that is generated by the activity of myeloperoxidase [46]. An array of reports have documented that they are robust biomarkers of oxidative stress [6, 47, 48].

Researchers report the high level of oxidized guanine nucleoside, 8-oxo-7, 8-dihydroguanosine (8-oxoGuo), and 8-hydroxy-2-deoxyguanosine (8-OHdG) in plasma and urine of prediabetic and T2DM individuals. These biomarkers, responsible for oxidative damage, oxidized guanine nucleoside and 8-oxoGuo for RNA, and 8-OHdG for DNA [49, 50]. Due to the single-stranded structure and without protective proteins, RNA is more susceptible to oxidative damage compared to DNA [50].

11.5 Polyphenols and Their Effects on Diabetes Management during Aging

Polyphenols are secondary metabolites naturally found in plants (vegetables, fruits, cereals, etc.) and plant-based beverages and are responsible for specific taste, color, and sometimes work as defense molecules against pathogens. Polyphenols are classified in several groups based on the presence of a number of phenolic rings and associated structural elements. Many studies including clinical works, report the anti-diabetic, antioxidative, anti-obesity, anti-hyperlipidemic, anti-inflammatory, anti-aging, anti-atherosclerosis, and anti-carcinogen roles of polyphenols such as catechins, luteolin, genistein, quercetin, resveratrol, gallic acid, and curcumin [51–53].

Catechins and its isoforms are a natural flavonoid class of polyphenols, present in green tea, red wine, and cocoa [54]. Catechins possess anti-aging, anti-diabetic, anti-obesity, anti-tumor, and anti-inflammatory properties [55, 56]. Evidence shows the strong antioxidative potential of catechins, which is useful in neutralizing oxidative load and thus preventing the associated complications in diabetic conditions, such as protecting IR by enhancing insulin sensitivity, alleviating endoplasmic reticulum (ER) stress, regulating mitochondrial function, anti-inflammatory activity, and regulating intestinal function [57, 58]. Catechins are also reported to prevent the oxidative modification of membrane transporters and activate the PMRS thereby restoring the redox homeostasis during aging [55].

Luteolin belongs to the flavone group of flavonoids. Many natural sources such as peppers, onion leaves, apples, carrots, cabbages, broccoli, celery, artichokes, and spices are rich in luteolin [59]. Luteolin is a strong antioxidant and has the ability to cure several health ailments where cellular oxidation mediates including inflammation, tumorigenesis, mutagenesis, apoptosis, allergic response, and hyperglycemia [59]. Luteolin has the potential to enhance insulin sensitivity by triggering the activity of Ak2 kinase. Ak2 prevents the attenuation of insulin signaling through the inhibition of dephosphorylation of insulin receptors [60]. In addition, luteolin's antioxidative activity provides therapeutic protection from the microvascular complications in aged diabetic patients [61].

The therapeutic role of genistein in reducing IR has been reported [62]. The main source and highest concentration of genistein is found in soybeans [63]. According to the experimental evidence, genistein has the ability to significantly reduce the fasting glucose in serum, glycated hemoglobin A1C, inflammatory protein expression, hepatocyte apoptosis, MDA and serum triglyceride, by mediating oxidative injury due to its strong antioxidant property [64, 65].

Quercetin is a bioactive compound largely present in onions, apples, red wine, and other citrus fruits [66]. Quercetin has the ability to induce the signaling molecules adenosine monophosphate kinase (AMPK) and CaMKII, which are responsible for the regulation of receptors like Akt and GLUT4 [67]. Quercetin is involved in the array of mechanisms towards the management of diabetes and related complications such as the promotion of insulin secretion through the activation of the intracellular Ca^{2+} signaling pathway [68], modification of IR through the modulation of the AMPK signaling pathway [67], maintenance of glucose homeostasis by regulating the activity of sirtuin 1 (SIRT1), Akt [69], and the inhibition of carbohydrate hydrolysis [70]. Quercetin exerts potent antioxidative activity through the inhibition of lipid peroxidation as evidenced by a decrease in MDA, TBARS, and via modulating the endogenous antioxidative enzymes' expression [71].

Another commonly studied bioactive compound for its health promoting properties is resveratrol, found in grapes, apples, pomegranates, and red wine. Several epidemiological studies have reported the preventive effect of resveratrol administration against chronic human diseases, and the major mechanism involved behind the action of resveratrol was found to be antioxidative [72, 73]. Resveratrol exerts beneficial effects against aging complications, inflammation, β-cell destruction, and IR through alleviating oxidative stress [73, 74]. Multiple associated mechanisms are involved against diabetic complications by resveratrol such as activation of sirtuin1 (SIRT1), AMPK, and affecting the phosphatidylinositol 3-kinase-Akt (PI3K-Akt) pathway, or protein kinase C-θ (PKC-θ) [74].

Gallic acid (GA), obtained from edible plant products such as red wine, green tea, and fruit, is the most potent antioxidant as evidenced by its use as standard for reporting the antioxidative activity of any compound [75]. GA exerts antioxidative, anti-aging, anti-inflammatory, anti-apoptotic, anti-obesity induced properties [76]. Hampered insulin sensitivity due to oxidative stress mediated by inhibition of signaling pathways such as PI3K-Akt, AMPK, and reduction in receptors has been reported [77]. GA has the ability to improve the activity and function of β-cells by the inhibition of caspase 9 related cellular apoptosis [78]. Ferk and co-workers have emphasized the possible benefits of GA during daily intake for prevention against oxidative DNA damage and associated complications such as cancer, cardiovascular diseases, and inflammation [79].

The bioactive compound curcumin, obtained from the rhizome of the turmeric plant (*Curcuma longa*), has been reported to be beneficial in preventing diabetic complications during aging [80] Among other reported properties of curcumin, its antioxidant effect is most effective in health promotion when administered [27, 81–83]. Curcumin induces insulin sensitivity, β-cell functions, prevention of β-cell death, and reduction in IR traits [82, 84, 85]. The antioxidant property mediates the possible mechanisms of curcumin involved in the control of age-induced diabetes, which are the attenuation of tumor necrosis factors-α (TNF-α), attenuation of plasma free fatty acid deposition, inhibition of the expression of nuclear factor-kappa B (NF-κB), reduction of the plasma protein carbonyls and lipid peroxidation, inhibition of activities of lysosomal enzyme (*N*-acetyl-β-D-glucosaminidase, β-D-glucuronidase, β-D-galactosidase), decrease in TBARS, and activation of peroxisome proliferator-activated receptor-gamma (PPAR-γ) [86–90].

More than 8000 different polyphenols have been discovered and yet their health promoting activities and the mechanisms involved are still under investigation. Besides those discussed, there are many other polyphenols that possess strong antioxidant properties and are beneficial in preventing the development and progression of diabetes during old age; however, due to the limitations of space it is not possible to describe them all. Chemical structures of the discussed polyphenols in the prevention of T2DM complications during aging are shown in Table 11.1.

Table 11.1 Chemical structures of different polyphenols beneficial in the prevention of oxidative mediated complications in type 2 diabetes during aging.

Catechin	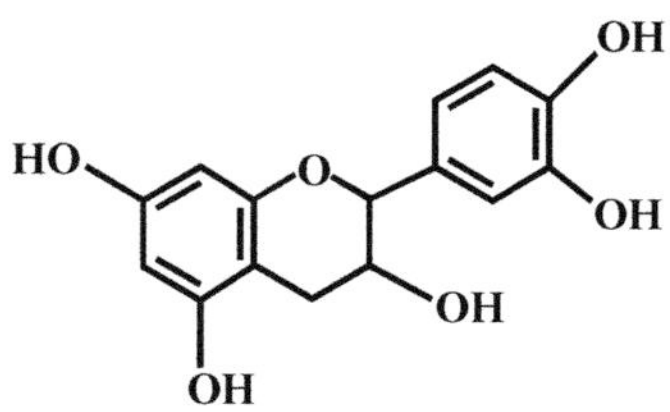
Resveratrol	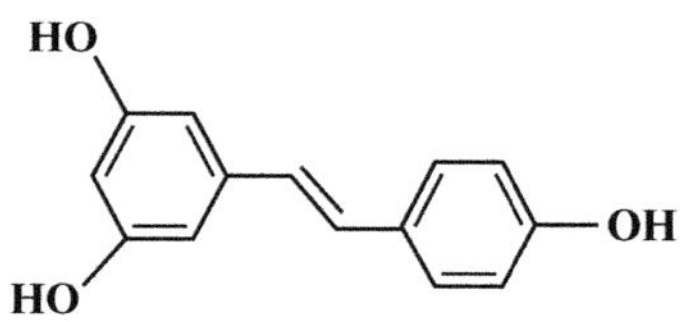
Luteolin	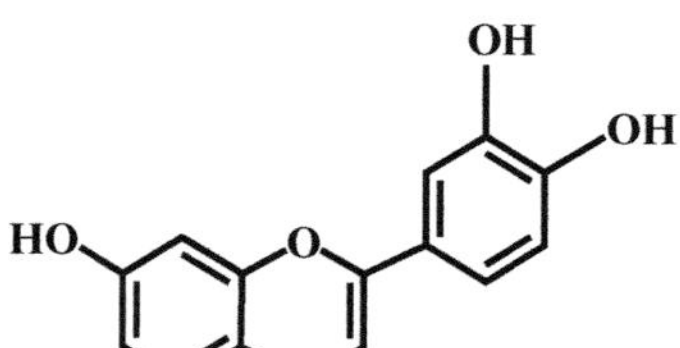
Quercetin	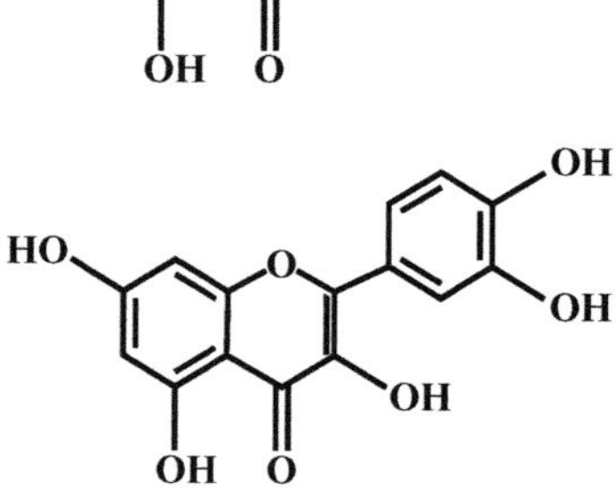
Gallic acid	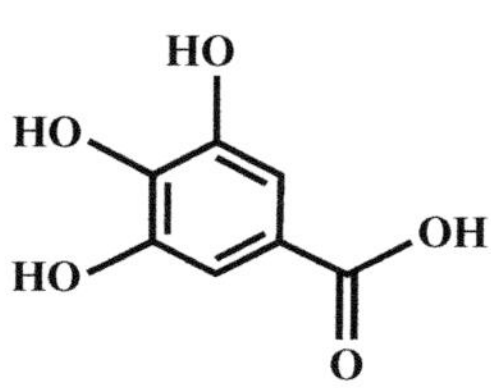

Table 11.1 (Continued)

Curcumin	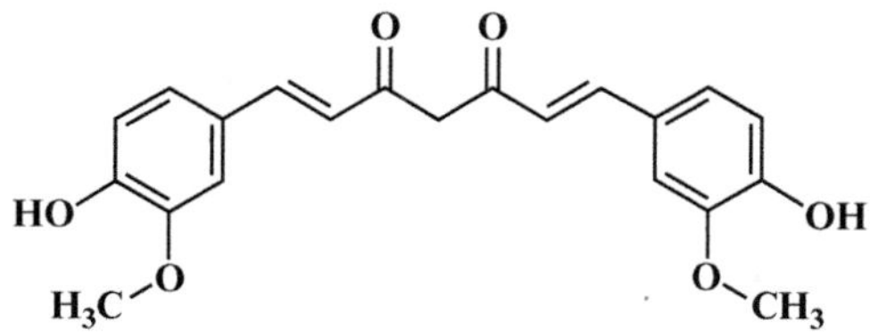
Genistein	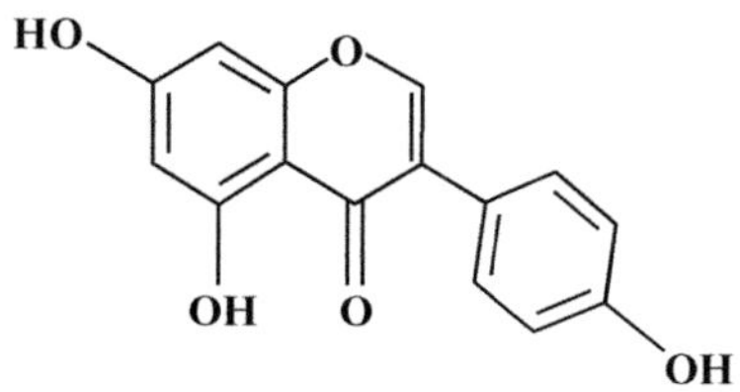

11.6 Future Prospective and Conclusion

With increasing age, susceptibility towards oxidative injury increases in T2DM patients and its improper management increases the chance of microvascular complications. The research outcomes of the aforementioned findings suggest that polyphenols such as phenolic acids, flavonoids, stilbenes, and lignins have the ability to combat the age-induced impairments in T2DM patients. It has been established that oxidative stress strongly mediates the diabetic complications in elderly people since the prevalence of oxidative stress is one of the many factors responsible for biological aging. Among the many mechanisms involved in the health-promoting effect of polyphenols, antioxidative action such as ROS scavenging, inducing the expression of antioxidant enzymes, inhibiting the oxidation/peroxidation of vital cellular biomolecules, preventing the apoptosis of β-cells, inhibiting MAPK/JNK pathway, chelating metal ions, etc. is significant, which is evidenced by the effect on biomarkers of oxidative stress.

Although the biological actions of polyphenols are still under investigation and the effects investigated still need broad clinical studies' interpretation, the data available to date strongly advocate their significant ameliorative effects in the management and prevention of oxidative stress mediated diabetic complications during aging. Further large group and computational studies are needed to understand the exact targets of polyphenols in order to develop drugs of natural origin, which is long awaited, especially in the case of T2DM.

Acknowledgement

KBP is thankful to the director, CSIR-CSMCRI, for his support and encouragement. PRIS.No. of the manuscript:259/2022.

References

1 IDA. (2021). *International Diabetes Federation*, 10th edn. IDF Diabetes Atlas. Brussels: Belgium.

2 Leiva, A.M., Martínez, M.A., Petermann, F. et al. (2018). Risk factors associated with type 2 diabetes in Chile. *Nutr. Hosp.* 35(2): 400–407.

3 Cowie, C.C., Rust, K.F., Ford, E.S. et al. (2009). Full accounting of diabetes and pre-diabetes in the U.S. population in 1988–1994 and 2005–2006. *Diabetes Care.* 32(2): 287–294.

4 Harman, D. (2003). The free radical theory of aging. *Antioxid. Redox Signal.* 5(5): 557–561.

5 Pandey, K.B. and Rizvi, S.I. (2011). Age-dependent oxidative stress biomarkers in type 2 diabetic patients. *J. Hum. Dis.* 1: 1–9.

6 Tiwari, B.K., Pandey, K.B., Abidi, A.B. et al. (2013). Markers of oxidative stress during diabetes mellitus. *J. Biomark.* 2013: 378790.

7 Choi, S.W. and Ho, C.K. (2018). Antioxidant properties of drugs used in type 2 diabetes management: Could they contribute to, confound or conceal effects of antioxidant therapy? *Redox. Rep.* 23(1): 1–24.

8 Sola, D., Rossi, L., Schianca, G.P.C. et al. (2015). Sulfonylureas and their use in clinical practice. *Arch. Med. Sci.* 11(4): 840–848.

9 Pandey, K.B. and Rizvi, S.I. (2012). Recent advances in health promoting effect of dietary polyphenols. *Curr. Nutr. Food Sci.* 8(4): 254–264.

10 Pandey, K.B. and Rizvi, S.I. (2017). Plant polyphenols in healthcare and aging. In: *Nutritional Antioxidant Therapies: Treatments and Perspectives.* (ed. K. Al-Gubory and I. Laher). 267–282. Cham: Springer. https://doi.org/10.1007/978-3-319-67625-8_11

11 Liu, K., Luo, M. and Wei, S. (2019). The bioprotective effects of polyphenols on metabolic syndrome against oxidative stress: Evidences and perspectives. *Oxid. Med. Cell Longev.* 2019: 1–16.

12 Tiwari, B.K. and Pandey, K.B. (2020). Role of natural polyphenols in oxidative stress: Prevention of diabetes. In: *Assessment of Medicinal Plants for Human Health.* 103–115. 1st edn. Apple Academic Press.

13 WHO. (2021). Ageing and health. WHO [Internet]. (Accessed January 30, 2023). Available from: https://www.who.int/news-room/fact-sheets/detail/ageing-and-health

14 Cui, H., Kong, Y. and Zhang, H. (2012). Oxidative stress, mitochondrial dysfunction, and aging. *J. Signal Transduct.* 2012: 1–13.

15 Nissanka, N. and Moraes, C.T. (2018). Mitochondrial DNA damage and reactive oxygen species in neurodegenerative disease. *FEBS Lett.* 592(5): 728–742.

16 Pandey, K.B. and Rizvi, S.I. (2015). Redox biology of aging: Focus on novel biomarkers. In: *Free Radicals in Human Health and Disease.* (ed. V. Rani and U. Yadav). 279–290. New Delhi: Springer. https://doi.org/10.1007/978-81-322-2035-0_18

17 Halim, M. and Halim, A. (2019). The effects of inflammation, aging and oxidative stress on the pathogenesis of diabetes mellitus (type 2 diabetes). *Diabetes Metab. Syndr.* 13(2): 1165–1172.

18 Kitada, M., Ogura, Y., Monno, I. et al. (2019). Sirtuins and type 2 diabetes: Role in inflammation, oxidative stress, and mitochondrial function. *Front. Endocrinol.* 10: 187.

19 Szoke, E., Shrayyef, M.Z., Messing, S. et al. (2008). Effect of aging on glucose homeostasis. *Diabetes Care.* 31(3): 539–543.

20 Goyal, R., Nguyen, M. and Jialal, I. (2022). Glucose intolerance. In: *StatPearls [Internet].* Treasure Island (FL): StatPearls Publishing. (cited 2022, Nov 11). (Accessed March 07, 2023). Available from: http://www.ncbi.nlm.nih.gov/books/NBK499910/.

21 Pandey, K.B., Mehdi, M.M., Maurya, P.K. et al. (2010). Plasma protein oxidation and its correlation with antioxidant potential during human aging. *Dis. Markers.* 29(1): 31–36.

22 Pandey, K.B. and Rizvi, S.I. (2010). Markers of oxidative stress in erythrocytes and plasma during aging in humans. *Oxid. Med. Cell Longev.* 3(1): 2–12.

23 Liguori, I., Russo, G., Curcio, F. et al. (2018). Oxidative stress, aging, and diseases. *Clin. Interv. Aging.* 13: 757–772.

24 Baynes, J.W. and Thorpe, S.R. (1990). Role of oxidative stress in diabetic complications: A new perspective on an old paradigm. *Diabetes.* 48(1): 1–9.

25 Ceriello, A. (2003). New insights on oxidative stress and diabetic complications may lead to a "causal" antioxidant therapy. *Diabetes Care.* 26(5): 1589–1596.

26 Geraldes, P. and King, G.L. (2010). Activation of protein kinase C isoforms and its impact on diabetic complications. *Circ. Res.* 106(8): 1319–1331.

27 Ighodaro, O.M. (2018). Molecular pathways associated with oxidative stress in diabetes mellitus. *Biomed. Pharmacother.* 108: 656–662.

28 Brownlee, M. (2001). Biochemistry and molecular cell biology of diabetic complications. *Nature.* 414(6865): 813–820.

29 Garg, S.S. and Gupta, J. (2022). Polyol pathway and redox balance in diabetes. *Pharmacol. Res.* 182: 106326.

30 Srikanth, K.K. and Orrick, J.A. (2022). Biochemistry, polyol or sorbitol pathways. In: *StatPearls [Internet].* Treasure Island (FL): StatPearls Publishing. (cited 2022

Nov 22). (Accessed March 07, 2023). Available from: http://www.ncbi.nlm.nih.gov/books/NBK576381.

31 Chaudhuri, J., Bains, Y., Guha, S. et al. (2018). The role of advanced glycation end products in aging and metabolic diseases: Bridging association and causality. *Cell Metab.* 28(3): 337–352.

32 Sharifi-Rad, M., Anil Kumar, N.V., Zucca, P. et al. (2020). Lifestyle, oxidative stress, and antioxidants: Back and forth in the pathophysiology of chronic diseases. *Front. Physiol.* 11: 694.

33 Singh, V.P., Bali, A., Singh, N. et al. (2014). Advanced glycation end products and diabetic complications. *Korean J. Physiol. Pharmacol.* 18(1): 1–14.

34 Vlassara, H. and Uribarri, J. (2014). Advanced glycation end products (AGE) and diabetes: Cause,effect, or both? *Curr. Diab. Rep.* 14(1): 453.

35 Kang, Q. and Yang, C. (2020). Oxidative stress and diabetic retinopathy: Molecular mechanisms, pathogenetic role and therapeutic implications. *Redox Biol.* 37: 101799.

36 Chistiakov, D.A., Sobenin, I.A., Orekhov, A.N. et al. (2014). Role of endoplasmic reticulum stress in atherosclerosis and diabetic macrovascular complications. *BioMed. Res. Int.* 2014: 1–14.

37 Weigert, C., Klopfer, K., Kausch, C. et al. (2003). Palmitate-induced activation of the hexosamine pathway in human myotubes. *Diabetes.* 52(3): 650–656.

38 Banik, S. and Ghosh, A. (2021). The association of oxidative stress biomarkers with type 2 diabetes mellitus: A systematic review and meta-analysis. *Health Sci. Rep. [Internet].* (cited 2022, Nov 22). Available from: https://onlinelibrary.wiley.com/doi/10.1002/hsr2.389

39 Góth, L. (2008). Catalase deficiency and type 2 diabetes. *Diabetes Care.* 31(12): e93–e93.

40 Kurutas, E.B. (2016). The importance of antioxidants which play the role in cellular response against oxidative/nitrosative stress: Current state. *Nutr. J.* 15(1): 71.

41 Barrera, G., Pizzimenti, S., Daga, M. et al. (2018). Lipid peroxidation-derived aldehydes, 4-Hydroxynonenal and Malondialdehyde in aging-related disorders. *Antioxidants (Basel).* 7(8): 102.

42 Ayala, A., Muñoz, M.F. and Argüelles, S. (2014). Lipid peroxidation: Production, metabolism, and signaling mechanisms of malondialdehyde and 4-hydroxy-2-nonenal. *Oxid. Med. Cell Longev.* 2014: 360438.

43 Martín-Gallán, P., Carrascosa, A., Gussinyé, M. et al. (2003). Biomarkers of diabetes-associated oxidative stress and antioxidant status in young diabetic patients with or without subclinical complications. *Free Radic. Biol. Med.* 34(12): 1563–1574.

44 Mas-Bargues, C., Escrivá, C., Dromant, M. et al. (2021). Lipid peroxidation as measured by chromatographic determination of malondialdehyde. Human plasma reference values in health and disease. *Arch. Biochem. Biophys.* 709: 108941.

45 Cakatay, U. (2005). Protein oxidation parameters in type 2 diabetic patients with good and poor glycaemic control. *Diabetes Metab.* 31(6): 551–557.

46 Pandey, K.B., Mishra, N. and Rizvi, S.I. (2010). Protein oxidation biomarkers in plasma of type 2 diabetic patients. *Clin. Biochem.* 43(4–5): 508–511.

47 Tiwari, B.K., Pandey, K.B., Jaiswal, N. et al. (2014). Anti-diabetic and anti-oxidative effect of composite extract of leaves of some Indian plants on alloxan induced diabetic wistar rats. *J. Pharm. Investig.* 44(3): 205–211.

48 Tiwari, B.K., Abidi, A.B., Rizvi, S.I. et al. (2018). Effect of oral supplementation of composite leaf extract of medicinal plants on biomarkers of oxidative stress in induced diabetic Wistar rats. *Arch. Physiol. Biochem.* 124(4): 361–366.

49 Urbaniak, S.K., Boguszewska, K., Szewczuk, M. et al. (2020). 8-Oxo-7,8-Dihydro-2′-Deoxyguanosine (8-oxodG) and 8-Hydroxy-2′-Deoxyguanosine (8-OHdG) as a potential biomarker for gestational diabetes mellitus (GDM) development. *Molecules.* 25(1): 202.

50 Chen, X., Yu, H., Li, Z. et al. (2022). Oxidative RNA damage in the pathogenesis and treatment of type 2 diabetes. *Front. Physiol.* 13: 725919.

51 Ganesan, K. and Xu, B. (2017). A critical review on polyphenols and health benefits of black soybeans. *Nutrients.* 9(5): 455.

52 Pandey, K.B. and Rizvi, S.I. (2009). Pleiotropic biological effects of resveratrol: Implications for human health. *Natl. Acad. Sci. Lett.* 32(11 & 12): 321–326.

53 Sharma, P., Hajam, Y.A., Kumar, R. et al. (2022). Complementary and alternative medicine for the treatment of diabetes and associated complications: A review on therapeutic role of polyphenols. *Phytomed. Plus.* 2(1): 100188.

54 Cheng, X., Zhang, L. and Lian, Y.J. (2015). Molecular targets in Alzheimer's disease: From pathogenesis to therapeutics. *Biomed. Res. Int.* 2015: 760758.

55 Pandey, K.B., Jha, R. and Rizvi, S.I. (2013). Erythrocyte membrane transporters during human ageing: Modulatory role of tea catechins. *Clin. Exp. Pharmacol. Physiol.* 40(2): 83–89.

56 Takahashi, M., Ozaki, M., Miyashita, M. et al. (2019). Effects of timing of acute catechin-rich green tea ingestion on postprandial glucose metabolism in healthy men. *J. Nutr. Biochem.* 73: 108221.

57 Cremonini, E., Fraga, C.G. and Oteiza, P.I. (2019). (-)-Epicatechin in the control of glucose homeostasis: Involvement of redox-regulated mechanisms. *Free Radic. Biol. Med.* 130: 478–488.

58 Xu, L., Li, W., Chen, Z. et al. (2019). Inhibitory effect of epigallocatechin-3-O-gallate on α-glucosidase and its hypoglycemic effect via targeting PI3K/AKT signaling pathway in L6 skeletal muscle cells. *Int. J. Biol. Macromol.* 125: 605–611.

59 Lin, Y., Shi, R., Wang, X. et al. (2008). Luteolin, a flavonoid with potential for cancer prevention and therapy. *Curr. Cancer Drug Targets.* 8(7): 634–646.

60 Ding, L., Jin, D. and Chen, X. (2010). Luteolin enhances insulin sensitivity via activation of PPARγ transcriptional activity in adipocytes. *J. Nutr. Biochem.* 21(10): 941–947.

61 Xiong, J., Wang, K., Yuan, C. et al. (2017). Luteolin protects mice from severe acute pancreatitis by exerting HO-1-mediated anti-inflammatory and antioxidant effects. *Int. J. Mol. Med.* 39(1): 113–125.

62 Gilbert, E.R. and Liu, D. (2013). Anti-diabetic functions of soy isoflavone genistein: Mechanisms underlying its effects on pancreatic β-cell function. *Food Funct.* 4(2): 200–212.

63 Bettaiah, A. and Prabhushankar, H.B. (2021). Screening of novel source for genistein by rapid and sensitive UPLC-APCI-TOF mass spectrometry. *Int. J. Food Sci.* 2021: 1–12.

64 Jain, R., Bolch, C., Al-Nakkash, L. et al. (2022). Systematic review of the impact of genistein on diabetes-related outcomes. *Am. J. Physiol. Regul. Integr. Comp. Physiol.* 323(3): R279–288.

65 Yang, R., Jia, Q., Mehmood, S. et al. (2021). Genistein ameliorates inflammation and insulin resistance through mediation of gut microbiota composition in type 2 diabetic mice. *Eur. J. Nutr.* 60(4): 2155–2168.

66 Batiha, G.E.S., Beshbishy, A.M., Ikram, M. et al. (2020). The pharmacological activity, biochemical properties, and pharmacokinetics of the major natural polyphenolic flavonoid: Quercetin. *Foods.* 9(3): 374.

67 Dhanya, R. (2022) Quercetin for managing type 2 diabetes and its complications, an insight into multitarget therapy. *Biomed. Pharmacother.* 146: 112560.

68 Bardy, G., Virsolvy, A., Quignard, J.F. et al. (2013). Quercetin induces insulin secretion by direct activation of L-type calcium channels in pancreatic beta cells: Quercetin increases L-type Ca currents in beta cells. *Br. J. Pharmacol.* 169(5): 1102–1113.

69 Peng, J., Li, Q., Li, K. et al. (2017). Quercetin improves glucose and lipid metabolism of diabetic rats: Involvement of Akt signaling and SIRT1. *J. Diabetes Res.* 2017: 1–10.

70 Oboh, G., Ademosun, A.O., Ayeni, P.O. et al. (2015). Comparative effect of quercetin and rutin on α-amylase, α-glucosidase, and some pro-oxidant-induced lipid peroxidation in rat pancreas. *Comp. Clin. Pathol.* 24(5): 1103–1110.

71 Yi, H., Peng, H., Wu, X. et al. (2021). The therapeutic effects and mechanisms of Quercetin on metabolic diseases: Pharmacological data and clinical evidence. *Oxid. Med. Cell Longev.* 2021: 1–16.

72 Pan, M.H., Wu, J.C., Ho, C.T. et al. (2018). Antiobesity molecular mechanisms of action: Resveratrol and pterostilbene: Antiobesity molecular mechanisms of action. *BioFactors.* 44(1): 50–60.

73 Tiwari, B.K. and Pandey, K.B. (2023). Mechanisms involved in prevention of dementia and promotion of healthy aging by resveratrol. In: *Plant Bioactives as Natural Panacea against Age-Induced Diseases.* (ed. K.B. Pandey and M Suttajit). 197–214. Elsevier.

74 Su, M., Zhao, W., Xu, S. et al. (2022). Resveratrol in treating diabetes and its cardiovascular complications: A review of its mechanisms of action. *Antioxidants (Basel).* 11(6): 1085.

75 Muir, R.M., Ibáñez, A.M., Uratsu, S.L. et al. (2011). Mechanism of gallic acid biosynthesis in bacteria (Escherichia coli) and walnut (Juglans regia). *Plant Mol. Biol.* 75(6): 555–565.

76 Bukhari, S.A., Yasmin, A., Rasul, A. et al. (2022).Identification of ascorbic acid and gallic acid as novel inhibitors of secreted frizzled-related protein for the treatment of obesity-induced type 2 diabetes. *Dose-Response*. 20(1): 155932582110697.

77 Xu, Y., Tang, G., Zhang, C. et al. (2021). Gallic acid and diabetes mellitus: Its association with oxidative stress. *Molecules*. 26(23): 7115.

78 Rahimifard, M., Baeeri, M., Bahadar, H. et al. (2020). Therapeutic effects of gallic acid in regulating senescence and diabetes; an in vitro study. *Molecules*. 25(24): 5875.

79 Ferk, F., Kundi, M., Brath, H. et al. (2018). Gallic acid improves health-associated biochemical parameters and prevents oxidative damage of DNA in type 2 diabetes patients: Results of a placebo-controlled pilot study. *Mol. Nutr. Food Res.* 62(4): 1700482.

80 Hettiarachchi, S.S., Dunuweera, S.P., Dunuweera, A.N. et al. (2021). Synthesis of curcumin nanoparticles from raw turmeric rhizome. *ACS Omega*. 6(12): 8246–8252.

81 Mirzaei, H., Shakeri, A., Rashidi, B. et al. (2017). Phytosomal curcumin: A review of pharmacokinetic, experimental and clinical studies. *Biomed. Pharmacother.* 85: 102–112.

82 Pivari, F., Mingione, A., Brasacchio, C. et al. (2019). Curcumin and type 2 diabetes mellitus: Prevention and treatment. *Nutrients*. 11(8): 1837.

83 Singh, P., Pandey, K.B. and Rizvi, S.I. (2016). Curcumin: The yellow molecule with pleiotropic biological effects. *Lett. Drug Des. Discov*. 13(2): 170–177.

84 Chuengsamarn, S., Rattanamongkolgul, S., Luechapudiporn, R. et al. (2012). Curcumin extract for prevention of type 2 diabetes. *Diabetes Care*. 35(11): 2121–2127.

85 Hodaei, H., Adibian, M., Nikpayam, O. et al. (2019). The effect of curcumin supplementation on anthropometric indices, insulin resistance and oxidative stress in patients with type 2 diabetes: A randomized, double-blind clinical trial. *Diabetol. Metab. Syndr.* 11(1): 41.

86 Badr, A.M., Sharkawy, H., Farid, A.A. et al. (2020). Curcumin induces regeneration of β cells and suppression of phosphorylated-NF-κB in streptozotocin-induced diabetic mice. *JoBAZ*. 81(1): 22.

87 Den Hartogh, D.J., Gabriel, A. and Tsiani, E. (2019). Antidiabetic properties of curcumin II: Evidence from in vivo studies. *Nutrients*. 12(1): 58.

88 El-Azab, M.F., Attia, F.M. and El-Mowafy, A.M. (2011). Novel role of curcumin combined with bone marrow transplantation in reversing experimental diabetes: Effects on pancreatic islet regeneration, oxidative stress, and inflammatory cytokines. *Eur. J. Pharmacol.* 658(1): 41–48.

89 Wojcik, M., Krawczyk, M., Wojcik, P. et al. (2018). Molecular mechanisms underlying curcumin-mediated therapeutic effects in type 2 diabetes and cancer. *Oxid. Med .Cell Longev*. 2018: 1–14.

90 Zhang, D.W., Fu, M., Gao, S.H. et al. (2013). Curcumin and diabetes: A systematic review. *Evid-based Complement. Altern. Med*. 2013: 1–16.

12

Bioactives from Clove Oil for Antibacterial Wound Dressings for the Treatment and Management of Wounds in Type 2 Diabetes Mellitus

Andleeb Shahzadi and Haktan Sonmez*

Department of Medical Pharmacology, Faculty of Medicine, Istanbul University-Cerrahpaşa, Istanbul, Turkey
** Corresponding author*

12.1 Introduction

Diabetes mellitus (DM) is a diverse metabolic condition followed by chronic hyperglycemia. Elevated homocysteine and DNA damage have been seen in type 2 diabetes mellitus (T2DM) patients that might cause endothelial damage [1]. Impaired wound healing in T2DM often leads to chronic lesions in these patients. The compromised wound restoration involves pathophysiology with an intricate interplay of vascular dysfunction, neuropathic abnormalities, immune irregularities, and various deviations in biochemical attributes [2]. Additionally, atherosclerosis and renal failure might be predisposing factors in poor wound healing [2, 3]. Therefore, patients with T2DM and minor wounds such as trauma related, to cutting a toenail, or stepping on a stone, later on, due to abnormalities in physiological functions linked to hyperglycemia, could often find these wounds becoming non-healing ulcers predisposed to infection [4]. It is not unusual for a chronic neuropathic ulcer to become gangrene and eventually require amputation [5]. Minor trauma in T2DM patients, such as closed puncture wounds, cuts, bruises, and burns, could also lead to a significant number of lower limb infections, although amputations are more rare [5]. T2DM complications such as diabetic foot ulcers, venous leg ulcers, pressure ulcers, and related infections are a common reason for morbidity and set a significant burden on the patient and society [6]. Almost 500 million individuals are suffering from DM, which is thought to further increase in the future. In the United States of America alone, above $300 billion is spent yearly on medical costs [7, 8].

Natural Products and their Bioactives in Antidiabetic Drug Discovery, First Edition.
Edited by Kanti Bhooshan Pandey, Maitree Suttajit and Pinar Atukeren.

Prevention and prophylactic care of wounds in T2DM have been encouraged to reduce patient illness, the consumption of costly resources, and the danger of amputation. These interventions, which comprise risk factor awareness, patient training, and intensive foot care have been shown to have a socioeconomical impact [9, 10]. Medicine based on medicinal plants has proven to be more cost effective, and has relatively few side effects compared to allopathic medicine. It has been shown in the literature that there has been a significant increase in the pharmaceutical industry's attention to the use of phytochemicals from medicinal plants. Phytochemicals have well established health benefits. As per the World Health Organization, 80% of patients from developing countries still depend on herbal or traditional medicines, and most of them are made from medicinal plants, to prevent or treat diseases [11].

The growth and infiltration of a diabetic wound is one of the main factors that can delay wound repair [12]. Aside from wound healing, another serious issue is to protect the lesion from bacterial contamination [13]. To speed up the restorative process of a wound, a bandage that delivers an antibacterial shield is required at the application site along with an appropriately humid setting. The role of phyto-compounds with numerous therapeutic applications, such as wound repair and antimicrobial properties has been shown in the literature [14].

Eugenia caryophyllata Thunberg, of the Myrtaceae family, universally identified as clove, is extensively consumed in food as an aromatic spice. The FDA classified clove oil as Generally Recognized as Safe (GRAS); nevertheless, it is lethal if taken above 3.75 g/kg per-oral. Exposure to light and moisture, and fluctuations in temperature turn clove oil yellow, which indicates it is chemically unstable [15]. Since ancient times, clove has been used conventionally to treat various diseases, not limited to the treatment of parasitic worms, respiratory disease (asthma, expectorant, decongestant), dental disorders (toothache, analgesic), pharyngitis (sore throat), digestive disorders (antiemetic, gastroesophageal reflux disease, dyspepsia, and diarrhea), as an antipyretic, aphrodisiac, anxiolytic and hypnotic, muscle relaxant, and also it is considered to be the most effective herb with anti-inflammatory and antibacterial properties [16–18].

Eugenol and β-caryophyllene, the chief components of clove essential oil, possess bactericidal properties against *Escherichia coli*, *Listeria monocytogenes*, and *Salmonella enterica*, foodborne and spoil-age microorganisms such as *Campylobacter jejuni*, *Salmonella enteritidis*, and *Staphylococcus aureus* [19, 20]. Clove essential oil produced as nanofibers with different polymers indicate robust antibacterial properties, specifically against *S. aureus*, *E. coli*, *Pseudomonas fluorescens*, and *Bacillus subtilis* [21, 22]. The wound healing properties of clove oil in nano-emulsion has been extensively investigated and it has been established that it has noticeable wound restorative capacity and also anti-inflammatory properties in comparison to pure clove oil [23].

Poly(ε-Caprolactone)/gelatin nanofiber clove oil exhibited notable antibacterial as well as discernable wound restorative properties [22]. The bioactivities of clove oil for further elaborating the antibacterial spectrum and its relation to the treatment and management of wounds caused by T2DM are discussed in this chapter.

12.2 Cellular Pathways of Wound Healing

The recovery of wounds is an intricate process and involves interrelated processes and synchronization of various cells with distinct roles in hemostasis overlapping with inflammation, proliferation, and remodeling phases. T2DM is generally associated with biochemical and vascular abnormalities due to hyperglycemia that leads to abnormal wound recovery, and ultimately the development of chronic ulcers. To understand the T2DM wound pathophysiologically it is important to understand the physiological cellular pathways that are involved in wound recovery (Figure 12.1).

12.2.1 Hemostatic and Inflammatory Response

After a vascularized response, hemostasis is the first phase that stops bleeding by vasoconstriction. The major cells that play a critical role in this phase are platelets

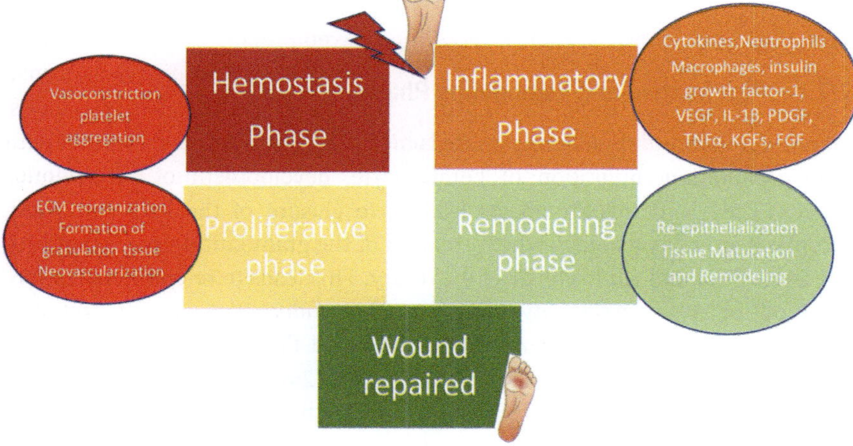

Figure 12.1 Different physiological pathways of wound healing. Vascular endothelial growth factor (VEGF), platelet-derived growth factor (PDGF), keratinocyte growth factors (KGFs), fibroblast growth factor (FGF), extracellular matrix (ECM), interleukin 1β (IL-1β), and tumor necrosis factor α (TNFα).

that adhere to the exposed collagen and activate the release of glycoproteins, which ultimately leads to the aggregation of platelets. Thrombin is activated by the release of intrinsic clotting factors; later, thrombin is involved in the formation of fibrin from fibrinogen [24]. The fibrin interacts with platelets aggregation to form a stable hemostatic plug [25]. Hemostasis is achieved within a few minutes of injury reducing the degree of hemorrhage [26].

Physiological recovery of wounds starts with the overlapping of hemostasis and the inflammatory phase. The inflammatory part of wound healing starts with vasodilation, which brings pro-inflammatory mediators at the site of injury and activates the phagocytosis process [27]. Aggregated platelets at the site of vascular damage are responsible for the recruitment of inflammatory mediators predominately the platelet-derived growth factor (PDGF), transforming growth factor β (TGFβ), and many other growth factors [28]. These pro-inflammatory mediators attract neutrophils and monocytes, which leads to the exudation of fluid (edema) at the wound site and are also responsible for the fibro-vascular scar formation. It has been found that wounds deprived of macrophages show delayed recovery [29]. Keratinocytes relocate in the wounded site and fibroblasts begin proliferating through the secretion of various growth factors; however, the change in full skin biology through keratinocyte-fibroblast interaction is not fully known [30]. These events appear early, called the acute phase, and may last for up to two weeks. Different recruited cell types are responsible for the granulation of tissue [28]. During the early phase of wound repair, macrophages are bactericidal and express tumor necrosis factor (TNF-α), interleukins (IL)-6, and interleukins (IL-1) β, following pro-inflammatory phagocytes that rapidly kill and engulf pathogens [31].

12.2.2 Proliferative and Remodeling Phase

The proliferative and remodeling phase is meant for the closure of a wound with tissue granulation, migration of keratinocyte, development of extracellular matrix (ECM), and myofibroblasts leading to closure of the lesion [12, 28]. Fibroblasts are responsible for the formation of the granulation tissues that produce fresh ECM and aid to constrict the injury. This leads to re-epithelialization, formation of new vessels involving vascular endothelial growth factor A (VEGF-A) in the wound, and restoration of blood flow that promotes wound healing with fresh epithelial cells and scar tissues [26]. Therefore, healing progresses to the next step, i.e. remodeling, and finally the type III collagen accumulation is substituted by type I collagen to improve tissue tensile strength and return to normal, and the wound repair follows the restoration of the physiology of the skin [32].

12.3 Various Cellular Components in Impaired Wound Healing in T2DM

Wound healing in T2DM is a multifactorial process involving cumbersome biochemical and cellular cascades (Figure 12.2). Besides, wound healing is under the influence of many other factors such as age, genetics, and comorbidities. Proper knowledge of these factors and how they influence wound repair is imperative. All of these factors are collectively important in the development and implementation of a phyto-compound for chronic wound repair [33].

The course of wound healing in diabetes is predominantly considered by prolongation of the inflammatory state, abnormal angiogenesis, reduction of endothelial progenitor cells, and an irregulation of extracellular matrix proteins [12]. Furthermore, an elevated response of chemotactic chemokines has been observed in diabetic wounds that leads to the migration of neutrophils and macrophages to the injured site [34]. Infiltration of the wounded area by these cells leads to the release of inflammatory cytokines such as IL-1 β, TNFα [35], whose levels remain elevated for a longer time thus maintaining the chronic inflammatory response [34].

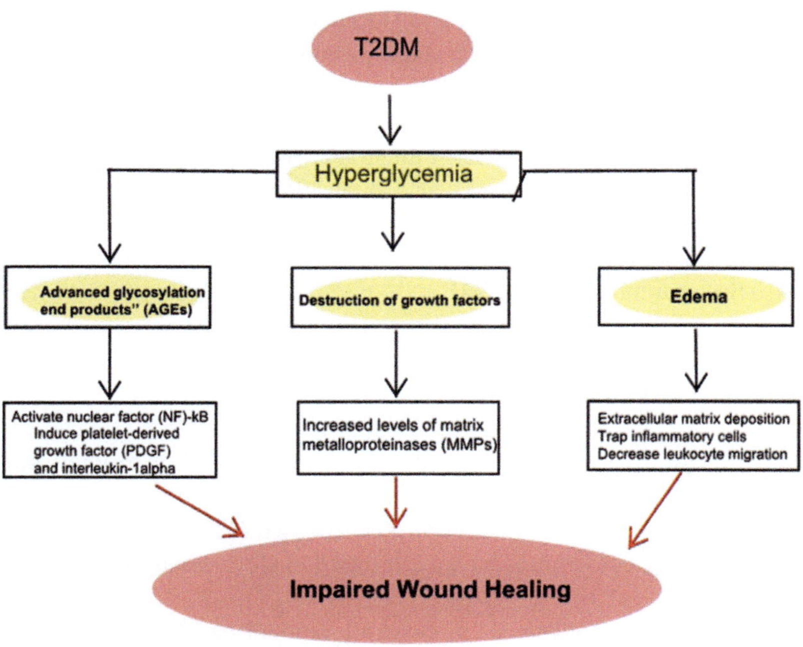

Figure 12.2 Pathophysiological processes involved in impaired wound repair linked with type 2 diabetes mellitus.

Evidence that diabetic wounds have a delayed influx of monocytes and activation of macrophages was also presented in the literature [34].

Furthermore, in T2DM, the release of growth factors, which play a vital role in initiating and maintaining the healing process, is impaired. It has been found that diabetic wounds have low concentrations of insulin-like growth factor-1 (IGF-1) and transforming growth factor-β (TGFβ). IGF-1 is required in cell-granulation and re-epithelialization of wounds, while TGFβ is involved in the migration of immune cells, keratinocytes, neoangiogenesis, and production of the ECM proteins [36]. The process of neoangiogenesis is abnormal in endothelial cells open to high glucose levels and with inadequate capillary density in the wounded area. Moreover, animal diabetic models showed macrophages with impaired release of VEGF and phagocytic activity [37]. Accordingly, in another experiment, diabetic ob/ob mice developed chronic wounds with low expression of VEGF-A mRNA in comparison to control mice [38]. It has been reported that local VEGF application is capable of repairing the wound by increasing growth factors involved in tissue repair [39].

Furthermore, interstitial edema is common in T2DM and can delay or complicate healing, predominantly accompanying other factors or comorbidities. The resulting increased pressure in the vessels impairs the supply of vital nutrients to the wounded area, thereby prolonging the normal wound repair course [40]. Edema also leads to decreased leukocyte migration and traps inflammatory cells. Matrix metalloproteinases (MMPs) are vital in wound repair and are capable of degrading ECM proteins [41]. They play a significant role in four stages of wound repair by eliminating impaired proteins in wound contraction and remodeling [42]. Abnormal levels of MMPs in T2DM might increase the duration of diabetic wound repair and lead to other vascular complications [43]. Impaired wound healing affects quality of life, leads to loss of function, and increases the morbidity risk in T2DM. It has been found that approximately 50% of patients with amputation have a five-year mortality rate [44].

12.4 Clove Essential Oil

The Myrtaceae family has more than 3000 species and *Syzygium aromaticum* L., frequently recognized as clove, is an aromatic flower. Numerous reports propose that *S. aromaticum* L. contains ~15–20% wt. of essential oil. High contents of phenolic compounds have been found in clove essential oil (CEO) [17, 45].

Eugenol is the major constituent of clove oil, accounting for approximately 50%. The 10–40% of clove oil contains eugenyl acetate, β-caryophyllene, and α-humulene. To the best of our knowledge, a total 36 mixtures from clove oil have been identified until now and the four most important ones are presented in Table 12.1 with their percentage content.

Table 12.1 Comparison of clove essential oil percentage with different analytical methods reported in various studies.

No.	Compound	Percentage (%)	References	Method
1	Eugenol	97.7573	[46]	GS-MS analysis
		88.58535	[21]	GS-MS analysis
		83.4	[45]	Microwave-assisted steam Distillation
		88.8	[45]	Microwave-assisted hydro-Distillation
		65.36 and 71.84	[47]	Microwave-assisted steam distillation outside and inside, respectively
		66.9	[48]	Coaxial microwave-assisted hydro-distillation
2	Eugenyl acetate	5.62	[21]	GS-MS analysis
		14.3	[45]	Microwave-assisted steam distillation
		7.46	[45]	Microwave-assisted hydro-distillation
		5.71 and 9.49	[47]	Microwave-assisted steam distillation outside and inside, respectively
		2.7	[48]	Coaxial microwave-assisted hydro-distillation
3	β-Caryophyllene	1.16	[46]	GS-MS analysis
		1.38	[21]	GS-MS analysis
		1.37	[45]	Microwave-assisted steam distillation
		2.65	[45]	Microwave-assisted hydro-distillation
		24.62 and 15.6	[47]	Microwave-assisted steam distillation outside and inside, respectively
		24.8	[48]	Coaxial microwave-assisted hydro-distillation

(Continued)

Table 12.1 (Continued)

No.	Compound	Percentage (%)	References	Method
4	α-Humulene	0.19	[21]	GS-MS analysis
		2.75	[49]	GS-MS analysis
		0.21	[45]	Microwave-assisted steam distillation
		0.4	[45]	Microwave-assisted hydro-distillation
		0.01	[47]	Microwave-assisted steam distillation inside
		3.1	[48]	Coaxial microwave-assisted hydro-distillation

Strong antibacterial properties of eugenol on gram negative bacteria (*E. coli, Salmonella, Pseudomonas aeruginosa, Klebsiella pneumoniae, Erwinia carotovora, Agrobacterium*), gram positive bacteria (*Staphylococcus, Streptococcus, Listeria*) *Aspergillus, Penicillium, Candida albicans*, and yeast have been reported in different studies [50, 51] as shown in Table 12.2. CEO inhibits gram-positive bacteria more significantly in comparison to gram-negative bacteria. It is the diffusible properties of the mucopeptide layer of gram-positive bacteria that renders them susceptible to antibacterial agents. In contrast to this, the gram-negative bacteria have a lipopolysaccharide layer in the outer cytoplasmic membrane that can considerably decrease the lipophilic diffusion of CEO across the cytoplasmic membrane [50]. A novel study augments the antimicrobial properties of CEO on an emerging multidrug-resistant *Streptococcus suis* that proposes the likelihood of considering it as a satisfactory substitute for the management of microbial diseases [46].

The lipophilic properties make the CEO easy to penetrate in the microbial cells. However, the antibacterial properties have been associated with the location of –OH groups at the meta and ortho positions. These groups link with the bacterial cell membrane, and damage leads to the escape of cellular content and that leads to cell death [50, 51, 59, 60]. A recent study has demonstrated that rats treated with CEO underwent similar healing to those treated with neomycin, an antibiotic used to fight bacteria, the inflammatory response of the body, and eventually repair the wounds. Thus, the untoward effects of antimicrobial agents can be reduced when used frequently [61].

Table 12.2 Antimicrobial properties of clove essential oil with mechanisms of action.

Compound	Chemical formal and structure	Microbes	Antibacterial mechanisms of action	References
Eugenol	$C_{10}H_{12}O_2$ 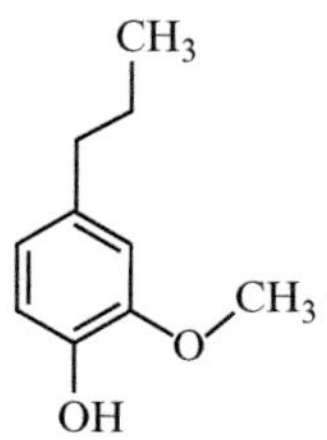	*Escherichia coli*	Inhibits bacterial migration and adhesion	[52]
			Inhibits biofilm formation and virulence	[53]
		Salmonella enteritidis	Decreases bacterial migration, inhibits bacterial adhesion and invasion, decreases integrity of cell wall and membrane, inhibits bacterial migration and invasion	[54, 55]
		Staphyloco-ccus aureu	Inhibits biofilm formation, decreases expression of virulence factors	[56]
		Streptococcus mutans	Reduces bacterial adhesion and biofilm formation	[57]
β-Caryo-phyllene	**(C15H24)** 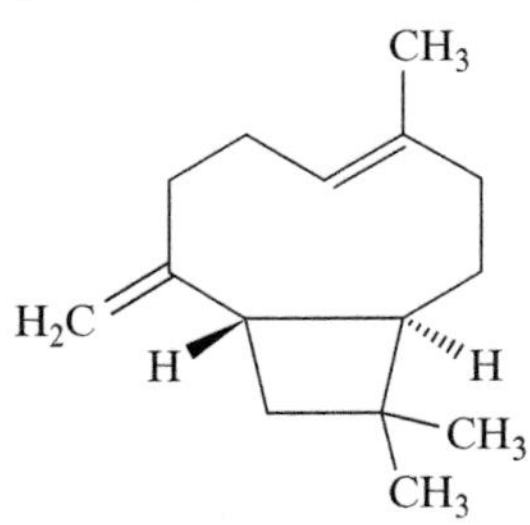	*Pseudomonas aeruginosa*	Decreases migration of virulence factors and biofilm formation	[58]
Eugenyl acetate	**($C_{12}H_{14}O_3$)** 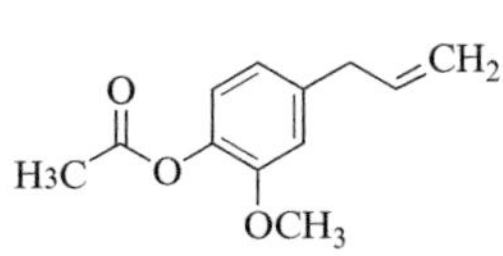			

12.5 Anti-inflammatory and Wound Healing Properties of CEO: Significance in T2DM

It has been reported that clove oil emulsion might be efficiently used as an anti-inflammatory and wound healer as an alternative to the commercially available synthetic preparations. It has been investigated that CEO and eugenol exhibit anti-inflammatory properties comparable to NSAIDs such as diclofenac gel and decrease inflammation by 60–20% after 3 hours of application. Similarly, wounds of rats treated with CEO showed a noteworthy contraction (95%) in 15 days [61].

CEO helped in decreasing the levels of inflammatory biomarkers; for example, vascular cell adhesion protein 1, interferon gamma-induced protein, and interferon–inducible T cell Alpha Chemoattractant-10 (I-TAC) at the wound site. CEO also inhibited proteins such as collagen I and collagen III, macrophage colony-stimulating factor, and the tissue inhibitor of metalloproteinase involved in tissue remodeling [23, 62]. Studies have reported that CEO can decrease the epidermis thickness and the expression of inflammatory cells. Eugenol inhibits COX-2 and ultimately inhibits the production of prostaglandins and other inflammatory mediators [63, 64]. The data from the latest study indicate that eugenol does not change IL-8 concentration in keratinocytes [61].

T2DM complications lead to pathophysiological abnormalities such as peripheral diabetic disorders, neuropathy, and diabetic foot ulcer (DFU) infection that prolong the inflammation duration with the induction of cytokines such as IL-1β [65], IL-6 and TNF-α [66], NF-κβ and MCP-1 [67], which leads to the abnormal proliferative phase, which is responsible for impaired diabetic wound restoration. Clove oil with various bioactives has the ability to switch the TNF-α inflammatory cellular pathway and activate the Nrf-2 pathway, which is thought to be the significant anti-inflammatory pathway [68].

Similarly, the results of a recent study showed that local application of clove flower extract hydrogel heals wound size in diabetic albino rats, restored wound healing markers such as MMP9 and TGFβ, growth factor signaling pathways (PPAR-α, EGF-β, VEGF-β, and FGF-β), inflammatory cytokine expression (IL8, TNFα, NFKβ and IL1β), and anti-inflammatory cytokines (IL4 and IL10). It could be determined that clove oil represents a reasonable antibiotic alternative for diabetic foot ulcer therapy as it may activate the diabetic wound healing process [69].

12.6 Conclusion

Clove oil bioactives offer a great deal of potential for clinical applications as alternative therapies for the management of T2DM related wounds due to their strong anti-inflammatory, antioxidant, and antimicrobial properties. However, evidence-based medical studies should be conducted before the inclusion of clove oil in clinical settings. It could be concluded that clove oil represents a true antibiotic substitute for T2DM wound healing since it exerts antibacterial and anti-inflammatory activities both in *in vitro* and *in vivo*. However, further work is required for the elaboration of novel mechanisms of the action of CEO and doses and experiments are required to be conducted. In addition, elaborative pharmacokinetics studies are also required to be conducted for CEO applications in clinical practices.

References

1 Bukhari SA, Javed S, Ali M, Shahzadi A, Rehman M. Serum haematological and biochemical indices of oxidative stress and their relationship with DNA damage and homocysteine in Pakistani type II diabetic patients. *Pak. J. Pharm. Sci.* 2015 May; 28(3):881–889.

2 Greenhalgh, D.G. Wound healing and diabetes mellitus. *Clin. Plast. Surg.* 2003 Jan; 30(1):37–45. doi:10.1016/s0094-1298(02)00066-4.

3 Mieczkowski M, Mrozikiewicz-Rakowska B, Kowara M, Kleibert M, Czupryniak L. The problem of wound healing in diabetes - From molecular pathways to the design of an animal model. *Int. J. Mol. Sci.* 2022 Jul 19; 23(14): doi:10.3390/ijms23147930.

4 Frykberg, R.G. and Banks, J. Challenges in the treatment of chronic wounds. *Adv. Wound Care (New Rochelle).* 2015 Sep 1; 4(9):560–582. doi:10.1089/wound.2015.0635.

5 East JM, Fray DA, Hall DE, Longmore CA. Chronic neuropathic ulcer is not the most common antecedent of lower limb infection or amputation among diabetics admitted to a regional hospital in Jamaica: Results from a prospective cohort study. *BMC Surg.* 2015 Sep 21; 15:104. doi:10.1186/s12893-015-0091-4.

6 Wu SC, Driver VR, Wrobel JS, Armstrong DG. Foot ulcers in the diabetic patient, prevention and treatment. *Vasc. Health Risk Manag.* 2007; 3(1):65–76.

7 Gao D, Zhang Y, Bowers DT, Liu W, Ma M. Functional hydrogels for diabetic wound management. *APL Bioeng.* 2021 Sep; 5(3):031503. doi:10.1063/5.0046682.

8 Glover K, Stratakos AC, Varadi A, Lamprou DA. 3D scaffolds in the treatment of diabetic foot ulcers: New trends vs conventional approaches. *Int. J. Pharm. Apr.* 2021 15; 599:120423. doi:10.1016/j.ijpharm.2021.120423.

9 Moreland ME, Kilbourne AM, Engelhardt JB, et al. Diabetes preventive care and non-traumatic lower extremity amputation rates. *J. Healthc. Qual.* 2004 Sep-Oct; 26(5):12–17. doi:10.1111/j.1945-1474.2004.tb00515.x.

10 Singh N, Armstrong DG, Lipsky BA. Preventing foot ulcers in patients with diabetes. *JAMA.* 2005 Jan 12; 293(2):217–228. doi:10.1001/jama.293.2.217.

11 Tran N, Pham B, Le L. Bioactive compounds in anti-diabetic plants: From herbal medicine to modern drug discovery. *Biology (Basel)*. 2020 Aug 28; 9(9): doi:10.3390/biology9090252.

12 Spampinato SF, Caruso GI, De Pasquale R, Sortino MA, Merlo S. The treatment of impaired wound healing in diabetes: Looking among old drugs. *Pharmaceuticals (Basel)*. 2020 Apr 1; 13(4): doi:10.3390/ph13040060.

13 Edwards R, Harding KG. Bacteria and wound healing. *Curr. Opin. Infect. Dis.* 2004 Apr; 17(2):91–96. doi:10.1097/00001432-200404000-00004.

14 Bittner Fialová S, Rendeková K, Mučaji P, Nagy M, Antibacterial SL. Activity of medicinal plants and their constituents in the context of skin and wound infections, Considering European legislation and folk medicine - A review. *Int. J. Mol. Sci.* 2021 Oct 4; 22(19): doi:10.3390/ijms221910746.

15 Hameed M, Rasul A, Waqas MK et al. Formulation and evaluation of a clove oil-encapsulated nanofiber formulation for effective wound-healing. *Molecules*. 2021 Apr 24; 26(9): doi:10.3390/molecules26092491.

16 Singh J, Baghotia A, Goel SP. Eugenia caryophyllata Thunberg (family Myrtaceae): A review. *Int. J. Res. Pharm.* 2012 12/20; 3:1469–1475.

17 Batiha GE, Alkazmi LM, Wasef LG, Beshbishy AM, Nadwa EH, Rashwan EK. Syzygium aromaticum L. (Myrtaceae): Traditional uses, bioactive chemical constituents, pharmacological and toxicological activities. *Biomolecules*. 2020 Jan 30; 10(2): doi:10.3390/biom10020202.

18 Otunola GA. Culinary spices in food and medicine: An overview of Syzygium aromaticum (L.) Merr. and L. M. Perry [Myrtaceae]. *Front. Pharmacol.* 2021; 12:793200. doi:10.3389/fphar.2021.793200.

19 Friedman M, Henika PR, Mandrell RE. Bactericidal activities of plant essential oils and some of their isolated constituents against Campylobacter jejuni, Escherichia coli, Listeria monocytogenes, and Salmonella enterica. *J. Food Prot.* 2002 Oct; 65(10):1545–1560. doi:10.4315/0362-028x-65.10.1545.

20 Kalemba D, Kunicka A. Antibacterial and antifungal properties of essential oils. *Curr. Med. Chem.* 2003 May; 10(10):813–829. doi:10.2174/0929867033457719.

21 Chaieb K, Hajlaoui H, Zmantar T, et al. The chemical composition and biological activity of clove essential oil, Eugenia caryophyllata (Syzigium aromaticum L. Myrtaceae): A short review. *Phytother. Res.* 2007 Jun; 21(6):501–506. doi:10.1002/ptr.2124.

22 Unalan I, Slavik B, Buettner A, Goldmann WH, Frank G, Boccaccini AR. Physical and antibacterial properties of peppermint essential oil loaded poly (ε-caprolactone) (PCL) electrospun fiber mats for wound healing. *Front. Bioeng. Biotechnol.* 2019; 7:346. doi:10.3389/fbioe.2019.00346.

23 Alam P, Ansari MJ, Anwer MK, Raish M, Kamal YK, Shakeel F. Wound healing effects of nanoemulsion containing clove essential oil. *Artif. Cells Nanomed. Biotechnol.* 2017 May; 45(3):591–597. doi:10.3109/21691401.2016.1163716.

24 Tennent GA, Brennan SO, Stangou AJ, O'Grady J, Hawkins PN, Pepys MB. Human plasma fibrinogen is synthesized in the liver. *Blood*. 2007 Mar 1; 109(5):1971–1974. doi:10.1182/blood-2006-08-040956.

25 Singer AJ, Clark RA. Cutaneous wound healing. *N. Engl. J. Med*. 1999 Sep 2; 341(10):738–746. doi:10.1056/nejm199909023411006.

26 Pastar I, Stojadinovic O, Yin NC, et al. Epithelialization in wound healing: A comprehensive review. *Adv. Wound Care (New Rochelle)*. 2014 Jul 1; 3(7):445–464. doi:10.1089/wound.2013.0473.

27 Ozgok Kangal MK, Regan JP. Wound Healing. StatPearls Publishing. Copyright © 2022, StatPearls Publishing LLC; 2022.

28 Falanga V. Wound healing and its impairment in the diabetic foot. *Lancet*. 2005 Nov 12; 366(9498):1736–1743. doi:10.1016/s0140-6736(05)67700-8.

29 Mirza R, DiPietro LA, Koh TJ. Selective and specific macrophage ablation is detrimental to wound healing in mice. *Am. J. Pathol*. 2009 Dec; 175(6):2454–2462. doi:10.2353/ajpath.2009.090248.

30 Fujiwara T, Kubo T, Kanazawa S, et al. Direct contact of fibroblasts with neuronal processes promotes differentiation to myofibroblasts and induces contraction of collagen matrix in vitro. *Wound Repair Regen*. 2013 Jul–Aug; 21(4):588–594. doi:10.1111/wrr.12059.

31 Slauch JM. How does the oxidative burst of macrophages kill bacteria? Still an open question. *Mol. Microbiol*. 2011 May; 80(3):580–583. doi:10.1111/j.1365-2958.2011.07612.x.

32 Han G, Ceilley R. Chronic wound healing: A review of current management and treatments. *Adv. Ther*. 2017 Mar; 34(3):599–610. doi:10.1007/s12325-017-0478-y.

33 Oguntibeju OO. Medicinal plants and their effects on diabetic wound healing. *Vet. World*. 2019 May; 12(5):653–663. doi:10.14202/vetworld.2019.653-663.

34 Wetzler C, Kämpfer H, Stallmeyer B, Pfeilschifter J, Frank S. Large and sustained induction of chemokines during impaired wound healing in the genetically diabetic mouse: Prolonged persistence of neutrophils and macrophages during the late phase of repair. *J. Invest. Dermatol*. 2000 Aug; 115(2):245–253. doi:10.1046/j.1523-1747.2000.00029.x.

35 Xiao T, Yan Z, Xiao S, Xia Y. Proinflammatory cytokines regulate epidermal stem cells in wound epithelialization. *Stem Cell Res. Ther*. 2020 2020/06/11; 11(1):232. doi:10.1186/s13287-020-01755-y.

36 Roberts AB. Transforming growth factor-beta: Activity and efficacy in animal models of wound healing. *Wound Repair Regen*. 1995 Oct-Dec; 3(4):408–418. doi:10.1046/j.1524-475X.1995.30405.x.

37 Khanna S, Biswas S, Shang Y, et al. Macrophage dysfunction impairs resolution of inflammation in the wounds of diabetic mice. *PLoS One*. 2010 Mar 4; 5(3):e9539. doi:10.1371/journal.pone.0009539.

38 Seitz O, Schürmann C, Hermes N, et al. Wound healing in mice with high-fat diet- or ob gene-induced diabetes-obesity syndromes: A comparative study. *Exp. Diabetes Res.* 2010; 2010:476969. doi:10.1155/2010/476969.

39 Galiano RD, Tepper OM, Pelo CR, et al. Topical vascular endothelial growth factor accelerates diabetic wound healing through increased angiogenesis and by mobilizing and recruiting bone marrow-derived cells. *Am. J. Pathol.* 2004 Jun; 164(6):1935–1947. doi:10.1016/s0002-9440(10)63754-6.

40 Armstrong DG, Nguyen HC. Improvement in healing with aggressive edema reduction after debridement of foot infection in persons with diabetes. *Arch. Surg.* 2000; 135(12):1405–1409. doi:10.1001/archsurg.135.12.1405.

41 Vitlianova K, Georgieva J, Milanova M, Tzonev S. Blood pressure control predicts plasma matrix metalloproteinase-9 in diabetes mellitus type II. *Arch. Med. Sci.* 2015 Mar 16; 11(1):85–91. doi:10.5114/aoms.2015.49208.

42 Serra R, Grande R, Buffone G, et al. Extracellular matrix assessment of infected chronic venous leg ulcers: Role of metalloproteinases and inflammatory cytokines. *Int. Wound J.* 2016 Feb; 13(1):53–58. doi:10.1111/iwj.12225.

43 Ayuk SM, Abrahamse H, Houreld NN. The role of matrix metalloproteinases in diabetic wound healing in relation to photobiomodulation. *J. Diabetes Res.* 2016; 2016:2897656. doi:10.1155/2016/2897656.

44 Fortington LV, Geertzen JH, van Netten JJ, Postema K, Rommers GM, Dijkstra PU. Short and long term mortality rates after a lower limb amputation. *Eur. J. Vasc. Endovasc. Surg.* 2013 Jul; 46(1):124–131. doi:10.1016/j.ejvs.2013.03.024.

45 Golmakani M-T, Zare M, Razzaghi S. Eugenol enrichment of clove bud essential oil using different microwave-assisted distillation methods. *Food Sci. Technol. Res.* 2017; 23(3):385–394. doi:10.3136/fstr.23.385.

46 Wongsawan K, Chaisri W, Tangtrongsup S, Mektrirat R. Bactericidal effect of clove oil against multidrug-resistant Streptococcus suis isolated from human patients and slaughtered pigs. *Pathogens.* 2019 Dec 21; 9(1): doi:10.3390/pathogens9010014.

47 Kennouche A, Benkaci-Ali F, Scholl G, Eppe G Chemical composition and antimicrobial activity of the essential oil of Eugenia caryophyllata cloves extracted by conventional and microwave techniques. *J. Biol. Act. Prod. Nat.* 2015 2015/01/02; 5(1):1–11. doi:10.1080/22311866.2014.961100.

48 González-Rivera J, Duce C, Falconieri D, et al. Coaxial microwave assisted hydrodistillation of essential oils from five different herbs (lavender, rosemary, sage, fennel seeds and clove buds): Chemical composition and thermal analysis. *Innov. Food Sci. Emerg. Technol.* 2016 2016/02/01/; 33:308–318. doi:10.1016/j.ifset.2015.12.011.

49 Wei MC, Xiao J, Yang YC. Extraction of α-humulene-enriched oil from clove using ultrasound-assisted supercritical carbon dioxide extraction and studies of its fictitious solubility. *Food Chem.* 2016 Nov 1; 210:172–181. doi:10.1016/j.foodchem.2016.04.076.

50 Alizadeh Behbahani B, Noshad M, Falah F Study of chemical structure, antimicrobial, cytotoxic and mechanism of action of Syzygium aromaticum essential oil on foodborne pathogens. *Potr. S. J. F. Sci.* 2019 11/28; 13:875–883. doi:10.5219/1226.

51 Shahbazi Y. Antioxidant, antibacterial, and antifungal properties of nanoemulsion of clove essential oil. *Nanomed. Res. J.* 2019; 4(4):204–208. doi:10.22034/nmrj.2019.04.001.

52 Baskaran SA, Kollanoor-Johny A, Nair MS, Venkitanarayanan K. Efficacy of plant-derived antimicrobials in controlling enterohemorrhagic Escherichia coli virulence in vitro. *J. Food Prot.* 2016 Nov; 79(11):1965–1970. doi:10.4315/0362-028x.Jfp-16-104.

53 Kim JS, Kuk E, Yu KN, et al. Antimicrobial effects of silver nanoparticles. Nanomedicine: Nanotechnology. *Biol. Med.* 2007 2007/03; 3(1):95–101. doi:10.1016/j.nano.2006.12.001.

54 Upadhyaya I, Upadhyay A, Kollanoor-Johny A, Darre MJ, Venkitanarayanan K. Effect of plant derived antimicrobials on Salmonella enteritidis adhesion to and invasion of primary chicken oviduct epithelial cells in vitro and virulence gene expression. *Int. J. Mol. Sci.* 2013 May 21; 14(5):10608–10625. doi:10.3390/ijms140510608.

55 Kollanoor-Johny A, Mattson T, Baskaran SA, et al. Reduction of Salmonella enterica serovar enteritidis colonization in 20-day-old broiler chickens by the plant-derived compounds trans-cinnamaldehyde and eugenol. *Appl. Environ. Microbiol.* 2012 Apr; 78(8):2981–2987. doi:10.1128/aem.07643-11.

56 Yadav MK, Chae SW, Im GJ, Chung JW, Song JJ. Eugenol: A phyto-compound effective against methicillin-resistant and methicillin-sensitive Staphylococcus aureus clinical strain biofilms. *PLoS One.* 2015; 10(3):e0119564. doi:10.1371/journal.pone.0119564.

57 Adil M, Singh K, Verma PK, Khan AU. Eugenol-induced suppression of biofilm-forming genes in Streptococcus mutans: An approach to inhibit biofilms. *J. Glob. Antimicrob. Resist.* 2014 Dec; 2(4):286–292. doi:10.1016/j.jgar.2014.05.006.

58 Houdkova M, Rondevaldova J, Doskocil I, Kokoska L. Evaluation of antibacterial potential and toxicity of plant volatile compounds using new broth microdilution volatilization method and modified MTT assay. *Fitoterapia.* 2017 Apr; 118:56–62. doi:10.1016/j.fitote.2017.02.008.

59 El-Darier S, Elahwany A, Elkenany E, Abdeldaim A. An in vitro study on antimicrobial and anticancer potentiality of thyme and clove oils. *Rendiconti Lincei Scienze Fisiche e Naturali.* 2018 02/07; 29: doi:10.1007/s12210-018-0672-0.

60 Marmouzi I, Karym EM, Alami R, et al. Modulatory effect of Syzygium aromaticum and Pelargonium graveolens on oxidative and sodium nitroprusside stress and inflammation. *Orient. Pharm. Exp. Med.* 2019 2019/06/01; 19(2):201–210. doi:10.1007/s13596-018-0335-9.

61 Banerjee K, Madhyastha H, Sandur VR, Manikandanath NT, Thiagarajan N, Thiagarajan P. Anti-inflammatory and wound healing potential of a clove oil emulsion. *Colloids Surf. B. Biointerfaces*. 2020 Sep; 193:111102. doi:10.1016/j.colsurfb.2020.111102.

62 Sugihartini N, Prabandari R, Yuwono T, Rahmawati D. The anti-inflammatory activity of essential oil of clove (Syzygium aromaticum) in absorption base ointment with addition of oleic acid and propylene glycol as enhancer. *Int. J. Appl. Pharm*. 2019 09/23; 106–109. doi:10.22159/ijap.2019.v11s5.T0081.

63 Sugihartini N, Lestari G, Yuliani S. Anti-inflammatory activity of essential oil of clove (Syzygium aromaticum) in O/W and W/O creams. *Pharmaciana*. 2019; 9(1): Pharmaciana. 2019.

64 Han X, Parker TL. Anti-inflammatory activity of clove (Eugenia caryophyllata) essential oil in human dermal fibroblasts. *Pharm. Biol*. 2017 Dec; 55(1):1619–1622. doi:10.1080/13880209.2017.1314513.

65 Yan C, Gao N, Sun H, et al. Targeting imbalance between IL-1β and IL-1 receptor antagonist ameliorates delayed epithelium wound healing in diabetic mouse corneas. *Am. J. Pathol*. 2016 Jun; 186(6):1466–1480. doi:10.1016/j.ajpath.2016.01.019.

66 Zubair M, Malik A, Ahmad J. Plasma adiponectin, IL-6, hsCRP, and TNF-α levels in subject with diabetic foot and their correlation with clinical variables in a North Indian tertiary care hospital. *Indian J. Endocrinol. Metab*. 2012 Sep; 16(5):769–776. doi:10.4103/2230-8210.100672.

67 Van Asten SA, Nichols A, La Fontaine J, Bhavan K, Peters EJ, Lavery LA. The value of inflammatory markers to diagnose and monitor diabetic foot osteomyelitis. *Int. Wound J*. 2017 Feb; 14(1):40–45. doi:10.1111/iwj.12545.

68 Ma L, Liu J, Lin Q, Gu Y, Yu W. Eugenol protects cells against oxidative stress via Nrf2. *Exp. Ther. Med*. 2021 Feb; 21(2):107. doi:10.3892/etm.2020.9539.

69 Ali R, Khamis T, Enan G, El-Didamony G, Sitohy B, Abdel-Fattah G. The healing capability of clove flower extract (CFE) in Streptozotocin-induced (STZ-induced) diabetic rat wounds infected with multidrug resistant bacteria. *Molecules*. 2022 Mar 31; 27(7): doi:10.3390/molecules27072270.

13

Nutritional Features and Bioactivities of Thymoquinone against Type 2 Diabetes Mellitus

Pınar Atukeren[1,2]

[1] *Department of Medical Biochemistry, Cerrahpaşa Faculty of Medicine, Istanbul University-Cerrahpaşa, Istanbul, Turkey*
[2] *Faculty of Pharmacy, Istanbul University-Cerrahpaşa, Istanbul, Turkey*

13.1 Introduction

Diabetes mellitus (DM) is one of the frequently seen chronic diseases that affects millions of people around the world with a strong negative influence on patients' health, and it is known to be a metabolic disorder affecting people of all ages and genders. The incidence of DM is increasing each year at a high rate and is predicted to be higher in the future [1]. Type 2 diabetes mellitus (T2DM) is characterized by an elevation in blood glucose levels because of a defect in the insulin action mechanism. Not only does blood glucose elevate in T2DM but also there is an enhancement in reactive oxygen species (ROS) production, which is also one of the main factors behind the damage in cells through impairing redox homeostasis, and this results in metabolic syndrome accompanied by various common diabetic complications such as neuropathy, nephropathy, and retinopathy [2, 3]. Diabetic neuropathy affects almost half of diabetic patients and is the major reason for morbidity causing neuropathic pain with susceptibility to injuries and foot ulceration [4]. T2DM is also associated with other diseases including obesity, cardiovascular diseases, hypertension, and also cancer of the pancreas, urinary tract, colorectal, liver, genitourinary system, etc. through an excess amount of oxidative stress, which plays an important role in the pathogenesis of the disease as it causes cells to be more prone to mutations [5].

Natural Products and their Bioactives in Antidiabetic Drug Discovery, First Edition.
Edited by Kanti Bhooshan Pandey, Maitree Suttajit and Pinar Atukeren.

13.2 Oxidative Stress and Redox Homeostasis

The prevalence of oxidative stress due to over-generation of ROS or free radicals induces damaging effects on macromolecules such as proteins, lipids, and nucleic acids in cells and tissues [6]. Polypeptide bonds with amino acids such as proline and lysine are easily affected by free radicals [7]. The oxidation of proteins, especially in amino acids such as histidine, tyrosine, phenylalanine, is in the formation of carbonyl groups. The aldehyde products of lipid peroxidation cause cross-linking in proteins by forming covalent bonds with sulfhydryl groups of cysteine, or with lysine and histidines [6, 7]. With all these mentioned effects of free radicals, various structural changes occur in the deoxyribonucleic acid (DNA) molecule (such as fragmentation of purine and pyrimidine bases, chain breaks, DNA denaturation) [6, 8]. As a result, free radicals cause breaks in DNA helices and disrupt DNA transcription and replication, leading to the death of the cell in the next mitotic division. If an imbalance occurs between the rate of formation of radicals derived from O_2 and the antioxidant capacity of the metabolism, oxidative stress and oxidative damage occur in the cells and redox homeostasis is disrupted [7]. In addition to the mitochondrial electron transport chain, free oxygen radicals are also formed in the organism during reactions catalyzed by many enzymes found in various fractions of cells [9]. Various external factors such as radiation, drugs, and harmful chemicals also lead to the formation of free radicals [7, 9].

As a result of lipid peroxidation in the cell membranes, the cellular transport systems are affected. The inevitable consequence of this is the disruption of both intracellular and extracellular ion balances. Thus, the intracellular concentration of calcium ion (Ca^{+2}) increases and, accordingly, proteases are activated. The autocatalytic mechanism triggered by lipid peroxidation leads to oxidative damage of the cell membranes. The toxic and reactive metabolites that result from this cause the death of the cell. Free radicals not only cause lipid peroxidation, but also oxidize them by forming covalent bonds with tissue proteins [9, 10]. Protein oxidation decreases enzyme activity and loss of function in proteins, which causes formation of protein deposits, disorders of proteolysis, cell wall degradation, proinflammation, etc. [11].

13.3 Antioxidants

Specific defense mechanisms have been developed in organisms to cope with the destructive effects of free radicals [6]. The main antioxidants are endogenous (enzymatic: superoxide dismutase (SOD), catalase (CAT), glutathione peroxidase (GSH-Px), glutathione reductase (GR), and mitochondrial cytochrome oxidase system; and non-enzymatic: melatonin, myoglobin, hemoglobin, ferritin, albumin, glutathione (GSH), ceruloplasmin, transferrin, uric acid, etc.) and exogenous antioxidants (α-tocopherol, β-carotene, ascorbic acid, folic acid, flavonoids, lycopene, etc.) [12].

13.4 Thymoquinone

Thymoquinone (TQ: 2-isopropyl-5-methylbenzo-1, 4-quinone), which is dark yellow, crystalline in appearance, with a molecular formula C10H12O2 and molecular weight 164.20 $gmol^{-1}$, is the most active biocomponent of the *Nigella sativa* (NS) plant [13–16]. It is known that TQ acts through antioxidant, anti-inflammatory, and immunomodulatory mechanisms to protect health [16].

The NS plant (black seed/black cumin), whose seeds have been used as spices and food additives for thousands of years and belong to the Ranunculaceae family, is known as "black seed" and is used during the production of bread, doughy foods, and varieties of cheese [13, 17].

This is a plant that is common in the Mediterranean region, grows up to about 20–30 cm and blooms every year in yellow, white, pink, or pale purple and with 5–10 petals. The fruit of the flower contains between 3 and 7 follicles and contains many seeds with a size of 0.2 cm inside each of them (Figure 13.1) [14].

The seeds of the plant are composed of 30–40% unsaturated fatty acids (arachidonic acid, linoleic and linolenic acid, oleic and palmitic acid, beta-sitosterol, sterol and sterol esters with cyclocalenol glucoside) and 0.5% saturated essential oils (TQ, thymohydroquinone (THQ), dithymoquinone, thymol, carvacrol, pinene, and citronellol) [13]. Apart from all these, the seeds also contain various vitamins, carbohydrates, essential and non-essential amino acids, iron (Fe), copper (Cu), potassium (K), zinc (Zn), calcium (Ca^{+2}), and phosphorus (P) [14, 15].

Figure 13.1 (a) *Nigella sativa* plant Credit: (by Franz Eugen Köhler, Köhler's Medizinal-Pflanzen – List of Koehler Images, Public Domain, https://commons.wikimedia.org/w/index.php?curid=255583). (b) Seeds Credit: (Picture Partners/Adobe Stock). (c) Chemical structure of thymoquinone.

Therapeutic biological effects of TQ, a quinine derivative, have been reported in many studies. TQ provides effective protection against many gram-negative and gram-positive bacteria with polyantibiotic resistance, as well as various types of fungi and protozoa [14]. TQ suppresses prostaglandins and leukotrienes, which are the main mediators of inflammatory processes. It also provides immune modulation by increasing TH and natural killer (NK) cells [14, 17, 18]. It is proposed that TQ in toxic agents' exposed tissues reduces DNA damage and prevents the onset of the carcinogenic process [19].

13.5 TQ against Type 2 Diabetes Mellitus

Recent studies have reported that TQ possesses strong antidiabetic properties that are more effective against T2DM complications. One of the other reported mechanisms, an antioxidative property, is documented as the most mediating action towards health promotion by TQ (Figure 13.2). In addition to increasing insulin secretion induced by glucose, it also has an effect on reducing glucose absorption from the intestinal mucosa [15, 20]. TQ, which is also known to regulate hepatic lipid peroxidation, protects the liver tissue against the degenerative effects of hyperglycemia. The fact that TQ specifically protects against liver damage induced

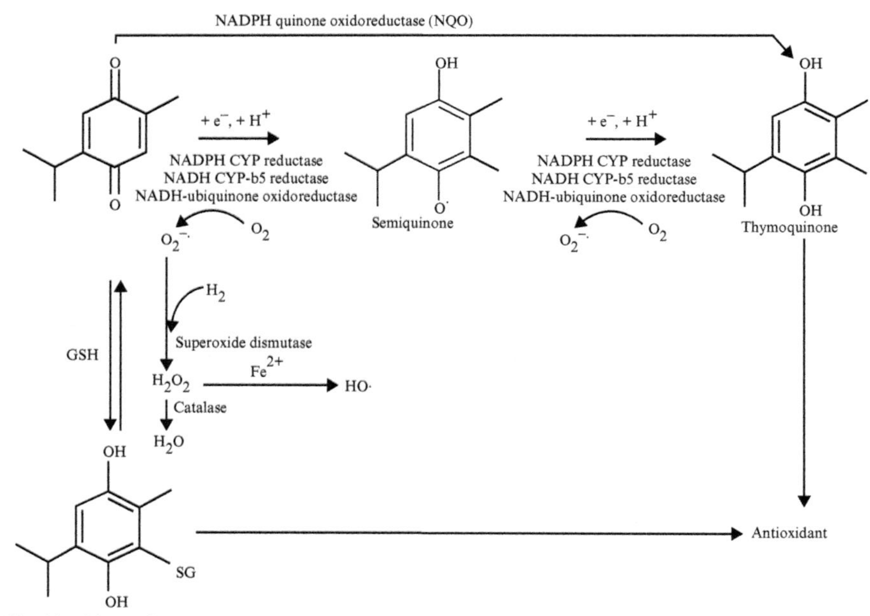

Figure 13.2 Schematic representation of the antioxidant effect elicited by thymoquinone.

by carbon tetrachloride (CCI4) is attributed to the antioxidant effect of this substance [21]. TQ is reported to decrease blood glucose level and enhance insulin level in experimental studies [22].

TQ decreases ROS and exhibits its antidiabetic properties protecting β-cell injury [23–25]. Over-production of ROS triggers various inflammatory signaling pathways leading to inflammation [26]. When the proinflammatory genes are activated and inflammation is triggered, this causes T2DM and apoptosis in β-cells [27, 28]. In diabetic rats, TQ is shown to lower the activity of COX-2 enzyme in β-cells, decrease lipid peroxidation levels, and increase SOD activity in pancreas [29].

Also, when experimental animals are supplemented with TQ during gestation, the offspring were protected from diabetes as the blood glucose levels were found to be low. Not only the glucose levels, but also the proinflammatory cytokines and ROS levels were found to be low too [30]. In another experimental diabetic study on gestational animals, TQ was shown to decrease blood glucose levels as it inhibits gluconeogenic enzymes' synthesis [31]. TQ was also reported to have reversing effects on streptozotocine's glycemic effects, by increasing insulin levels and lowering glucose levels via enhancing glucose utilization. Administration of TQ (80 mg/kg b.w) for 45 days to induced diabetic rats significantly protected the decrease in the activities of antioxidant enzymes such as glutathione peroxidase (GPx), catalase (CAT), and antioxidants such as vitamin E and GSH. So, it can be concluded that TQ protects β-cells via exhibiting an antioxidant property [32].

TQ is known to be the main antioxidant component of NS [33]. It is thought to diminish oxidative stress and maintain insulin secretion via different mechanisms [34, 35]. Also, TQ is known to downregulate the expression of gluconeogenic enzymes and lower the absorption of glucose in the gut [36]. Also, TQ activates the adenosine monophosphate-activated protein, and this inhibits gluconeogenesis [37]. Diabetic nephropathy, which results in renal failure, is frequently seen due to hyperglycemia, which provokes oxidative stress in renal tubules where the elevated levels of free radical molecules cause damage, so fibrosis is seen. When TQ-rich NS is supplemented in addition to other polyphenols, it is reported to exhibit impairing effects in diabetic nephropathy [38, 39]. However, further clinical studies are highly recommended for TQ.

13.6 Conclusion

T2DM is one of the most frequently seen metabolic disorders affecting both genders worldwide. The associated side effects of synthetic medicines have triggered the discovery of antidiabetic therapy with few or minimal side effects. The natural bioactive compound TQ has shown promising effects against many T2DM-induced complications. Antioxidant and other biological activities are reported to mediate the effects elicited by TQ; however, in depth studies on its mechanisms of action, doses, iteration with other biological compounds, and clinical studies are still anticipated in order to define TQ-based drug discovery against T2DM.

References

1 Whiting DR, Guariguata L, Weil C, Shaw J. IDF diabetes atlas: Global estimates of the prevalence of diabetes for 2011 and 2030. *Diabetes Res Clin Pract.* 2011; 94(3):311–321.
2 Fiorentino TV, Prioletta A, Zuo P, Folli F. Hyperglycemia-induced oxidative stress and its role in diabetes mellitus related cardiovascular diseases. *Curr Pharm Des.* 2013; 19:5695–5703.
3 Aronson D. Hyperglycemia and the pathobiology of diabetic complications. *Adv Cardiol.* 2008; 45:1–16.
4 Tesfaye S, Selvarajah D. Advances in the epidemiology, pathogenesis and management of diabetic peripheral neuropathy. *Diabetes Metab Res Rev.* 2012; 28(S1):8–14.
5 Monnier L, Colette C, Dunseath GJ, Owens DR. The loss of postprandial glycemic control precedes stepwise deterioration of fasting with worsening diabetes. *Diabetes Care.* 2007; 30:263–269.
6 Pisoschi AM, Pop A. The role of antioxidants in the chemistry of oxidative stress: A review. *Eur J Med Chem.* 2015; 97:55–74.
7 Nakazawa H, Genka C, Fujishima M. Pathological aspects of active oxygens/free radicals. *Japn J Physiol.* 1996; 46(1):15–32.
8 Reiter RJ, Tan D, Manchester LC, Qi W. Biochemical reactivity of melatonin with reactive oxygen and nitrogen species: a review of the evidence. *Cell Biochem Biophys.* 2001; 34:237–256.
9 Fedorova M, Bollineni RC, Hoffmann R. Protein carbonylation as a major hallmark of oxidative damage: Update of analytical strategies. *Mass Spectrom.* 2009; 33(2):79–97.
10 Catalá A. Lipid peroxidation of membrane phospholipids generates hydroxy-alkenals and oxidized phospholipids active in physiological and/or pathological conditions. *Chem Phys Lipids.* 2009;157:1–11.
11 Shacter E. Protein oxidative damage. *Methods Enzym.* 2000; 319:428–436.
12 Aguilar TAF, Navarro BCH, Pérez JAM. Endogenous antioxidants: A review of their role in oxidative stress. In: Morales-Gonzalez JA, editor. *A Master Regulator of Oxidative Stress - the Transcription Factor Nrf2.* InTech Open, 2016; doi:10.5772/65715.
13 Ahmad A, Husain A, Mujeeb M, Khan SA, Najmi AK, Siddique NA, et al. A review on therapeutic potential of Nigella sativa: A miracle herb. *Asian Pac J Trop Biomed.* 2013; 3(5):337–352.

14 Forouzanfar F, Fazly Bazzaz BS, Hosseinzadeh H. Black cumin (Nigella sativa) and its constituent (thymoquinone): A review on antimicrobial effects. *Iran J Basic Med Sci.* 2014; 17(12):929–938.

15 Ali BH, Blunden G. Pharmacological and toxicological properties of Nigella sativa. *Phyther Res.* 2003; 17(4):299–305.

16 Darakhshan S, Bidmeshki Pour A, Hosseinzadeh Colagar A, Sisakhtnezhad S. Thymoquinone and its therapeutic potentials. *Pharmacol Res.* 2015; 95–96:138–158.

17 Salem ML. Immunomodulatory and therapeutic properties of the Nigella sativa L. seed. *Int Immunopharmacol.* 2005; 5(13–14):1749–1770.

18 Majdalawieh AF, Fayyad MW. Immunomodulatory and anti-inflammatory action of Nigella sativa and thymoquinone: A comprehensive review. *Int Immunopharmacol.* 2015; 28(1):295–304.

19 Kundu J, Chun KS, Aruoma OI, Kundu JK. Mechanistic perspectives on cancer chemoprevention/chemotherapeutic effects of thymoquinone. *Mutat Res - Fundam Mol Mech Mutagen.* 2014; 768(C):22–34.

20 Ismail M, Yaheya M. Therapeutic role of prophetic medicine Habbat El Baraka (Nigella sativa L.) - A review. *World Appl Sci J.* 2009; 7(9):1203–1208.

21 Mollazadeh H, Hosseinzadeh H. The protective effect of Nigella sativa against liver injury: A review. *Iran J Basic Med Sci.* 2014; 17(12):958–966.

22 Abdelrazek HMA, Kilany OE, Muhammad MAA, Tag HM, Abdelazim AM. Black seed Thymoquinone improved insulin secretion, hepatic glycogen storage, and oxidative stress in streptozotocin-induced diabetic male Wistar rats. *Oxidative Med Cell Longev.* 2018; 2018:10.

23 Al-Trad B, Al-Batayneh K, El-Metwally S, Alhazimi A, Ginawi I, Alaraj M, et al. Nigella sativa oil and thymoquinone ameliorate albuminuria and renal extracellular matrix accumulation in the experimental diabetic rats. *Eur Rev Med Phamacol Sci.* 2016; 20(12):2680–2688.

24 Pei X, Li X, Chen H, Han Y, Fan Y. Thymoquinone inhibits angiotensin II-induced proliferation and migration of vascular smooth muscle cells through the AMPK/PPARg/PGC-1a pathway. *DNA Cell Biol.* 2016; 35(8):426–433.

25 Fouad AA, Alwadani F. Ameliorative effects of thymoquinone against eye lens changes in streptozotocin diabetic rats. *Environ Toxicol Pharmacol.* 2015; 40(3):960–965.

26 Samarghandian S, Farkhondeh T, Samini F, Borji A. Protective effects of carvacrol against oxidative stress induced by chronic stress in rat's brain, liver, and kidney. *Biochem Res Int.* 2016; 2016:2645237.

27 Zheng Y, Wang X, Ji S, Tian S, Wu H, Luo P et al. Mepenzolate bromide promotes diabetic wound healing by modulating inflammation and oxidative stress. *Am J Transl Res.* 2016; 8(6):2738–2747.

28 Dias JA, Fredrikson ON, Ericson U, Gullberg B, Hedblad B, Engstrom G, et al. Low-grade inflammation, oxidative stress and risk of invasive post-menopausal breast cancer - a nested case - control study from the Mälmo diet and cancer cohort. *PLoS One*. 2016; 11(7):e0158959.

29 Al Wafai RJ. Nigella sativa and thymoquinone suppress cyclooxygenase-2 and oxidative stress in pancreatic tissue of streptozotocin-induced diabetic rats. *Pancreas*. 2013; 42(5):841–849.

30 Badr G, Mahmoud MH, Farhat K, Waly H, AI-Abdin OZ, Rabah OM. Maternal supplementation of diabetic mice with thymoquinone protects their offspring from abnormal obesity and diabetes by modulating their lipid profile and free radical production and restoring lymphocyte proliferation via PI3K/AKT signaling. *Lipids Health Dis*. 2013; 12:37.

31 Fararh KM., Shimizu Y, Shiina T, Nikami H, Ghanem MM, Takewaki T. Thymoquinone reduces hepatic glucose production in diabetic hamsters. *Res Vet Sci*. 2005; 79(3):219–223.

32 Sankaranarayanan C, Pari L. Thymoquinone ameliorates chemical induced oxidative stress and ß-cell damage in experimental hyperglycemic rats. *Chem Biol Interact*. 2011; 190(2–3):148–154.

33 Abdelmeguid NE, Fakhomy R, Kamal SM, AI Wafai RJ. Effects of Nigella sativa and thymoquinone on biochemical and subcellular changes in pancreatic β-cells of streptozotocin-induced diabetic rats. *J Diabetes*. 2010; 2(4):256–266.

34 Randhawa MA, Alghamdi MS, Maulik SK. The effect of thymoquinone, an active component of Nigella sativa, on isoproterenol induced myocardial injury. *Pak J Pharm Sci*. 2013; 26(6):1215–1219.

35 Mohmoud Saleh Mansi K. Effects of oral administration of water extract of Nigella sativa on serum concentrations of insulin and testosterone in alloxan-induced diabetic rats. *Pak J Biol Sci*. 2005; 8:1152–1156.

36 Meddah B, Ducroc R, El Abbes Faouzi M, Eto B, Mahraoui L, Benhaddou-Andaloussi A, Martineau LC, Cherrah Y, Haddad PS. Nigella sativa inhibits intestinal glucose absorption and improves glucose tolerance in rats. *J Ethnopharmacol*. 2009; 121:419–424.

37 Coughlan KA, Valentine RJ, Ruderman NB, Saha AK. AMPK activation: A therapeutic target for type 2 diabetes? *Diabetes Metab Syndr Obes Targets Ther*. 2014; 7:241–253.

38 Sayed A. Thymoquinone and proanthocyanidin attenuation of diabetic nephropathy in rats. *Eur Rev Med Pharmacol Sci*. 2012; 16(6):808–815.

39 Morcos M, Sayed AA, Bierhaus A, Yard B, Waldherr R, Merz W, et al. Activation of tubular epithelial cells in diabetic nephropathy. *Diabetes*. 2002; 51(12):3532–3544.

14

Effect of Resveratrol and Catechins in Maintaining Redox Homeostasis during Type 2 Diabetes Mellitus

Prachee Dubey[1,*] *and Geeta Watal*[2]

[1] *Government G.I. College, Malwan, Fatehpur, UP-212601, India*
[2] *Department of Chemistry, University of Allahabad, Prayagraj, UP-211002, India*
* *Corresponding author*

14.1 Introduction

Oxidation-reduction or “redox” is the vital mechanism of life that involves essential transfer of reducing equivalents to an acceptor from a donor in the system. Cellular redox biology comprises a complicated network of reactive species, signaling, antioxidant enzymes and non-enzymes, which are the prerequisites for the occurrence of cellular metabolisms, diseases, and therapeutic inventions. Reactive oxygen species (ROS) are the by-products of physiological reactions in the body generated in many organelles including the electron transport chain located in the mitochondria [1, 2].

These ROS are highly specific in action and play a vital role in cell signaling, which is necessary for a successful physiology. Although ROS are strictly regulated by different antioxidant systems in the cell, since, as usual, no system is perfect, sometimes their generation supersedes their neutralization and a state prevails known as oxidative stress. There may be various reasons, including weakened inherent defense mechanisms, exposure to radiation, pollutants, toxic chemicals, and/or malnutrition [3]. To a certain extent, the oxidative load benefits the system; it upregulates defense potential; however, excessive generation of ROS, thereby persistent oxidative stress, damages the vital biomolecules of the cell including proteins, lipids, and nucleic acids, and thus alters the redox homeostasis of the cellular system [1, 4].

Type 2 diabetes mellitus (T2DM) is a metabolic disorder that comprises about 90% of all types of diabetes and is characterized by insulin resistance (IR) and associated hyperglycemia [5, 6]. Prevalence of T2DM has increased significantly

Natural Products and their Bioactives in Antidiabetic Drug Discovery, First Edition.
Edited by Kanti Bhooshan Pandey, Maitree Suttajit and Pinar Atukeren.

in the last few years and it is predicted that it may be an epidemic in the coming years. Long-term uncontrolled T2DM culminates in micro and macro vascular complications such as retinopathy, neuropathy, nephropathy, cardiac and cerebral diseases, even premature death [5, 7, 8]. It has been proposed that developing countries are prone to T2DM and may contribute about 77% of total diabetics globally by the year 2030 [9, 10].

Epidemiological studies demonstrate the urgent need for alternative therapies to control the development and progression of T2DM due to the long-term associated serious side effects of synthetic medicines and their high cost [10]. Data from clinical studies have provided enough evidence that changes in lifestyle and dietary interventions are the most effective strategies to delay and/or prevent the onset of this chronic human health impairment with minimal side effects [11, 12]. Polyphenols are a group of bioactive compounds largely present in many plants and their products. Recently, polyphenols received a great deal of interest in clinical and epidemiological studies due to their effective therapeutic biological effects on many of the chronic human diseases including cancer, cardiovascular diseases, and neuro disorders [13–15]. This chapter provides a short summary of the key players that are involved in the redox homeostasis in the cellular system and their implementation in T2DM along with the possible therapeutic potentials of two major polyphenols; resveratrol and catechins, on restoring the redox state thereby preventing the associated complications.

14.2 Cellular Redox Homeostasis

14.2.1 Cellular Reactive Species

Reactive species in the cell that take part in redox homeostasis are different radicals including ROS, reactive nitrogen species (RNS), and reactive sulfur species (RSS), in which ROS are dominant since oxygen transforms in the enzymatic reactions in living organisms and becomes part of the biomass and releases energy on oxidation of glucose. This may be the reason why ROS are the most studied in the redox homeostasis system and their signaling is better understood than other reactive species [3]. ROS represents a group of oxygen-derived compounds including hydrogen peroxide (H_2O_2), hydroxyl radicals (OH^o), singlet oxygen ($^1O_2$), superoxide anion ($^oO_2^-$), peroxyl radical (ROO^o), ozone (O_3), etc. ROS can be produced in the cell during intrinsic metabolic pathways (intracellular) or by triggers received from external factors such as smoke, metals, toxins, chemical fertilizers, pollutants, and radiation [16, 17].

In the cell, nitric oxide (NO) acts as a vasodilator and is biosynthesized from NADPH, oxygen, and L-arginine by mediation of nitric oxide synthases (NOS).

NO is a free radical that quickly reacts with superoxide anions to form peroxynitrite ($ONOO^-$). $ONOO^-$ is highly reactive, and modifies lipids, nucleic acids, proteins, and thiols followed by its accumulation in the cell. The most common modifications by RNS are tyrosine nitration and oxidized cysteines [1, 18]. Gaseous H_2S is generated in the cytoplasm and in mitochondria from amino acids, cysteine and homocysteine, by involving different specific enzymes. Like ROS, the transfer of a single electron from H_2S results in the generation of thiyl radical (HS^o), hydrogen persulfide (H_2S_2), supersulfide radical ($^oS_2^-$), and ultimately in elemental sulfur. These reactive species can alter the membrane bound transport channels and the cellular redox balance [1].

14.2.2 Cellular Antioxidants

To neutralize the deleterious effects of reactive species, cells are naturally equipped with inherent defense systems that comprise a group of enzymatic and non-enzymatic molecules. Reduced glutathione (GSH) is the most studied antioxidative molecule that provides the first line of defense against reactive oxygen/nitrogen/sulfur species. To reduce these reactive species, GSH quickly oxidizes into disulfide (GSSG) in the presence of glutathione peroxidase (GPX), which in favorable condition can be reversed back to GSH by the enzyme glutathione reductase (GR) by receiving the reducing equivalents from NADPH. GSH occurs in all mammalian cells and is an authentic measure of oxidative stress that the GSH/GSSG ratio is calculated to display in the extent of cellular oxidative load [19, 20].

Superoxide dismutase (SOD) and catalase (CAT) are the most well studied antioxidant enzymes in the biological system. SOD catalyzes superoxide dismutation and converts superoxide into O_2 and H_2O_2. Different organelles of the cell possess different subtypes of SOD such as SOD_1 in the mitochondrial intermembrane space and cytosol, SOD_2 in the mitochondrial matrix, and SOD3, which occurs in intracellular space [21]. SOD also gains importance because it generates H_2O_2, which engages in redox signaling [22]. H_2O_2 generated by SOD is detoxified by CAT, which is a tetrameric protein. CAT is predominantly located in peroxisomes and breaks H_2O_2 into water and O_2 and thus maintains the redox balance in the cell. H_2O_2 can also be reduced by GPX in the cellular system [23].

In addition, the nuclear factor E2-related factor (Nrf_2), which is the central transcription factor of the antioxidant defense, controls many genes that are actually antioxidative proteins and are directly involved in nullifying the hazardous actions of reactive oxygen/nitrogen/sulfur species [1]. The cellular antioxidant systems involved in neutralizing the deleterious effects of oxidative stress are well regulated in their expression of miRNAs, which creates an extensive network of tight regulation that allows the cellular system to take all of the actions on an immediate basis that are critically required to maintain the redox balance [24].

14.3 Cellular Redox Imbalance and Type 2 Diabetes Mellitus

Redox regulation in the cellular system is extremely important for a successful physiology. Since all the cells, tissues, and organelles are interconnected, a redox imbalance in any of them affects the overall physiology and invites multiple pathophysiological events. T2DM is one of the most studied pathological human events; however, the mechanisms behind its development and cause has still not been fully explored. Mounting evidence on recent *in vivo* and clinical studies has shown that redox imbalance is one of the critical players in T2DM and widely influences the signaling of insulin [25, 26]. Elevated levels of markers of oxidative stress and diminished antioxidant potential have been reported in different biological fluids of diabetic patients [27]. An array of studies has documented an increased lipid peroxidation/protein/nucleotides oxidation product and improper GSH/GSSg ratio, diminished expression of antioxidant enzymes, low levels of vitamins, and a compromised plasma membrane redox system in T2DM patients [27–30]. Increased oxidative stress due to over generation of reactive oxygen/nitrogen species is accompanied by dyslipidemia and hyperglycemia, the common symptoms of diabetes [31]. The untoward influence of redox imbalance on key regulators and insulin signaling such as the serine/threonine kinase (Akt), forkhead box class O family member proteins (FOXO), protein tyrosine phosphatase 1B (PTP1B), have been reported in T2DM [32]. Nevertheless, a high oxidative stress/imbalance redox state is directly associated with IR, inflammation, and non-enzymatic glycation [33]. FOXO transcription factors incorporate cysteine residues, which are conductive to increase antioxidants thereby diminishing the oxidative overload, are highly sensitive to redox stress. Critically compromised redox homeostasis triggers FOXO transcription factors and may hyperactivate them resulting in hyperglycemia, hypertriglyceridemia, followed by IR [25, 34]. Considering the significance of redox homeostasis in T2DM, its restoration by natural bioactives may be an effective therapy against this multifactorial metabolic error.

14.4 Polyphenols as Dietary Antioxidants in Restoring/Maintaining Redox Homeostasis during T2DM

Polyphenols are secondary metabolites in plants and are synthesized to provide them with protection from biotic and abiotic stress [35]. Recent epidemiological, clinical, and animal studies have provided evidence that polyphenols may be strong curative agents against many chronic human diseases due to their potent antioxidative, antimicrobial, selective apoptosis, cell signaling, and wound healing activities [35, 36]. Polyphenols, due to reducing properties, act as electron donors and may act as redox maintaining/restoring agents in cellular systems, which is crucial in the prevention of development of an array of health impairments and a successful physiology [36, 37]. More than 10 000 polyphenols have been investigated; however, only a few have been explored for their health beneficial activities and mechanisms involved behind their activities. In this chapter, the redox balancing activities of resveratrol and catechins, relevant to T2DM protection, have been discussed.

14.4.1 Resveratrol

Resveratrol is a stilbene class of polyphenolic compound, and occurs naturally in a variety of plants and their products such as grapes, berries, peanuts, and red wine. Resveratrol is synthesized in plants against infections in unfavorable conditions [38]. Serious studies on the health promotive effects of resveratrol were started in the 1990s when it was reported that resveratrol may be responsible for the reduced cardiovascular diseases in French people despite evidence of unhealthy life styles and smoking habits [39]. Recent studies have reported that resveratrol may provide therapeutic support against a variety of health impairments via multiple mechanism(s) of action; however, being antioxidative is the major factor [3, 40]. Many researchers have reported that resveratrol possesses a strong antioxidative property and its administration very effectively maintains the redox homeostasis and reduces the oxidative stress load [40, 41]. Administration of resveratrol enhances glucose uptake, its metabolism, protects pancreatic β-cells, improves insulin sensitivity, and reduces insulin resistance [42–44]. Resveratrol offers redox balancing effects via a variety of mechanisms such as reducing ROS generation, scavenging free radicals, inducing endogenous antioxidative enzymes, recycling ascorbic acid, and others [41, 45]. It was shown that 0.3% resveratrol fed to T2DM model *db/db* mice for eight weeks showed improved glycolipid metabolism, and the effect was mediated by Mn-SOD [46]. In another study performed by Ramadori and colleagues, it was reported that central administration of resveratrol for 14 weeks improved diet-induced diabetes (hyperglycemia) by hypothalamic NF-kB inflammatory signal transduction [47]. C57BL/6 mice on a high fructose diet, fed on 400 mg/kg resveratrol orally for 16 weeks reduced the IR, enhanced mitochondrial biogenesis and oxidative phosphorylation; and the proposed mechanisms were activation of sirtuin1 (SIRT1) and peroxisome proliferator-activated receptor-gamma coactivator 1alpha (PGC-1α) [48]. Hussein and Mahfouz (2016) reported that administration of resveratrol 5 mg/kg orally for eight weeks to streptozotocin (STZ)-induced diabetic Wistar rats significantly reduced the blood glucose levels and improved the antioxidant status; the mechanisms proposed were induction of TGF-β1, fibronectin, NF-Kb/p65, Nrf2, SIRT1, and FOXO1 in the kidney [49].

Another study has reported that administration of 20 mg/kg resveratrol for eight weeks to STZ-induced diabetes rats very effectively decreased hyperglycemia, lipid peroxidation by the mechanism, which incorporates reduction in oxidative stress and induces potential of extrahepatic tissues to absorb glucose [50]. All these results establish resveratrol as a potential agent that may be explored as an anti-diabetic drug with natural origin.

14.4.2 Catechins

Catechins are the flavonoid group of polyphenols and are found naturally in many plants in abundant amounts such as fresh tea leaves, grapes, berries, beans, and red wine. The tea plant (*Camellia sinensis*) possesses catechins in high concentration; it has been reported that catechins account for 70% of the total polyphenols in tea leaves [51]. Catechins represent a group of similar compounds that differ only due to esterification. Many different catechins are reported to occur; however, epicatechin (EC), epigallocatechin (EGC), epicatechin-3-gallate (ECG), and epigallocatechin-3-gallate (EGCG) are the major catechins whose therapeutic applications have been well explored [52] (Figure 14.1).

Catechins are reported as potential anti-diabetic agents due to their strong antioxidant properties. An array of studies including *in vitro*, *in vivo*, animal, and clinical studies have provided enough evidence in favor of the anti-diabetic action of catechins [53, 54]. Catechins are reported to upregulate the PMRS via donating reducing equivalents that are ultimately utilized to recycle ascorbate in erythrocytes, and operate in the cell to establish redox homeostasis, which is reported as compromised in T2DM patients [30, 55]. Interestingly, catechins elicit their redox restoring effects by scavenging free radicals (antioxidant) as well as by generating them (pro-oxidant) as per requirement [52]. Catechins exert their antioxidative effects through chelating metal ions, neutralizing ROS/RNS, inducing expression on antioxidative enzymes, and producing phase II detoxifying enzymes [56]. EGCG, ECG, and EGC have been reported to protect peroxidation of lipids and oxidation of proteins and to reduce the oxidative load by preventing degradation of GSH and thiols in T2DM patients [57, 58]. Malfunctioned membrane transporters have been documented in T2DM, and administration of catechins significantly protected the oxidative stress induced deviations in membrane bound ion channels and maintained the cellular pH [59].

The family of NADPH oxidase is the major source of intracellular superoxide and H_2O_2 and thus induces oxidative stress. It has been reported that metabolites of catechins and EC have abilities to suppress the expression of NADPH oxidase by reducing redox-sensitive kinases (JNK and IKK), thereby enhancing insulin sensitivity [60]. Bettaieb and co-workers have reported impaired activation of insulin signaling proteins such as insulin receptors (IR), insulin receptor

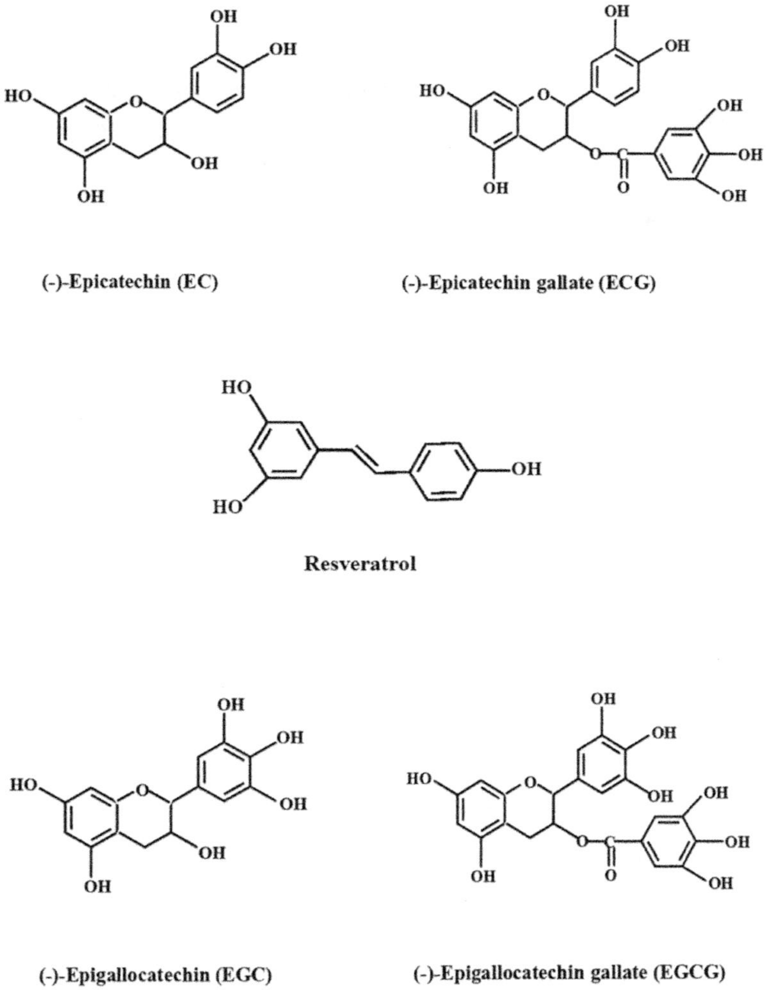

Figure 14.1 The chemical structure of the main tea catechins: EC ((-)-Epicatechin), EGC ((-)-Epigallocatechin), ECG ((-)-Epicatechin gallate), EGCG ((-)-Epigallocatechin gallate), and resveratrol.

substrate 1 (IRS-1), etc. in high fructose diet induced insulin resistant rats, and the impairments were reversed by administration of EC [61]. Supplementation of catechins also reduced the oxidative burden, suppressed the IR, and diminished the overexpression of redox-sensitive signals such as IKK/NF-kB, JNK1/2, as well as PTP1B [62]. Altogether, the results report that catechins have the ability to protect the suppression of insulin sensitivity, IR via redox-sensitive signaling, which has a documented regulatory influence on diabetes.

14.5 Future Trends and Conclusion

Cellular redox-imbalance is one of the major impairments in T2DM that mediates diabetic complications. Results obtained from different studies on various model systems and clinical trials on humans have provided enough evidence of the potent anti-diabetic effects of catechins and resveratrol. However, large-scale randomized clinical trials, bioavailability, dose, and exposure are other important issues related to catechins and resveratrol that are warranted to establish these polyphenolic bioactive compounds as potential therapeutic natural agents against T2DM.

Acknowledgment

PD acknowledges support of her college and DSE, Govt. of UP, India

References

1 Tretter V, Hochreiter B, Zach ML, Krenn K, Klein KU. Understanding cellular redox homeostasis: A challenge for precision medicine. *Int J Mol Sci.* 2021 Dec 22; 23(1):106. doi: 10.3390/ijms23010106. PMID: 35008532; PMCID: PMC8745322.

2 Sies H, Jones DP. Reactive oxygen species (ROS) as pleiotropic physiological signalling agents. *Nat Rev Mol Cell Biol.* 2020 Jul; 21(7):363–383. doi: 10.1038/s41580-020-0230-3. Epub 2020 Mar 30. PMID: 32231263.

3 Sies H. Oxidative stress: Concept and some practical aspects. *Antioxidants (Basel).* 2020 Sep 10; 9(9):852. doi: 10.3390/antiox9090852. PMID: 32927924; PMCID: PMC7555448.

4 Holmstrom KM., Finkel T. Cellular mechanisms and physiological consequences of redox-dependent signalling. *Nat Rev Mol Cell Biol.* 2014; 15:411–421.

5 Brunton S. Pathophysiology of type 2 diabetes: The evolution of our understanding. *J Fam Pract.* 2016 Apr; 65(4 Suppl): supp_az_0416.PMID: 27262256.

6 Gomez-Guijarro MD, Álvarez-Bueno C, Saz-Lara A, Sequí-Domínguez I, Lucerón-Lucas-Torres M, Cavero-Redondo I. Association between severe hypoglycemia and risk of dementia in patients with type 2 diabetes mellitus: A systematic review and meta-analysis. *Diabetes Metab Res Rev.* 2023 Jan 17: e3610. doi: 10.1002/dmrr.3610. Epub ahead of print. PMID: 36649373.

7 Henning RJ. Type-2 diabetes mellitus and cardiovascular disease. *Future Cardiol.* 2018 Nov; 14(6):491–509. doi:10.2217/fca-2018-0045. Epub 2018 Nov 9. PMID: 30409037.

8 Damanik J, Yunir E. Type 2 Diabetes mellitus and cognitive impairment. *Acta Med Indones*. 2021 Apr; 53(2):213–220. PMID: 34251351.

9 Animaw W, Seyoum Y. Increasing prevalence of diabetes mellitus in a developing country and its related factors. *PLoS One*. 2017 Nov 7; 12(11):e0187670. doi: 10.1371/journal.pone.0187670. PMID: 29112962; PMCID: PMC5675402.

10 Wu Y, Ding Y, Tanaka Y, Zhang W. Risk factors contributing to type 2 diabetes and recent advances in the treatment and prevention. *Int J Med Sci*. 2014 Sep 6; 11(11):1185–1200. doi: 10.7150/ijms.10001. PMID: 25249787; PMCID: PMC4166864.

11 Bekele H, Asefa A, Getachew B, Belete AM. Barriers and strategies to lifestyle and dietary pattern interventions for prevention and management of TYPE-2 diabetes in Africa, systematic review. *Journal of Diabetes Research*. 2020; 2020 (Article ID 7948712):14. doi:10.1155/2020/7948712.

12 Tariq MNM, Stojanovska L, Dhaheri ASA, Cheikh Ismail L, Apostolopoulos V, Ali HI. Lifestyle interventions for prevention and management of diet-linked non-communicable diseases among adults in Arab countries. *Healthcare*. 2023; 11(1):45. doi:10.3390/healthcare11010045.

13 Rana A, Samtiya M, Dhewa T, Mishra V, Aluko RE. Health benefits of polyphenols: A concise review. *J Food Biochem*. 2022 Oct; 46(10):e14264. doi: 10.1111/jfbc.14264. Epub 2022 Jun 13. PMID: 35694805.

14 Mishra J, Srivastava SK, Pandey KB. Dietary polyphenols in the intervention of gestational diabetes. *Curr Trad Med*. 2021; 7(6):8–15(8). doi:10.2174/2215083807666211007094515.

15 Pandey KB, Rizvi SI. (2017). Plant polyphenols in healthcare and aging. In: *Nutritional Antioxidant Therapies: Treatments and Perspectives*. (ed. K. Al-Gubory and I. Laher). 267–282. Cham: Springer. https://doi.org/10.1007/978-3-319-67625-8_11

16 de Almeida AJPO, de Oliveira JCPL, da Silva Pontes LV, de Souza Júnior JF, Gonçalves TAF, Dantas SH, de Almeida Feitosa MS, Silva AO, de Medeiros IA. ROS: Basic concepts, sources, cellular signaling, and its implications in aging pathways. *Oxid Med Cell Longev*. 2022 Oct 19; 2022:1225578. doi: 10.1155/2022/1225578. PMID: 36312897; PMCID: PMC9605829.

17 Brieger K, Schiavone S, Miller FJ, Jr, Krause KH. Reactive oxygen species: From health to disease. *Swiss Med Wkly*. 2012 Aug 17; 142:w13659. doi: 10.4414/smw.2012.13659. PMID: 22903797.

18 Janssen-Heininger YM, Persinger RL, Korn SH, Pantano C, McElhinney B, Reynaert NL, Langen RC, Ckless K, Shrivastava P, Poynter ME. Reactive nitrogen species and cell signaling: Implications for death or survival of lung epithelium. *Am J Respir Crit Care Med*. 2002 Dec 15; 166(12 Pt 2):S9–S16. doi: 10.1164/rccm.2206008. PMID: 12471083.

19 Lu SC. Glutathione synthesis. *Biochim Biophys Acta*. 2013 May; 1830(5):3143–3153. doi: 10.1016/j.bbagen.2012.09.008. Epub 2012 Sep 17. PMID: 22995213; PMCID: PMC3549305.

20 Liu T, Sun L, Zhang Y, Wang Y, Zheng J. Imbalanced GSH/ROS and sequential cell death. *J Biochem Mol Toxicol*. 2022 Jan; 36(1):e22942. doi: 10.1002/jbt.22942. Epub 2021 Nov 2. PMID: 34725879.

21 McCord JM, Edeas MA. SOD, oxidative stress and human pathologies: A brief history and a future vision. *Biomed Pharmacother*. 2005 May; 59(4):139–142. doi: 10.1016/j.biopha.2005.03.005. Epub 2005 Mar 22. PMID: 15862706.

22 Wang Y, Branicky R, Noë A, Hekimi S. Superoxide dismutases: Dual roles in controlling ROS damage and regulating ROS signaling. *J Cell Biol*. 2018 Jun 4; 217(6):1915–1928. doi: 10.1083/jcb.201708007. Epub 2018 Apr 18. PMID: 29669742; PMCID: PMC5987716.

23 Nandi A, Yan LJ, Jana CK, Das N. Role of catalase in oxidative stress- and age-associated degenerative diseases. *Oxid Med Cell Longev*. 2019 Nov 11; 2019:9613090. doi: 10.1155/2019/9613090. PMID: 31827713; PMCID: PMC6885225.

24 Ciesielska S, Slezak-Prochazka I, Bil P, Rzeszowska-Wolny J. Micro RNAs in regulation of cellular redox homeostasis. *Int J Mol Sci*. 2021; 22(11):6022. doi:10.3390/ijms22116022.

25 Zuo J, Zhang Z, Luo M, Zhou L, Nice EC, Zhang W, Wang C, Huang C. Redox signaling at the crossroads of human health and disease. *MedComm*. 2022 Mar 31; 3(2):e127. doi: 10.1002/mco2.127. PMID: 35386842; PMCID: PMC8971743.

26 Yan LJ. Redox imbalance stress in diabetes mellitus: Role of the polyol pathway. *Animal Model Exp Med*. 2018 Mar; 1(1):7–13. doi: 10.1002/ame2.12001. Epub 2018 Apr 19. PMID: 29863179; PMCID: PMC5975374.

27 Tiwari BK, Pandey KB, Abidi AB, Rizvi SI. Markers of oxidative stress during diabetes mellitus. *J Biomark*. 2013; 2013:378790. doi:10.1155/2013/378790. Epub 2013 Dec 17. PMID: 26317014; PMCID: PMC4437365.

28 Pandey KB, Mishra N, Rizvi SI. (2009). Myricetin can mitigate altered redox status in type 2 diabetic erythrocytes. *Kuwait J Sci Eng*. 36(2A): 115–124.

29 Asmat U, Abad K, Ismail K. Diabetes mellitus and oxidative stress-A concise review. *Saudi Pharm J*. 2016 Sep; 24(5):547–553. doi: 10.1016/j.jsps.2015.03.013. Epub 2015 Mar 21. PMID: 27752226; PMCID: PMC5059829.

30 Rizvi SI, Srivastava N. Erythrocyte plasma membrane redox system in first degree relatives of type 2 diabetic patients. *Int J Diabetes Mellitus*. 2010 Aug 1; 2(2):119–121.

31 Nishikawa T, Araki E. Impact of mitochondrial ROS production in the pathogenesis of diabetes mellitus and its complications. *Antioxid Redox Signal*. 2007; 9(3):343–353.

32 Lennicke C, Cocheme HM. Redox regulation of the insulin signalling pathway. *Redox Biol*. 2021; 42:101964.

33 Kaneto H, Katakami N, Matsuhisa M, Matsuoka TA. Role of reactive oxygen species in the progression of type 2 diabetes and atherosclerosis. *Mediators Inflamm.* 2010; 2010:453892.

34 Barthel A, Schmoll D, Unterman TG. FoxO proteins in insulin action and metabolism. *Trends Endocrinol Metab.* 2005; 16(4):183–189.

35 Pandey KB, Rizvi SI. Recent advances in health promoting effect of dietary polyphenols. *Curr Nutr Food Sci.* 2012; 8(4):254–264(11).

36 Fraga CG, Croft KD, Kennedy DO, Tomás-Barberán FA. The effects of polyphenols and other bioactives on human health. *Food Funct.* 2019 Feb 20; 10(2):514–528. doi: 10.1039/c8fo01997e. PMID: 30746536.

37 Rizvi SI, Jha R, Pandey KB. Activation of erythrocyte plasma membrane redox system provides a useful method to evaluate antioxidant potential of plant polyphenols. In: Armstrong D, editor. *Advanced Protocols in Oxidative Stress II. Methods in Molecular Biology*. vol. 594. Totowa, NJ: Humana Press, 2010; doi:10.1007/978-1-60761-411-1_24.

38 Pandey KB, Rizvi SI. Pleiotropic biological effects of resveratrol: Implications for human health. *Natl Acad Sci Lett.* 2009; 32(11/12):321–326.

39 Biagi M, Bertelli AA. Wine, alcohol and pills: What future for the French paradox? *Life Sci.* 2015 Jun 15;131:19–22. doi: 10.1016/j.lfs.2015.02.024. Epub 2015 Apr 1. PMID: 25841977.

40 Planinc M, Jovanović IN, Rašić D, Peraica M, Sutlić Ž. Resveratrol as antioxidant in cardiac surgery: Is there potential for clinical application? *Arh Hig Rada Toksikol.* 2023 Jan 7; 73(4):256–259. doi: 10.2478/aiht-2022-73-3643. PMID: 36607724.

41 Rizvi SI, Pandey KB. Activation of the erythrocyte plasma membrane redox system by resveratrol: A possible mechanism for antioxidant properties. *Pharmacol Rep.* 2010; 62:726–732.

42 Bo S, Gambino R, Ponzo V, Cioffi I, Goitre I, Evangelista A, Ciccone G, Cassader M, Procopio M. Effects of resveratrol on bone health in type 2 diabetic patients. A double-blind randomized-controlled trial. *Nutr Diabetes.* 2018 Sep 20; 8(1):51. doi: 10.1038/s41387-018-0059-4. PMID: 30237505; PMCID: PMC6147949.

43 Sattarinezhad A, Roozbeh J, Shirazi Yeganeh B, Omrani GR, Shams M. Resveratrol reduces albuminuria in diabetic nephropathy: A randomized double-blind placebo-controlled clinical trial. *Diabetes Metab.* 2019 Jan; 45(1):53–59. doi: 10.1016/j.diabet.2018.05.010. Epub 2018 Jun 8. PMID: 29983230.

44 Su M, Zhao W, Xu S, Weng J. Resveratrol in treating diabetes and its cardiovascular complications: A review of its mechanisms of action. *Antioxidants (Basel).* 2022 May 30; 11(6):1085. doi: 10.3390/antiox11061085. PMID: 35739982; PMCID: PMC9219679.

45 Meng X, Zhou J, Zhao CN, Gan RY, Li HB. Health benefits and molecular mechanisms of resveratrol: A narrative review. *Foods.* 2020 Mar 14; 9(3):340. doi: 10.3390/foods9030340. PMID: 32183376; PMCID: PMC7143620.

46 Kitada M, Kume S, Imaizumi N, Koya D. Resveratrol improves oxidative stress and protects against diabetic nephropathy through normalization of Mn-SOD dysfunction in AMPK/SIRT1-independent pathway. *Diabetes*. 2011 Feb; 60(2):634–643. doi: 10.2337/db10-0386. PMID: 21270273; PMCID: PMC3028365.

47 Ramadori G, Gautron L, Fujikawa T, Vianna CR, Elmquist JK, Coppari R. Central administration of resveratrol improves diet-induced diabetes. *Endocrinology*. 2009; 150:5326–5333.

48 Lagouge M, Argmann C, Gerhart-Hines Z, Meziane H, Lerin C, Daussin F, Messadeq N, Milne J, Lambert P, Elliott P, et al. Resveratrol improves mitochondrial function and protects against metabolic disease by activating SIRT1 and PGC-1. *Cell*. 2006; 127:1109–1122.

49 Hussein MM, Mahfouz MK. Effect of resveratrol and rosuvastatin on experimental diabetic nephropathy in rats. *Biomed Pharmacother*. 2016; 82:685–692.

50 Gencoglu H, Tuzcu M, Hayirli A, Sahin K. Protective effects of resveratrol against streptozotocin-induced diabetes in rats by modulation of visfatin/sirtuin-1 pathway and glucose transporters. *Int J Food Sci Nutr*. 2015; 66:314–320.

51 Bae J, Kim N, Shin Y, Kim SY, Kim YJ. Activity of catechins and their applications. *Biomed Dermatol*. 2020; 4(1):8. doi:10.1186/s41702-020-0057-8. Epub 2020 Feb 26. PMCID: PMC7149075.

52 Bernatoniene J, Kopustinskiene DM. The role of catechins in cellular responses to oxidative stress. *Molecules*. 2018 Apr 20; 23(4):965. doi: 10.3390/molecules23040965. PMID: 29677167; PMCID: PMC6017297.

53 Brimson JM, Prasanth MI, Kumaree KK, Thitilertdecha P, Malar DS, Tencomnao T, Prasansuklab A. Tea plant (Camellia sinensis): A current update on use in diabetes, obesity, and cardiovascular disease. *Nutrients*. 2022 Dec 21; 15(1):37. doi: 10.3390/nu15010037. PMID: 36615695; PMCID: PMC9823498.

54 Wen L, Wu D, Tan X, Zhong M, Xing J, Li W, Li D, Cao F. The role of catechins in regulating diabetes: An update review. *Nutrients*. 2022 Nov 4; 14(21):4681. doi: 10.3390/nu14214681. PMID: 36364943; PMCID: PMC9654920.

55 Pandey KB, Rizvi SI. Upregulation of erythrocyte ascorbate free radical reductase by tea catechins: Correlation with their antioxidant properties. *Food Res Int*. 2012; 1(46):46–49.

56 Rosen T. Green tea catechins: Biologic properties, proposed mechanisms of action, and clinical implications. *J Drugs Dermatol*. 2012 Nov; 11(11):e55–60. PMID: 23135094.

57 Rizvi SI, Zaid MA, Anis R, Mishra N. Protective role of tea catechins against oxidation-induced damage of type 2 diabetic erythrocytes. *Clin Exp Pharmacol Physiol*. 2005 Jan-Feb; 32(1–2):70–75. doi: 10.1111/j.1440-1681.2005.04160.x. PMID: 15730438.

58 Mostafa T, Sabry D, Abdelaal AM, Mostafa I, Taymour M. Cavernous antioxidant effect of green tea, epigallocatechin-3-gallate with/without sildenafil citrate intake in aged diabetic rats. *Andrologia*. 2013 Aug; 45(4):272–277. doi: 10.1111/and.12005. Epub 2012 Aug 29. PMID: 22928786.

59 Rizvi SI, Zaid MA. Impairment of sodium pump and Na/H exchanger in erythrocytes from non-insulin dependent diabetes mellitus patients: Effect of tea catechins. *Clin Chim Acta*. 2005 Apr; 354(1–2):59–67. doi: 10.1016/j.cccn.2004.11.008. Epub 2004 Dec 24. PMID: 15748600.

60 Cremonini E, Oteiza PI. (-)-Epicatechin and its metabolites prevent palmitate-induced NADPH oxidase upregulation, oxidative stress and insulin resistance in HepG2 cells. *Arch Biochem Biophys*. 2018; 646:55–63.

61 Bettaieb A, Vazquez Prieto MA, Rodriguez Lanzi C, Miatello RM, Haj FG, Fraga CG, Oteiza PI. (-)-Epicatechin mitigates high-fructose-associated insulin resistance by modulating redox signaling and endoplasmic reticulum stress. *Free Radic Biol Med*. 2014 Jul; 72:247–256. doi: 10.1016/j.freeradbiomed.2014.04.011. Epub 2014 Apr 16. PMID: 24746618; PMCID: PMC4077617.

62 Daveri E, Cremonini E, Mastaloudis A, Hester SN, Wood SM, Waterhouse AL, Anderson M, Fraga CG, Oteiza PI. Cyanidin and delphinidin modulate inflammation and altered redox signaling improving insulin resistance in high fat-fed mice. *Redox Biol*. 2018 Sep; 18:16–24. doi: 10.1016/j.redox.2018.05.012. Epub 2018 May 30. PMID: 29890336; PMCID: PMC6035912.

15

Cannabis

Action Mechanisms and Potential Roles in the Management of Type 2 Diabetes Mellitus

Karolin Yanar[1,] and Zeynep Mine Coşkun Yazıcı[2]*

[1] *Department of Medical Biochemistry, Cerrahpaşa Faculty of Medicine, Istanbul University-Cerrahpaşa, Istanbul, Turkey*
[2] *Department of Molecular Biology and Genetics, Faculty of Arts and Sciences, Demiroglu Bilim University, Istanbul Turkey*
[*] *Corresponding author*

15.1 Introduction

Cannabis sativa L. has been used in many fields; from industry to medicine since ancient times. This plant is an herbaceous species originating from Central Asia [1, 2]. While the total number of natural compounds identified from *C. sativa* was 483 in 1995, this number increased to 490 by 2005 [3, 4]. Recently, it has been reported that a total of 565 compounds have been identified in the *Cannabis* plant, including cannabinoids and non-cannabinoids [5].

Cannabinoids are classified into three groups; endocannabinoids (produced in humans and animals), phytocannabinoids (produced in plants), and synthetic cannabinoids (manufactured in the laboratory) [6]. The pharmacological effects of cannabinoids occur as a result of interactions between these compounds and cannabinoid receptors (CBRs), which are members of the G-protein coupled receptor family [1, 7]. Thus, endocannabinoids (endogenous cannabinoids), CBRs, and additionally enzymes (responsible for the synthesis and degradation of endocannabinoids) form the endocannabinoid system [8]. Interest in the endocannabinoid system, which is thought to be a potential target in the treatment of various diseases in drug discovery and development and medical research, has increased recently [9]. This chapter elucidates the potential role and mechanisms of action of the *Cannabis* plant and its components in the treatment of type 2 diabetes mellitus (T2DM), a metabolic disease.

Natural Products and their Bioactives in Antidiabetic Drug Discovery, First Edition.
Edited by Kanti Bhooshan Pandey, Maitree Suttajit and Pinar Atukeren.

15.2 Endocannabinoid System Elements

The term "cannabinoids" is known as a family of compounds that act on CBRs. Evidence shows that phytocannabinoids and endocannabinoids acting on CBRs, in other words, the endocannabinoid system, regulate different aspects of human metabolic, immunological, physiological, and behavioral functions [6, 10]. One of the two important compounds of *Cannabis* is cannabidiol (CBD), and the other is delta-9-tetrahydrocannabinol (Δ9-THC). Although CBD was first isolated from *Cannabis* in 1940, its structure was not reported until 1963. However, CBD did not come to the fore, as it was not psychoactive [11], while the structure of Δ9-THC, a major psychoactive phytocannabinoid, was reported in 1964 by an Israeli research team [12, 13]. Later, the first CBR (CB1R) was identified. Subsequently, the first endocannabinoid, *N*-arachidonoylethanolamine (AEA), was found and is also called anandamide, derived from ananda, which means happiness and pleasure in Sanskrit [14, 15]. After AEA, another endocannabinoid, 2-arachidonoylglycerol (2-AG), was identified in 1995 [16]. In the following years, various endocannabinoids such as 2-arachidonyl glyceryl ether [17], O-arachidonoyl-ethanolamine [18], and *N*-arachidonoylethanolamine [19, 20] were identified. However, these remained the putative endocannabinoids [21].

As mentioned above, the first identified CBR is CB1R, which is the most abundant. CB2Rs are expressed at much lower levels in the central nervous system (CNS) than CB1Rs [22, 23]. Orphan G protein-coupled receptor 55 (GPR55), first isolated from CNS regions by Sawzdargo et al. in 1999, was presented as a major candidate to be considered a "third" CBR, as it shows binding with some cannabinoids and non-cannabinoid ligands [24]. Apart from these receptors, cannabinoids exert their therapeutic effects through various targets, including ligand-gated ion channels, transient receptor potential (TRP) channels, metabolic enzymes, and nuclear receptors [25]. The other components are diacylglycerol lipase (DAGL) isozymes α and β, fatty acid amide hydrolase (FAAH), monoacylglycerol lipase (MAGL), and *N*-acyl-phosphatidylethanolamine-selective phospholipase D (NAPE-PLD), which are the enzymes responsible for the synthesis and degradation of endocannabinoids [9, 26].

15.3 Cannabinoid Receptors (CBRS)

15.3.1 Distribution in Tissues

Researchers initially suggested that due to the lipophilic nature of cannabinoids, these compounds act by nonspecifically disrupting the cell membrane. However, mapping and pharmacological characterization of cannabinoid binding sites in the brain following the emergence of THC and other synthesized cannabinoids revealed the presence of CBRs [15, 27–29].

In a review written by Howlett, it is reported that CB1Rs are mostly present in the CNS and peripheral nervous system (PNS), while CB2Rs are found in immune cells [30]. In addition to the known form of CB1R encoded by the CNR1 gene, two additional isoforms have been reported. It has been suggested that the CB1Rb form shows a higher expression level in pancreatic islet cells and hepatocytes and may be associated with metabolism [7, 31–33]. On the other hand, compared to CB1R, CB2R has more species differences between humans and rodents. It has been reported that CB2R, encoded by the CNR2 gene, has two isoforms (CB2A and CB2B), and in humans CB2A is mostly found in the testes and brain, and CB2B is expressed especially in other peripheral tissues [34]. Yang et al. (2016) specified that although GPR55 exhibits low homology with CBRs, it shares several cannabinoid ligands with CBRs [35]. In 1999, when GPR55 was first isolated, it was detected to be expressed in different parts of the CNS [24]. It was later determined that GPR55 was also found in peripheral tissues, including the adrenal glands, parts of the gastrointestinal tract, and endothelial cells [36]. Recent reports advocate that GPR55 is also expressed in the endometrium, adipocytes, and liver cells [37, 38].

15.3.2 Signaling

When activated by G-protein coupled receptor family members CB1R and CB2R endocannabinoids, CB1/CB2 receptors bind to the inhibitory G proteins (Gi/o) and inhibit adenylyl cyclase (AC) activity, activate voltage-gated calcium channels (VGCC) and mitogen-activated protein kinases (MAPK), and also initiate multiple gene signaling pathways [39]. In addition to Gi/o-mediated signaling, these two receptors are able to mediate signaling pathways in a G protein-independent manner through beta arrestins (β-arrestins) [40–43]. The signal transduction pathway of GPR55 is dissimilar to that of CB1R and CB2R. It has been shown that GPR55 only binds to $G\alpha 12{,}13$ proteins and activates ras homologous gene family members A (RhoA) and Rho-associated protein kinase (ROCK). Then, ROCK opens the phospholipase C (PLC) pathway, thereby increasing the cytosolic concentration of Ca^{2+}. The increase in intracellular Ca^{2+} causes extracellular regulated kinase (ERK) phosphorylation and gene expression [36, 37, 44].

15.4 Type 2 Diabetes Mellitus

T2DM is a heterogeneous, chronic, and progressive metabolic disorder characterized by chronic hyperglycemia together with impairment of the macromolecular metabolism resulting from defective action and qualitative and quantitative secretion of insulin and its possible long-term complications [45–47]. T2DM is a global burden, and is the most common form accounting for 90–95% of all diabetes [47, 48]. T2DM affects approximately 462 million individuals worldwide, and the number of affected individuals is rising daily in socioeconomically developed and developing countries [49]. Epidemiologists predict that the prevalence of diabetes mellitus could increase by 7862 per 100 000 by 2040 worldwide [49, 50]. The medical expenses of individuals with diabetes are more than double than they would be without diabetes [48]. In order to decrease the economic burden of T2DM and provide a better quality, healthy life by using the most effective treatments for individuals with T2DM, the risk factors, underlying pathophysiology of T2DM, and potential mechanisms of therapeutics must be clearly elucidated.

15.5 Risk Factors, Complications, and Pathophysiology of T2DM

Lifestyle, demographic and psychosocial factors, environmental factors, metabolic status, epigenetic alterations, hereditary factors and their interrelations contribute to the development of T2DM [51]. Physical inactivity is defined as exercising less than 30–60 minutes three or four times a week, leading to diminished insulin sensitivity and progressive loss of pancreatic β-cells. Consequently, it results in glucose intolerance, one of the characteristics of T2DM. Physical inactivity and a sedentary life are also causes of obesity [51]. Smoking and alcohol consumption are preventable risk factors for T2DM. According to epidemiological studies and the 2014 Surgeon General's Report, smoking increases the development risk of T2DM by approximately 30–40%. Cigarettes contain nicotine, which binds to nicotinic acetylcholine receptors. These receptors involve signaling inside the CNS and PNS, as well as in various metabolic organs, such as the pancreatic islets, adipose tissue, macrophages, liver, and skeletal muscle [52]. Nicotine alters glucoregulation and energy homeostasis. High-fat diets affect insulin action via reduced mitochondrial oxidative capacity. The reduction of mitochondrial oxidative capacity results in a decrease in the number of insulin receptors without altering receptor affinity, changes the autophosphorylation and IRS-1 and Akt phosphorylation, and activation of PI3K and Akt. Furthermore, a high-fat diet stimulates uncoupling proteins and inflammatory cytokines, which are associated with mitochondrial dysfunction. In addition, this diet changes the phospholipid content in the muscle cells' membrane and affects the functions of GLUT4 or binding of insulin. In hepatic tissue, a high fat diet is associated with insulin resistance (IR) [53].

A complex neuroendocrine structure is the primary underlying mechanism of chronic stress in the development of T2DM. Chronic stress, a consequence of metabolic failure, induces an increased concentration of glucocorticoids and catecholamines. Elevated concentrations of these hormones increase insulin demand and lead to IR [54]. Besides this, low quality and quantity of sleep and disrupted circadian rhythm also result in IR and glucose intolerance via hypoxia-induced sympathetic activity, oxidative stress, obesity, and inflammation [51, 55, 56]. Sleep deprivation activates the orexin system, stimulates the sympathetic system, and causes overeating [57]. Furthermore, enhanced serum ghrelin levels, IL-6, body mass index (BMI), diminished leptin levels, and impaired lipid metabolism are seen in obstructive sleep apnea, a sleep-related breathing disorder [55, 58]. Aging is another important risk factor for T2DM [59]. Between 2017 and 2045, the global population of adults aged ≥65 years with diabetes mellitus is projected to grow from 122 million to 253 million, in parallel with an estimated increase in the number of adults aged 65–99 years from 652 million to 1.42 billion [50]. Apart from the described factors, ethnicity, sex hormones, obesity, hypertension, and epigenetic factors also play a very crucial role in the development and progression of T2DM. When we evaluate all of the risk factors of T2DM we conclude that T2DM is associated with IR and dysfunctional pancreatic β cells, inflammatory processes, and impaired redox status [60].

15.6 Association of *Cannabis* with T2DM and its Complications

The relationship between *Cannabis* and diabetes dates back to the 1970s. Research from that time to the present appears to be on cannabinoids and their associated receptors. Studies show that endocannabinoids play a role in controlling glucose utilization and energy homeostasis by regulating pancreatic hormone secretion. It has also been reported that endocannabinoids are associated with cell proliferation and classification during pancreatic islet formation [61, 62]. An irregularity in the endocannabinoid system, which plays a role in lipid and glucose metabolism, causes visceral fat accumulation in various organs involved in energy homeostasis and decreased adiponectin secretion from this tissue. This may contribute to the onset of various cardiometabolic risk factors associated with obesity and T2DM [63]. A study on patients with T2DM receiving insulin suggested that Rimonabant, a selective CB1R antagonist, modulated glycemic control and cardiometabolic risk factor [64]. However, it has been reported that the European Medicines Agency (EMEA) called for its withdrawal from the market as a result of the association of CB1R antagonist group drugs, including Rimonabant, with increased risk of anxiety, depression, and suicide [65]. In addition, it has been noted that controlled blocking of CB1R can ameliorate the endoplasmic reticulum (ER) and mitochondrial stress that induce IR and T2DM [66]. On the other hand,

WIN 55,212–2, a non-selective cannabinoid agonist, has a curative effect on T2DM complications through activation of CB1R [67]. Similarly, in a study on T2DM rats, it was found that CB1R mRNA expression was decreased in the liver [68]. Moreover, Δ9-THC, a partial CB1R agonist, shows therapeutic effects on impaired redox status and inflammation in the rat kidney with T2DM [69, 70].

There is evidence that increased activation of another receptor, CB2R, may be associated with decreased inflammation and oxidative stress, which play a role in the pathophysiology of T2DM and its complications [71, 72]. In a study on T2DM rats, ß-Caryophyllene, a CB2R agonist, decreased the progression and risk of IR and T2DM by increasing the insulin receptor, insulin receptor-substrate-1 (IRS-1), Akt, and glucose transporter-4 (GLUT4) expression in the skeletal muscle [73].

According to the data from epidemiological studies, regular chronic use of marijuana (one of the other names of *Cannabis*) is related to a decreased prevalence of T2DM, fasting insulin, IR, and lower waist circumference [74, 75]. Tuduri et al. (2017) suggested that the presence of GRP55 in β-cells may have a role in the regulation of insulin secretion [76]. It has been reported that GPR55 agonists applied to rat islets increase intracellular Ca^{2+} concentration and cAMP levels, resulting in increased insulin secretion [77, 78].

15.7 Conclusion

Despite studies showing that the endocannabinoid system is associated with the prevention of T2DM and its complications, there is no consensus on whether cannabinoid receptor agonists or antagonists are more beneficial in treating this metabolic disease. An array of health beneficial effects of *Cannabis* and its bioactive compounds has been reported in many studies; however, there are only a few, or compromised, T2DM specific studies. Clinical studies on large groups of populations of different ethnicity are required to finalize the results on its antidiabetic properties.

References

1 Breijyeh Z, Jubeh B, Bufo SA, Karaman R, Scrano L. Cannabis: A toxin-producing plant with potential therapeutic uses. *Toxins (Basel)*. 2021; 13(2):117.

2 Andre CM, Hausman JF, Guerriero G. Cannabis sativa: The plant of the thousand and one molecules. *Front Plant Sci*. 2016; 7:1–17.

3 Ross SA, ElSohly MA. Constituents of cannabis sativa L. XXVIII. A review of the natural constituents: 1980–1994. *Zagazig J Pharm Sci*. 1995; 4(1):150–160.

4 ElSohly MA, Slade D. Chemical constituents of marijuana: The complex mixture of natural cannabinoids. *Life Sci.* 2005; 78(5):539–548.
5 ElSohly MA, Radwan MM, Gul W, Chandra S, Galal A. Phytochemistry of Cannabis sativa L. *Prog Chem Org Nat Prod.* 2017; 103:1–36.
6 Fraguas-Sánchez AI, Fernández-Carballido A, Torres-Suárez AI. Phyto-, endo- and synthetic cannabinoids: Promising chemotherapeutic agents in the treatment of breast and prostate carcinomas. *Expert Opin Investig Drugs.* 2016; 25(11):1311–1323.
7 Zou S, Kumar U. Cannabinoid receptors and the endocannabinoid system: Signaling and function in the central nervous system. *Int J Mol Sci.* 2018; 19(3):833.
8 Lu HC, Mackie K. Review of the endocannabinoid system. *Biol Psychiatry Cogn Neurosci Neuroimaging.* 2021; 6(6):607–615.
9 Lowe H, Toyang N, Steele B, Bryant J, Ngwa W. The endocannabinoid system: A potential target for the treatment of various diseases. *Int J Mol Sci.* 2021; 22(17):9472.
10 Joshi N, Onaivi ES. Endocannabinoid system components: Overview and tissue distribution. *Adv Exp Med Biol.* 2019; 1162:1–12.
11 Adams R, Hunt M, Clark JH. Structure of cannabidiol, a product isolated from the marihuana extract of Minnesota wild hemp.I. *J Am Chem Soc.* 1940; 2:196–200.
12 Mechoulam R, Gaoni Y. A total synthesis of Dl-Delta-1-Tetrahydrocannabinol, the active constituent of hashish. *J Am Chem Soc.* 1965; 87:3273–3275.
13 Pertwee RG. Cannabinoid pharmacology: The first 66 years. *Br J Pharmacol.* 2006; 147(S1):S163–S171.
14 Devane WA, Dysarz FA, Johnson MR, Melvin LS, Howlett AC. Determination and characterization of a cannabinoid receptor in rat brain. *Mol Pharmacol.* 1988; 34:605–613.
15 Devane WA, Hanus L, Breuer A, Pertwee RG, Stevenson LA, Griffin G, Gibson D, Mandelbaum A, Etinger A, Mechoulam R. Isolation and structure of a brain constituent that binds to the cannabinoid receptor. *Science.* 1992; 258(5090):1946–1949.
16 Mechoulam R, Ben-Shabat S, Hanus L, Ligumsky M, Kaminski NE, Schatz AR, Gopher A, Almog S, Martin BR, Compton DR, Pertwee RG, Griffin G, Bayewitch M, Barg J, Vogel Z. Identification of an endogenous 2-monoglyceride, present in canine gut, that binds to cannabinoid receptors. *Biochem Pharmacol.* 1995; 50:83–90.
17 Hanus L, Abu-Lafi S, Fride E, Breuer A, Vogel Z, Shalev DE, Kustanovich I, Mechoulam R. 2- Arachidonyl glyceryl ether, an endogenous agonist of the cannabinoid CB1 receptor. *Proc Natl Acad Sci USA.* 2001; 98:3662–3665.
18 Porter AC, Sauer JM, Knierman MD, Becker GW, Berna MJ, Bao J, Nomikos GG, Carter P, Bymaster FP, Leese AB, Felder CC. Characterization of a novel endocannabinoid, virodhamine, with antagonist activity at the CB1 receptor. *J Pharmacol Exp Ther.* 2002; 301:1020–1024.

19 Bisogno T, Melck D, Bobrov MYu, Gretskaya NM, Bezuglov VV, De Petrocellis L, Di Marzo V. *N*-acyl-dopamines: Novel synthetic CB(1) cannabinoid-receptor ligands and inhibitors of anandamide inactivation with cannabimimetic activity in vitro and in vivo. *Biochem J*. 2000; 3:817–824.

20 Huang SM, Bisogno T, Trevisani M, Al-Hayani A, De Petrocellis L, Fezza F, Tognetto M, Petros TJ, Krey JF, Chu CJ, Miller JD, Davies SN, Geppetti P, Walker JM, Di Marzo V. An endogenous capsaicin like substance with high potency at recombinant and native vanilloid VR1 receptors. *Proc Natl Acad Sci USA*. 2002; 99:8400–8405.

21 Pertwee RG. Endocannabinoids and their pharmacological actions. *Handb Exp Pharmacol*. 2015; 231:1–37.

22 Mackie K. Distribution of cannabinoid receptors in the central and peripheral nervous system. *Handb Exp Pharmacol*. 2005; 168:299–325.

23 Lu HC, Mackie K. An introduction to the endogenous cannabinoid system. *Biol Psychiatry*. 2016; 79(7):516–525.

24 Sawzdargo M, Nguyen T, Lee DK, Lynch KR, Cheng R, Heng HH, George SR, O'Dowd BF. Identification and cloning of three novel human G protein-coupled receptor genes GPR52, PsiGPR53 and GPR55: GPR55 is extensively expressed in human brain. *Brain Res Mol Brain Res*. 1999; 64:193–198.

25 Lago-Fernandez A, Zarzo-Arias S, Jagerovic N, Morales P. Relevance of peroxisome proliferator activated receptors in multitarget paradigm associated with the endocannabinoid system. *Int J Mol Sci*. 2021; 22(3):1001.

26 Di Marzo V, Piscitelli F. The endocannabinoid system and its modulation by phytocannabinoids. *Neurotherapeutics*. 2015; 12(4):692–698.

27 Gaoni Y, Mechoulam R. Isolation, structure, and partial synthesis of an active constituent of hashish. *J Am Chem Soc*. 1964; 86:1646–1647.

28 Matsuda LA, Lolait SJ, Brownstein MJ, Young AC, Bonner TI. Structure of a cannabinoid receptor and functional expression of the cloned cDNA. *Nature*. 1990; 346:561–564.

29 Pertwee RG, Howlett AC, Abood ME, Alexander SP, di Marzo V, Elphick MR, Greasley PJ, Hansen HS, Kunos G, Mackie K, Mechoulam R, Ross RA.International union of basic and clinical pharmacology. LXXIX. Cannabinoid receptors and their ligands: Beyond CB1and CB2. *Pharmacol Rev*. 2010; 62:588–631.

30 Howlett AC. The cannabinoid receptors. *Prostag Oth Lipid M*. 2002; 68–69:619–631.

31 Ryberg E, Vu HK, Larsson N, Groblewski T, Hjorth S, Elebring T, Sjorgren S, Greasley PJ. Identification and characterisation of a novel splice variant of the human CB1 receptor. *FEBS Lett*. 2005; 579:259–2564.

32 Shire D, Carillon C, Kaghad M, Calandra B, Rinaldicarmona M, Lefur G, Caput D, Ferrara P. An amino-terminal variant of the central cannabinoid receptor resulting from alternative splicing. *J Biol Chem*. 1995; 270:3726–3731.

33 Gonzalez-Mariscal I, Krzysik-Walker SM, Doyle ME, Liu QR, Cimbro R, Calvo SSC, Ghosh S, Ciesla L, Moaddel R, Carlson OD, Witek RP, O'Connell JF, Egan JM. Human CB1 receptor isoforms, present in hepatocytes and β-cells, are involved in regulating metabolism. *Sci Rep.* 2016; 6:33302.

34 Liu QR, Pan CH, Hishimoto A, Li CY, Xi ZX, Llorente-Berzal A, Viveros MP, Ishiguro H, Arinami T, Onaivi ES, Uhl GR. Species differences in cannabinoid receptor 2 (CNR2 gene): Identification of novel human and rodent CB2 isoforms, differential tissue expression and regulation by cannabinoid receptor ligands. *Genes Brain Behav.* 2009; 8(5):519–530.

35 Yang H, Zhou J, Lehmann C. GPR55 - a putative "type 3" cannabinoid receptor in inflammation. *J Basic Clin Physiol Pharmacol.* 2016; 27(3):297–302.

36 Ryberg E, Larsson N, Sjögren S, Hjorth S, Hermansson NO, Leonova J, Elebring T, Nilsson T, Drmota T, Greasley PJ. The orphan receptor GPR55 is a novel cannabinoid receptor. *Br J Pharmacol.* 2007; 152:1092–1101.

37 Lipina C, Walsh SK, Mitchrll SE, Speakman JR, Wainwright CL, Hundal HS. GPR55 deficiency is associated with increased adiposity and impaired insulin signaling in peripheral metabolic tissues. *FASEB.* 2019; 33:1299–1312.

38 Ayakannu T, Taylor AH, Konje JC. Expression of the putative cannabinoid receptor GPR55 is increased in endometrial carcinoma. *Histochem Cell Biol.* 2021; 156:449–460.

39 Howlett AC, Barth F, Bonner TI, Cabral G, Casellas P, Devane WA, Felder CC, Herkenham M, Mackie K, Martin BR, Mechoulam R, Pertwee RG. International union of pharmacology. XXVII. Classification of cannabinoid receptors. *Pharmacol Rev.* 2002; 54(2):161–202.

40 Turu G, Soltész-Katona E, Tóth AD, Juhász C, Cserző M, Misák Á, Balla A, Caron MG, Hunyady L. Biased coupling to β-arrestin of two common variants of the CB_2 cannabinoid receptor. *Front Endocrinol (Lausanne).* 2021; 2:714561.

41 Nogueras-Ortiz C, Yudowski GA. The multiple waves of cannabinoid 1 receptor signaling. *Mol Pharmacol.* 2016; 90(5):620–626.

42 Breivogel CS, Puri V, Lambert JM, Hill DK, Huffman JW, Razdan RK. The influence of beta-arrestin2 on cannabinoid CB1 receptor coupling to G-proteins and subcellular localization and relative levels of beta-arrestin1 and 2 in mouse brain. *J Recept Signal Transduct Res.* 2013; 33(6):367–379.

43 Howlett AC, Blume LC, Dalton GD. CB (1) cannabinoid receptors and their associated proteins. *Curr Med Chem.* 2010; 17(14):1382–1393.

44 Henstridge CM, Balenga NA, Ford LA, Ross RA, Waldhoer M, Irving AJ. The GPR55 ligand L-alpha-lysophosphatidylinositol promotes RhoA-dependent Ca^{2+} signaling and NFAT activation. *FASEB.* 2009; 23(1):183–193.

45 Landgraf R, Aberle J, Birkenfeld AL, Gallwitz B, Kellerer M, Klein H, Müller-Wieland D, Nauck MA, Reuter HM, Siegel E. Therapy of type 2 diabetes. *Exp Clin Endocrinol Diabetes.* 2019; 127(S 01):S73–S92.

46 Reed J, Bain S, Kanamarlapudi V. A review of current trends with type 2 diabetes epidemiology, aetiology, pathogenesis, treatments and future perspectives. *Diabetes, Metab Syndr and Obes: Targets Ther*. 2021; 14:3567–3602.
47 Servan PR. Obesity and diabetes. *Nutr Hosp*. 2013; 28(5):138–143.
48 American Diabetes Association; 2. Classification and diagnosis of diabetes: Standards of medical care in diabetes—2020. *Diabetes Care*. 2020; 43(1):S14–31.
49 Khan MAB, Hashim MJ, King JK, Govender RD, Mustafa H, Al Kaabi J. Epidemiology of type 2 diabetes - Global burden of disease and forecasted trends. *J Epidemiol Glob Health*. 2020; 10(1):107–111.
50 Cho NH, Shaw JE, Karuranga S, Huang Y, da Rocha Fernandes JD, Ohlrogge AW, Malanda B. IDF diabetes atlas: Global estimates of diabetes prevalence for 2017 and projections for 2045. *Diabetes Res Clin Pract*. 2018; 138:271–281.
51 Ismail L, Materwala H, Al Kaabi J. Association of risk factors with type 2 diabetes: A systematic review. *Comput Struct Biotechnol J*. 2021; 19:1759–1785.
52 Maddatu J, Anderson-Baucum E, Evans-Molina C. Smoking and the risk of type 2 diabetes. *Trans Res*. 2017; 184:101–107.
53 Gheibi S, Kashfi K, Ghasemi A. A practical guide for induction of type-2 diabetes in rat: Incorporating a high-fat diet and streptozotocin. *Biomed Pharmacother*. 2017; 95:605–613.
54 Ingrosso DMF, Primavera M, Samvelyan S, Tagi VM, Chiarelli F. Stress and diabetes mellitus: pathogenetic mechanisms and clinical outcome. *Horm Res Paediatr*. 2023; 96(1):34–43.
55 Stamatakis KA, Punjabi NM. Effects of sleep fragmentation on glucose metabolism in normal subjects. *Chest*. 2010; 137:95–101.
56 Khandelwal D, Dutta D, Chittawar S, Kalra S. Sleep disorders in type 2 diabetes. *Indian J Endocrinol Metab*. 2017; 21(5):758–761.
57 Taheri S, Lin L, Austin D, Young T, Mignot E. Short sleep duration is associated with reduced leptin, elevated ghrelin, and increased body mass index. *PLoS Med*. 2004; 1:e62.
58 Nannapaneni S, Ramar K, Surani S. Effect of obstructive sleep apnea on type 2 diabetes mellitus: A comprehensive literature review. *World J Diabetes*. 2013; 4:238–244.
59 Nanayakkara N, Curtis AJ, Heritier S, Gadowski AM, Pavkov ME, Kenealy T, Owens DR, Thomas RL, Song S, Wong J, Chan JC, Luk AO, Penno G, Ji L, Mohan V, Amutha A, Romero-Aroca P, Gasevic D, Magliano DJ, Teede HJ, Chalmers J, Zoungas S. Impact of age at type 2 diabetes mellitus diagnosis on mortality and vascular complications: Systematic review and meta-analyses. *Diabetologia*. 2021; 64:275–287.
60 Coskun ZM, Bolkent S. Biochemical and immunohistochemical changes in delta-9-tetrahydrocannabinol-treated type 2 diabetic rats. *Acta Histochem*. 2014; 116(1):112–116.

61 Malenczyk K, Keimpema E, Piscitelli F, Calvigioni D, Björklund P, Mackie K, Di Marzo V, Hökfelt TG, Dobrzyn A, Harkany T. Fetal endocannabinoids orchestrate the organization of pancreatic islet microarchitecture. *Proc Natl Acad Sci USA*. 2015; 112(45):E6185–194.

62 Rahman SMK, Uyama T, Hussain Z, Ueda N. Roles of endocannabinoids and endocannabinoid-like molecules in energy homeostasis and metabolic regulation: A nutritional perspective. *Annu Rev Nutr*. 2021; 41:177–202.

63 Di Marzo V. The endocannabinoid system in obesity and type 2 diabetes. *Diabetologia*. 2008; 51(8):1356–1367.

64 Hollander PA, Amod A, Litwak LE, Chaudhari U. ARPEGGIO study group. Effect of rimonabant on glycemic control in insulin-treated type 2 diabetes: The ARPEGGIO trial. *Diabetes Care*. 2010; 33(3):605–607.

65 Le Foll B, Gorelick DA, Goldberg SR. The future of endocannabinoid-oriented clinical research after CB1 antagonists. *Psychopharmacology (Berl)*. 2009; 205(1):171–174.

66 Lu D, Dopart R, Kendall DA. Controlled downregulation of the cannabinoid CB1 receptor provides a promising approach for the treatment of obesity and obesity-derived type 2 diabetes. *Cell Stress Chaperones*. 2016; 21(1):1–7.

67 Vera G, López-Miranda V, Herradón E, Martín MI, Abalo R. Characterization of cannabinoid-induced relief of neuropathic pain in rat models of type 1 and type 2 diabetes. *Pharmacol Biochem Behav*. 2012; 102(2):335–343.

68 Coskun ZM, Beydogan AB, Bolkent S. Changes in the expression levels of CB1 and GLP-1R mRNAs and microRNAs 33a and 122 in the liver of type 2 diabetic rats treated with ghrelin. *J Biochem Mol Toxicol*. 2019; 33(10):e22388.

69 Pertwee RG. The diverse CB1 and CB2 receptor pharmacology of three plant cannabinoids: Delta9-tetrahydrocannabinol, cannabidiol and delta9-tetrahydrocannabivarin. *Br J Pharmacol*. 2008; 153(2):199–215.

70 Yanar K, Coskun ZM, Beydogan AB, Aydin S, Bolkent S. The effects of delta-9-tetrahydrocannabinol on Krüppel-like factor-4 expression, redox homeostasis, and inflammation in the kidney of diabetic rat. *J Cell Biochem*. 2019; 120(9):16219–16228.

71 Kumawat VS, Kaur G. Therapeutic potential of cannabinoid receptor 2 in the treatment of diabetes mellitus and its complications. *Eur J Pharmacol*. 2019 5; 862:172628.

72 Oguntibeju OO. Type 2 diabetes mellitus, oxidative stress and inflammation: Examining the links. *Int J Physiol Pathophysiol Pharmacol*. 2019; 11(3):45–63.

73 Mani V, Badrachalam R, Shanmugam SN, Balraj M, Kasthuri R, Danavel A, Babu S. Effect of β-Caryophyllene on insulin resistance in skeletal muscle of high fat diet and fructose-induced type-2 diabetic rats. *Bioinformation*. 2021; 17(8):741–747.

74 Rajavashisth TB, Shaheen M, Norris KC, Pan D, Sinha SK, Ortega J, Friedman TC. Decreased prevalence of diabetes in marijuana users: Cross-sectional data

from the National Health and Nutrition Examination Survey (NHANES) III. *BMJ Open*. 2012; 2(1):e000494.

75 Penner EA, Buettner H, Mittleman MA. The impact of marijuana use on glucose, insulin, and insulin resistance among US adults. *Am J Med*. 2013; 126(7):583–589.

76 Tudurí E, Imbernon M, Hernández-Bautista RJ, Tojo M, Fernø J, Diéguez C, Nogueiras R. GPR55: A new promising target for metabolism? *J Mol Endocrinol*. 2017; 58(3):R191–R202.

77 McKillop AM, Moran BM, Abdel-Wahab YH, Flatt PR. Evaluation of the insulin releasing and antihyperglycaemic activities of GPR55 lipid agonists using clonal beta-cells, isolated pancreatic islets and mice. *Br J Pharmacol*. 2013; 170(5):978–990.

78 Romero-Zerbo SY, Rafacho A, Díaz-Arteaga A, Suárez J, Quesada I, Imbernon M, Ross RA, Dieguez C, Rodríguez de Fonseca F, Nogueiras R, Nadal A, Bermúdez-Silva FJ. A role for the putative cannabinoid receptor GPR55 in the islets of Langerhans. *J Endocrinol*. 2011; 211(2):177–185.

Part IV

Gestational Diabetes

Prevention and Management by Natural Compounds

16

Epidemiology of Gestational Diabetes Mellitus

Preventive Significance of Dietary Pattern

Burcu Yeşilkaya

Department of Nutrition and Dietetics, Faculty of Health Science, Okan University, Istanbul, Turkey

16.1 Introduction

During pregnancy, a woman experiences many metabolic changes to meet her nutritional needs for both her and the fetus. Pregnancy is a period of hormonal stress. The fetus uses glucose to grow. This causes a serious change and reaction in the mother's glucose and energy metabolism. The first of these reactions may be the development of insulin resistance (IR), and hyperglycemia, and consequently, gestational diabetes mellitus (GDM). In cases where the complications of GDM are not prevented and treated, the development of type 2 diabetes mellitus (T2DM) is inevitable [1, 2]. Diabetes is a chronic disease requiring continuous medical care and glycemic control [1]. GDM is defined by the American Diabetes Association (ADA) as glucose intolerance that accompanies or is first noticed during pregnancy [3]. According to the International Diabetes Federation (IDF) 2022 Diabetes Atlas, the global prevalence of GDM is 14.0% and it varies geographically [4].

GDM may be a predictable complication in many cases. It is also linked with body mass index (BMI), high body fat, and blood glucose levels. A glycemic test can prove that there may be a problem like GDM during pregnancy. Glycemic tests are not routinely administered to every woman of reproductive age. This, unfortunately, means the hyperglycemia that occurs when women become pregnant is a situation that cannot be predicted. The severity of hyperglycemia is directly linked to the severity of GDM. The increase in this equation directly causes an increase in fetal risk. Diagnosis of hyperglycemia before pregnancy may prevent complications such as premature births and stillbirths [1, 5, 6]. Diagnostic tests should be performed to

Natural Products and their Bioactives in Antidiabetic Drug Discovery, First Edition.
Edited by Kanti Bhooshan Pandey, Maitree Suttajit and Pinar Atukeren.

identify the risk. It would be more accurate to prioritize the groups to be tested according to risk factors. Women who are found to be at risk in glucose tests should be followed up with both nutritional counseling and regular tests. Although fasting glucose tests and Hba1c tests are low-cost, Hba1c tests may not give accurate results after the 15th week of pregnancy. Increased red blood cell turnover and other blood values (hemoglobinopathies, anemia) can overestimate Hba1c [1, 7].

GDM is a β-cell disorder with a high risk of developing T2DM. For this reason, women diagnosed with GDM should have medical and nutritional support to reduce the risk of T2DM, and lifelong screening is recommended [3].

16.2 Dietary Patterns and Recommendations for GDM

Many maternal adaptations take place throughout the pregnancy process. These are both psychological and physiological adjustments. It occurs in many systemic areas, including respiratory, digestive, cardiovascular, and metabolic [8–10].

Glucose metabolism adapts itself differently. As a part of gestational glucose metabolism, from the beginning of pregnancy, women can develop IR because of many maternal hormones and factors. The purpose of maternally adapted IR is to provide adequate nutrients to the growing fetus. Women who develop IR have a higher risk of developing GDM [10, 11].

Management of GDM requires many changes, both medical and non-medical interventions. "Lifestyle changes" is at the top of the ADA's recommendation list for the management of GDM. It includes Medical Nutritional Therapy (MNT) with weight management and physical activity (PA) [12]. Medical Nutritional Therapy is defined as "designing meals with controlled carbohydrate levels for nutritional adequacy with normal nutrition, normal sugar levels, and prevention of Ketosis" [13, 14]. It is necessary to control blood glucose during GDM. The MNT for women with GDM should include sufficient macronutrients and micronutrients to support the growth of the fetus. It should also promote appropriate maternal gestational weight gain by keeping postprandial glucose levels at the limits [15].

Gestational weight gain (GWG) has a high priority for both the mother and fetus. Sufficient energy and nutrient intake should be considered during pregnancy [16]. Insufficient or excessive GWG can affect maternal and newborn health. High GWG can mean hypertension, diabetes, cesarean, postpartum weight retention, and obesity for the mother. Insufficient GWG, on the other hand, affects the fetus most and is seen as infant deaths, premature birth, and fetal growth retardation [17].

The Institute of Medicine (IOM) has a recommendation for weight gain according to the pre-pregnancy BMI that can be explained as seen below [16]:

- Underweight (<18.5 kg/m^2) expected weight gain: 12.5–18 kg
- Normal weight (18.5–24.9 kg/m^2) expected weight gain: 11.5–16 kg

- Overweight (25.0–29.9 kg/m^2) expected weight gain: 7–11.5 kg
- Obese (≥30 kg/m^2) expected weight gain: 5–9 kg.

Healthy GWG can be supported with a healthy diet. Research showed that the "Western Type Diet Pattern", which is high in processed food, saturated fats, high in salt sauces, and alcoholic drinks, may cause obesity during pregnancy and high GWG may result in GDM [17, 18].

According to IOM (US) National Research Council (US) Committee recommendations, approximately a 350–400 kcal increase in daily energy intake is sufficient for a normal GWG. This amount of energy is recommended for pregnant women without obesity [16]. There are many studies about the kcal intake of obese-pregnant women. Half of them recommend no additive energy during the second and third trimesters, and the others recommend at least 200 kcal/day. For a healthy GWG, the energy intake should be equal to the energy spent and enough for a growing fetus [17], because unbalanced weight gain can cause unbalanced blood glucose levels that result in GDM. The recommendations for weight gain and the calculation of energy requirements for normal glucose tolerance pregnancies (NGTP) are also appropriate for women with GDM [15]. With or without obesity, the daily energy intake should not be below 1600 kcal [19, 20].

Sufficient energy intake is important, but the type of energy source may be the most important area. Carbohydrate intake has a vital role during GDM because it is directly linked with blood glucose and insulin secretion. The type and the way of consuming carbohydrates is a different story that should be handled well. IOM states that 45–60% [16, 21] of energy sources should come from carbohydrates. The authorities also find the restriction of carbohydrates suitable for women with GDM. According to the authorities, the carbohydrate energy recommendations are between 35–45% of the daily energy needs [22–24]. There is another and more recent meta-analysis about how much carbohydrate is optimum for normal fetal growth, which states that the optimum carbohydrate is 47–70% of energy intake (EI) according to the woman's medical status. The authors add that the lowest percentage (47%) is based on 100 000 women included in the study [25]. The American Diabetes Association's (ADA) report stated that the Dietary Reference Intakes (DRI) guide should be taken as a reference guide when planning a nutrition program. Although not specific to women with GMD, all pregnant women reported that it is appropriate to consume a minimum of 175g of CHO and 28g of fiber. The importance of the amount and type of CHO for women with GDM has been emphasized and it has been reported that it effects postprandial blood sugar directly [26]. The American Academy of Nutrition and Dietetics advocates that all women diagnosed with GDM should receive medical nutrition therapy from a dietitian. For carbohydrate intake, not specific to GDM, all pregnant women stated that they should consume at least 157 g of CHO and 28 g of fiber and that the type and amount of carbohydrates should be evaluated according to the medical condition of the women. In addition, carbohydrate consumption should be divided into main meals and snacks. Thus, it helps to regulate blood sugar levels and ensure that postprandial blood sugar levels do not increase excessively [14, 27].

There is another perspective that low-carbohydrate intake can lead to high protein and fat consumption. Low-carbohydrate and high-fat diets may increase maternal free fatty acids and maternal insulin resistance [28]. In a study in which women were followed up for 12 years, it was reported that low carbohydrate diets are linked with high protein and fat content diet patterns and may cause GDM while fiber-rich diets may prevent GDM [29].

The type of carbohydrates matters a lot. Carbohydrates should mainly consist of foods with a low glycemic index that are rich in fiber, such as vegetables, legumes, fruits, and whole grains; these foods possess large amounts of bioactive compounds [14]. The high-quality, nutrient-dense carbohydrates included in nutrition programs benefit controlled fasting/postprandial glucose, lower free fatty acids, and enhanced insulin action. Consumption of simple carbohydrates can cause rapid and unstable blood sugar levels [28].

The daily protein requirement of a woman with GDM is recommended as 71 g of protein or 1.1 g/kg/day by the Academy of Nutrition and Dietetics. Generally, the type of protein is not found to be associated with glycemic control, GWG, birth weight, or fetal growth [30, 31]. A meta-analysis showed that animal and plant-based protein consumption is no different in most cases. For example, women with a vegetarian or vegan diet can deliver infants without any abnormalities if nutrient supplementation (iron, B12) is added to the diet. Because of the poverty in some rural areas, women with vegan/vegetarian diets as a culture may not be able to consume sufficient energy or micronutrients. Low birth weight is an outcome of this. It may explain the association between vegetarian/vegan diets and low birth weight. As a result, it has been shown that protein intake from vegetables, lean meat, and fish has positive effects on the treatment of GDM and can increase insulin sensitivity. Plant-based proteins may not have a direct healing effect on the GDM process. This improvement may be caused by plant-based foods being more included in the diet, and fewer carbohydrates and saturated fat in the diet may have a healing effect on the GDM process. In addition, the results cannot be generalized to all ethnicities, as many studies focused on Asian and Middle Eastern women [15].

Authorities recommend nearly the same amount of fat consumption in nutrition plans for women with GDM as for women with normal blood glucose levels. A fat intake of 20–40% of daily energy meets the required amount. Saturated fat should be less than 10% of fat energy. A 350 g/week fatty fish can meet the requirements of PUFA, which has a high priority for the development of the fetus. High fat consumption causes abnormal GWG and increased oxidative stress [15, 16, 32]. In a meta-analysis it is reported that the lipid profile of GDM women is significantly associated with age, BMI, OGTT, and GDM diagnosis criteria. According to these findings, the fat content of the nutrition program can be changed according to the status of women with GDM [33].

During pregnancy, the vitamin and mineral requirements increase. Recommendations for vitamin and mineral supplements do not differ for women with normal blood sugar levels and women with GDM. Recommendations agree that the supplementation of folic acid, vitamin D, and iron is essential [15, 16, 32]. There is a study that reported that a lack of vitamin D and folic acid causes a higher GDM risk [34].

When the evidence comes together, it has been shown that dietary intake and food selection before and during pregnancy is associated with the risk of GMD. In many studies, it is reported that processed food, red meat, and food containing eggs, animal protein, cholesterol, and saturated fat contents as nutrients were observed as risk factors for GDM. Mediterranean-type diets and vegetarian-type diets are more protective diets for GDM [35–38]. A study examining the diets of women with GDM concluded that high amounts of processed grains, fats, and added sugar consumed during pregnancy may cause GDM formation, while high fruit and vegetable consumption may be protective [39, 40]. A systematic review has shown that a dietary pattern rich in fruit, vegetables, whole grains, and fish is beneficial. Processed food, high-fat dairy, and refined sugars are associated with GDM [41].

GDM is also associated with pregnancy obesity [42]. Another study showed that in women who were overweight before pregnancy, a high protein-low starch diet was associated with a lower risk of GDM [43].

However, since eating habits differ ethnically and geographically, it is more beneficial to evaluate each ethnic group separately. The pre-pregnancy BMI and GWG link also has differences ethnically and geographically [44]. As the prevalence of GDM differs, the food groups that increase or trigger the risk of GDM may also be different. In some Asian groups, seafood consumption is seen as a risk factor for GDM, while in some other groups, it is called a GDM risk reducer with this adverse effect. In a study of Chinese women's diet patterns, it is reported that a vegetarian diet and a traditional diet (including consumption of cereals, potatoes, legumes, veggies, fruits, eggs, and meats) are linked with low birth weight [45]. Both diet patterns have a high consumption of fruits and vegetables, which seem mostly "healthy", but the consumption style and additives like soy sauce may trigger adverse outcomes. The other reason for association with low birth weight may be the lack of protein/low protein content diet patterns. It is also important to know that some studies report that low birth weight is highly linked to high body fat percentages and obesity in later years [19].

Few studies investigate the pre-pregnancy dietary patterns and GDM interactions. These studies target pre-pregnancy BMI and diet habits to the effectiveness of pregnancy complications. Healthy diet habits of non-pregnant women may prevent complications like gestational diabetes or gestational hypertension during pregnancy. The healthy diet recommendation for non-pregnant women is balanced and sufficient nutrition. A balanced and sufficient diet includes high consumption of vegetables, fruits, fish, lean meats, and whole grains, and lower consumption of added sugars, processed foods, and fats [46].

The Academy of Nutrition and Dietetics 2018 report shows two types of dietary patterns that can affect GDM. The first dietary pattern "Dietary Approaches to Stop Hypertension (DASH)" is known to be high in carbohydrates and fibers, and lower in sugars, fats, cholesterol, and sodium. DASH is effective in improving outcomes such as IR, blood glucose levels, blood pressure, and lipid profile [27]. In a meta-analysis, results showed that it also has an effect on neonates such as lower incidence of preeclampsia, LGA, and newborn head circumference [47].

The second dietary pattern "Low-Moderate Glycemic Index (GI)" has improvements in glycemic control. GI explains how foods affect blood glucose levels over time [27]. A GI score ranges from 0 to 100. Pure glucose has a GI of 100 and other nutrients are scored based on pure glucose. For the classification, a GI of 55 or lower means low GI, 56–69 means moderate GI, and 70 or higher means high GI [48]. Some starchy foods, fruits, vegetables, dairy, nuts, legumes, whole grains, and bread generally have a low GI. The high content of protein, fat, and fiber can decrease the GI of the food [49, 50]. Glycemic load (GL) is another parameter that we shouldn't skip. It provides information about how it affects blood sugar with one portion of food (quality/amount of carbohydrate) because it calculates the available carbohydrates in food. For the classification, a GL of 10 or lower means low GL, 11–19 means moderate GL, and 20 or more means high GL [48]. Both GI and GL provide quantitative and qualitative perceptions for the management of carbohydrate-containing foods. A study of Chinese women shows that dietary GI, GL, and fiber intake before and during pregnancy affect glucose homeostasis. For women with a high GI and high GL diet before the pregnancy, their first and second trimesters were more likely to develop GDM than women with a low GI and GL diet. It has also been reported that women who eat a high-fiber diet have a lower risk of GDM than women who eat less fiber. The same study also reported that sugar-sweetened beverages are highly associated with the development of GDM [51].

An intermittent fasting (IF) diet can be preferred regionally or culturally for its health benefits. IF aims to stimulate IGF-1, which primarily regulates glucose metabolism. A recent review article reported that dietary measures including IF, calorie restriction, and exercise can improve insulin sensitivity, reduce body fat percentage, and correct fetal growth abnormalities associated with hyperglycemia with essential micronutrient supplements [52]. Another study showed that a low-carb diet (LCD) that contains less than 35–45% carbohydrates and a low glycemic index diet (LGID) that contains food GI less than 55 have positive effects on postprandial glycemia. LCD also has a lower large for gestational age (LGA), while LGID can increase insulin sensitivity. The same study also showed that a caloric restriction diet (30–50% caloric restriction) may lead to better glycemic control and a Mediterranean diet (MedDiet) can result in better cardiovascular outcomes [53].

The MedDiet is found to be highly safe for pregnancy. It is high in vegetables, olive/olive oil, fruit, legumes, nuts, and whole grains, with moderate animal and dairy products. It has an anti-inflammatory effect, so it is protective against chronic inflammation, including metabolic syndrome, diabetes, and obesity. This diet has high PUFA and MUFA, so it supplies healthy growth and development for the fetus [53–56]. In a study it was found that a MedDiet-based medical nutrition therapy as part of GDM management is associated with similar outcomes as women with normal glucose levels [35]. A study that has a different perspective reported that the women who followed a MedDiet in early pregnancy and carried the rs7903146 T-allele had a lower risk of developing GDM than CC carriers. This rs7903146 variant is a part of the TCF7L2 gene and refers to people at high risk of developing diabetes. Accordingly, it was stated that determining genetic predisposition and making nutritional interventions would give healthier results [57, 58].

Ketogenic diets are diets that are high in protein and fat, and very low in carbohydrates (<20–50 g/day or <10% kcal/day). Ketogenic diets stimulate the use and production of ketones, acetoacetate, and β-hydroxybutyrate instead of glucose as energy sources. Its safety for pregnancy has not been established [59].

The Alternative Healthy Eating Index-2010 (AHEI-2010) is a food-based measure of diet quality that predicts major chronic disease risk [60]. It emphasizes high intakes of legumes, nuts, whole grains, vegetables, fruits, omega-3 fats, and PUFA while limiting red meats, processed meats, sodium, and sugary and alcoholic beverages. The effect of (AHEI-2010) on reducing GDM was found to be between 19–46%. In addition, it has been observed that this diet can provide 83% risk reduction together with the reduction of other risk factors (PA, smoking, and BMI). It has been reported that the risk of GDM is 41% lower in people who consume foods according to the AHEI-2010. The same study also compared three different dietary patterns and their protective effects on GDM. The result showed that the Alternative Healthy Eating Index had a more protective effect compared to the DASH diet and the MedDiet [61].

16.3 Conclusion

Studies show that eating habits before and during pregnancy can trigger the development of GDM. Eating habits are shaped by geographical, ethnic, economic, and individual preferences. These parameters should be considered when assessing the diet of individuals. The nutrition plans should be realistic. Nutritional plans that people do not feel are suitable for themselves or have difficulty getting involved in are not sustainable. Screening the glycemic profile before pregnancy can prevent the development of GDM and its damage. Eating a diet rich in fiber and low glycemic index before and during pregnancy has been found to be both healthy and protective in the development of GDM. In addition, low-carbohydrate nutrition, and consuming carbohydrates by incorporating them into meals were also found to be important. Vitamin D and folic acid supplements are recommended before and during pregnancy. MUFA and PUFA have come to the fore with their preventive effects against GDM. The Mediterranean-type diet, which mostly includes all these dietary recommendations, was found to be beneficial in preventing GDM. To apply the Mediterranean-type diet, it is essential to make a lifestyle change as soon as possible.

References

1 American Diabetes Association Professional Practice Committee. 2. Classification and diagnosis of diabetes: Standards of medical care in diabetes 2022. *Diabetes Care*. 2022; 45(1):17–38.

2 Al Subhi SK, Al Kindi RM, Al Rawahi A, Al Seyabi IS, Al Mukhaini A. Prevalence of gestational diabetes mellitus using the latest World Health Organization diagnostic criteria among Omani women in Muscat, Oman. *Oman Med J*. 2021; 36(1):215. doi:10.5001/omj.2021.08.

3 American Diabetes Association. Gestational diabetes mellitus. ddfinition, detection, and diagnosis. *Diabetes Care*. 2002; 25(1):94–96. doi:10.2337/diacare.25.2007.S94.

4 Wang H, Li N, Chivese T, et al. IDF Diabetes Atlas: Estimation of global and regional gestational diabetes mellitus prevalence for 2021 by International Association of Diabetes in pregnancy study group's criteria. *Diabetes Res Clin Pract*. 2022; 183:109050. doi:10.1016/j.diabres.2021.109050.

5 Ye W, Luo C, Huang J, et al. Gestational diabetes mellitus and adverse pregnancy outcomes: A systematic review and meta-analysis. *BMJ*. 2022; 377:067946. doi:10.1136/bmj-2021-067946.

6 LeBlanc ES, Hillier TA. The impact of gestational weight gain on glucose and insulin physiology in pregnancy - does timing matter? *J Clin Endocrinol Metab*. 2022; 107(3):1303–e1304. doi:10.1210/clinem/dgab745.

7 International Diabetes Federation. IDF guide for diabetes epidemiology studies. [cited 10 07 2022]. Available from: https://diabetesatlas.org/idfawp/resource-files/2021/11/IDF-guide-diabetes-epidemiology-studies.pdf.

8 Bjelica A, Kapor-Stanulović N. Trudnoća kao psiholoski dogadaj [Pregnancy as a psychological event]. *Med Pregl*. 2004; 57(3–4):144–148. doi:10.2298/mpns0404144b.

9 Soma-Pillay P, Nelson-Piercy C, Tolppanen H, Mebazaa A. Physiological changes in pregnancy. *Cardiovasc J Afr*. 2016; 27(2):89–94. doi:10.5830/CVJA-2016-021.

10 Angueira AR, Ludvik AE, Reddy TE, Wicksteed B, Lowe WL Jr, Layden BT. New insights into gestational glucose metabolism: Lessons learned from 21st century approaches. *Diabetes*. 2015; 64(2):327–334. doi:10.2337/db14-0877.

11 Sonagra AD, Biradar SM, Dattatreya K, Murthy DSJ. Normal pregnancy - a state of insulin resistance. *J Clin Diagn Res.* 2014; 8(11):CC01–CC3. doi:10.7860/JCDR/2014/10068.5081.

12 American Diabetes Association. Standards of medical care in diabetes 2022. Abridged for primary care providers. *Clin Diabetes.* 2022; 40(1):10–35. doi:10.2337/cd22-as01.

13 Jovanovic-Peterson L, Peterson CM. Nutritional management of the obese gestational diabetic pregnant woman. *J Am Coll Nutr.* 1992; 11:246–250. doi:10.1080/07315724.1992.10718223.

14 Dolatkhah N, Hajifaraji M, Shakouri SK. Nutrition therapy in managing pregnant women with gestational diabetes mellitus: A literature review. *J Family Reprod Health.* 2018; 12(2):57–72.

15 Rasmussen L, Poulsen CW, Kampmann U, Smedegaard SB, Ovesen PG, Fuglsang J. Diet and healthy lifestyle in the management of gestational diabetes mellitus. *Nutrients.* 2020; 12(10):3050. Published 2020 Oct 6. doi:10.3390/nu12103050.

16 Institute of Medicine (US) and National Research Council (US) Committee to Reexamine IOM Pregnancy Weight Guidelines. *Weight Gain during Pregnancy: Reexamining the Guidelines.* (ed. K.M. Rasmussen et. al.). National Academies Press (US); 2009. doi:10.17226/12584.

17 Maugeri A, Barchitta M, Favara G, La Rosa MC, La Mastra C, Magnano San Lio R, Agodi A. Maternal dietary patterns are associated with pre-pregnancy body mass index and gestational weight gain: Results from the "mamma & bambino" cohort. *Nutrients.* 2019; 11(6):1308. doi:10.3390/nu11061308.

18 Itani L, Radwan H, Hashim M et al. Dietary patterns and their associations with gestational weight gain in the United Arab Emirates: Results from the MISC cohort. *Nutr J.* 2020; 19:36. doi:10.1186/s12937-020-00553-9.

19 Most J, Amant MS, Hsia DS, Altazan AD, et al. Evidence-based recommendations for energy intake in pregnant women with obesity. *J Clin Invest.* 2019; 129(11):4682–4690. doi:10.1172/JCI130341.

20 Parrettini S, Caroli A, Torlone E. Nutrition and metabolic adaptations in physiological and complicated pregnancy: Focus on obesity and gestational diabetes. *Front Endocrinol (Lausanne).* 2020; 11:611929. Published 2020 Nov 30. doi:10.3389/fendo.2020.611929.

21 Mustad VA, Huynh DTT, López-Pedrosa JM, Campoy C, Rueda R. The role of dietary carbohydrates in gestational diabetes. *Nutrients.* 2020; 12(2):385. doi:10.3390/nu12020385.

22 Tsirou E, Grammatikopoulou MG, Theodoridis X, et al. Guidelines for medical nutrition therapy in gestational diabetes mellitus: Systematic review and critical appraisal. *J Acad Nutr Diet.* 2019; 119(8):1320–1339. doi:10.1016/j.jand.2019.04.002.

23 Blumer I, Hadar E, Haddan DR, et al. Diabetes and pregnancy: An endocrine society clinical practice guideline. *J Clin Endo Metab.* 2013; 98:4227–4249. doi:10.1210/jc.2013-2465.

24 Hod M, Kapur A, Sacks DA, et al. The International Federation of Gynecology and Obstetrics (FIGO). Initiative on gestational diabetes mellitus: A pragmatic guide for diagnosis, management, and care. *Int J Gynaecol Obstet.* 2015; 131(3):173–211. doi:10.1016/S0020-7292(15)30033-3.

25 Sweeting A, Mijatovic J, Brinkworth GD, et al. The carbohydrate threshold in pregnancy and gestational diabetes: How low can we go? *Nutrients.* 2021; 13(8):2599. doi:10.3390/nu13082599.

26 American Diabetes Association. 14. Management of diabetes in pregnancy: Standards of medical care in diabetes. *Diabetes Care.* 2019; 42(1):165–172. doi:10.2337/dc19-S014.

27 Gardea D, et al. Academy of nutrition and dietetics. Gestational diabetes evidence-based nutrition practice guideline. *J Acad Nutr Diet.* 2018; 118(9): 1719–1742. doi:10.1016/j.jand.2018.03.014.

28 American Diabetes Association. Professional Practice Committee; 15. Management of diabetes in pregnancy: Standards of medical care in diabetes 2022. *Diabetes Care.* 2022; 45(1):232–243. doi:10.2337/dc22-S015.

29 Looman M, Schoenaker D, Soedamah-Muthu S, et al. Pre-pregnancy dietary carbohydrate quantity and quality, and risk of developing gestational diabetes: The Australian longitudinal study on women's health. *Br J Nutr.* 2018; 120(4):435–444. doi:10.1017/S0007114518001277.

30 Duarte-Gardea MO, Gonzales-Pacheco DM, Reader DM et al. Academy of nutrition and dietetics gestational diabetes evidence-based nutrition practice guideline. *J Acad Nutr Diet.* 2018; 118:1719–1742. doi:10.1016/j.jand.2018.03.014.

31 Eshak ES, Okada C, Baba S, et al. Maternal total energy, macronutrient and vitamin intakes during pregnancy associated with the offspring's birth size in the Japan environment and children's study. *Br J Nutr.* 2020; 124(6):558–566. doi:10.1017/S0007114520001397.

32 Nordic Council of Ministers. Nordic Nutrition Recommendations 2012: Integrating Nutrition and Physical Activity. 2012. Copenhagen, Denmark. 2014:627. [Cited 10.08.2022]. Available at: Nordic Nutrition Recommendations 2012: Integrating nutrition and physical - Nordic Council of Ministers - Google Books.

33 Hu J, Gillies CL, Lin S, et al. Association of maternal lipid profile and gestational diabetes mellitus: A systematic review and meta-analysis of 292 studies and 97 880 women. *EClinicalMedicine.* 2021; 34:100830. doi:10.1016/j.eclinm.2021.100830.

34 Meinilä J, Koivusalo SB, Valkama A, et al. Nutrient intake of pregnant women at high risk of gestational diabetes. *Food Nutr Res.* 2015; 59:26676. doi:10.3402/fnr.v59.26676.

35 Assaf-Balut C, Garcia de la Torre N, Durán A, et al. Medical nutrition therapy for gestational diabetes mellitus based on Mediterranean diet principles: A subanalysis of the St Carlos GDM prevention study. *BMJ Open Diabetes Res Care.* 2018; 6(1):e000550. doi:10.1136/bmjdrc-2018-000550.

36 Yong HY, Mohd Shariff Z, Mohd Yusof BN, et al. The association between dietary patterns before and in early pregnancy and the risk of gestational diabetes mellitus (GDM): Data from the Malaysian SECOST cohort. *PLoS One*. 2020; 15(1):e0227246. doi:10.1371/journal.pone.0227246.

37 Sedaghat F, Akhoondan M, Ehteshami M, Aghamohammadi V, Ghanei N, Mirmiran P, Rashidkhani B. Maternal dietary patterns and gestational diabetes risk: A case-control study. *J Diabetes Res*. 2017; 2017:5173926. doi:10.1155/2017/5173926.

38 Shin D, Lee KW, Song WO. Dietary patterns during pregnancy are associated with risk of gestational diabetes mellitus. *Nutrients*. 2015; 7(11):9369–9382. doi:10.3390/nu7115472.

39 Chen X, Zhang Z, Yang H, et al. Consumption of ultra-processed foods and health outcomes: A systematic review of epidemiological studies. *Nutr J*. 2020; 19:86. doi:10.1186/s12937-020-00604-1.

40 Wen L, Ge H, Qiao J, et al. Maternal dietary patterns and risk of gestational diabetes mellitus in twin pregnancies: A longitudinal twin pregnancies birth cohort study. *Nutr J*. 2020; 19:13. doi:10.1186/s12937-020-00529-9.

41 Schoenaker DA, Mishra GD, Callaway LK, Soedamah-Muthu SS. The role of energy, nutrients, foods, and dietary patterns in the development of gestational diabetes mellitus: A systematic review of observational studies. *Diabetes Care*. 2016; 39(1):16–23. doi:10.2337/dc15-0540.

42 Goldstein RF, Abell SK, Ranasinha S, et al. Gestational weight gain across continents and ethnicity: Systematic review and meta-analysis of maternal and infant outcomes in more than one million women. *BMC Med*. 2018; 16(1):153. doi:10.1186/s12916-018-1128-1.

43 Mak JKL, Pham NM, Lee AH, et al. Dietary patterns during pregnancy and risk of gestational diabetes: A prospective cohort study in Western China. *Nutr J*. 2018; 17:107. doi:10.1186/s12937-018-0413-3.

44 Yan H, Dang S, Zhang Y, et al. Dietary patterns of Chinese women of childbearing age during pregnancy and their relationship to the neonatal birth weight. *Nutr J*. 2020; 19:89. doi:10.1186/s12937-020-00607-y.

45 Jarman M, Mathe N, Ramazani F, et al. Dietary patterns prior to pregnancy and associations with pregnancy complications. *Nutrients*. 2018; 10(7):914. doi:10.3390/nu10070914.

46 Nutrition Working Group. O'Connor DL, Blake J, et al. Canadian consensus on female nutrition: Adolescence, reproduction, menopause, and beyond [published correction appears in. *J Obstet Gynaecol Can*. 2018 Feb;40(2):268]. *J Obstet Gynaecol Can*. 2016; 38(6):508–554.e18. doi:10.1016/j.jogc.2016.01.001.

47 Li S, Gan Y, Chen M, et al. Effects of the dietary approaches to stop hypertension (DASH) on pregnancy/neonatal outcomes and maternal glycemic control: A systematic review and meta-analysis of randomized clinical trials. *Complement Ther Med*. 2020; 54:102551. doi:10.1016/j.ctim.2020.102551.

48 Kim D. Obesity. In: *Chapter 14 - Glycemic Index.* (ed. R. Mehrzad). Elsevier, 2020; pp. 183–189. 10.1016/B978-0-12-818839-2.00014-4. [Cited 15.08.2022]. Available at: https://www.sciencedirect.com/science/article/pii/B9780128188392000144.

49 The British Diabetic Association. Glycaemic index and diabetes. *What is the glycaemic index?* [Cited: 17.08.2022]. Available at: https://www.diabetes.org.uk/guide-to-diabetes/enjoy-food/carbohydrates-and-diabetes/glycaemic-index-and-diabetes.

50 Atkinson FS, Brand-Miller JC, Foster-Powell K. International tables of glycemic index and glycemic load values 2021: A systematic review. *Am J Clin Nutr.* 2021; 114(5):1625–1632. doi:10.1093/ajcn/nqab233.

51 Zhang X, Gong Y, Della Corte K, et al. Relevance of dietary glycemic index, glycemic load and fiber intake before and during pregnancy for the risk of gestational diabetes mellitus and maternal glucose homeostasis. *Clin Nutr.* 2021; 40(5):2791–2799. doi:10.1016/j.clnu.2021.03.041.

52 Ali AM, Kunugi H. Intermittent fasting, dietary modifications, and exercise for the control of gestational diabetes and maternal mood dysregulation: A review and a case report. *Int J Environ Res.* 2020; 17(24):9379. doi:10.3390/ijerph17249379.

53 Parrettini S, Caroli A, Torlone E. Nutrition and metabolic adaptations in physiological and complicated pregnancy: Focus on obesity and gestational diabetes. *Front Endocrinol.* 2020; 11:611929. doi:10.3389/fendo.2020.611929.

54 Fedullo AL, Schiattarella A, Morlando M, Raguzzini A, Toti E, De Franciscis P, Peluso I. Mediterranean diet for the prevention of gestational diabetes in the Covid-19 era: Implications of IL-6 in diabesity. *Int J Mol Sci.* 2021; 22(3):1213. doi:10.3390/ijms22031213.

55 de la Torre NG, Assaf-Balut C, Jiménez Varas I, et al. Effectiveness of following Mediterranean diet recommendations in the real world in the incidence of gestational diabetes mellitus (GDM) and adverse maternal-fetal outcomes: A prospective, universal, interventional study with a single group. The St Carlos study. *Nutrients.* 2019; 11(6):1210. doi:10.3390/nu11061210.

56 Mahjoub F, Ben Jemaa H, Ben Sabeh F, Ben Amor N, Gamoudi A, Jamoussi H. Impact of nutrients and Mediterranean diet on the occurrence of gestational diabetes. *Libyan J Med.* 2021; 16(1):1930346. doi:10.1080/19932820.2021.1930346.

57 Barabash A, Valerio JD, Garcia de la Torre N, et al. TCF7L2 rs7903146 polymorphism modulates the association between adherence to a Mediterranean diet and the risk of gestational diabetes mellitus. *Metabol Open.* 2020; 8:100069. doi:10.1016/j.metop.2020.100069.

58 Cropano C, Santoro N, Groop L, et al. The rs7903146 variant in the TCF7L2 gene increases the risk of prediabetes/type 2 diabetes in obese adolescents by impairing β-cell function and hepatic insulin sensitivity. *Diabetes Care.* 2017; 40(8):1082–1089. doi:10.2337/dc17-0290.

59 Mahajan A, Donovan LE, Vallee R, Yamamoto JM. Evidenced-based nutrition for gestational diabetes mellitus. *Curr Diab Rep.* 2019; 19(10):94. doi:10.1007/s11892-019-1208-4.

60 Al-Ibrahim AA, Jackson RT. Healthy eating index versus alternate healthy index in relation to diabetes status and health markers in U.S. adults: NHANES 2007–2010. *Nutr J.* 2019; 18:26. doi:10.1186/s12937-019-0450-6.

61 Altemani AH, Alzaheb RA. The prevention of gestational diabetes mellitus (the role of lifestyle): A meta-analysis. *Diabetol Metab Syndr.* 2022; 14:83. doi:10.1186/s13098-022-00854-5.

17

Biomarkers of Gestational Diabetes Mellitus, Dietary Polyphenols, and Drug Discovery

Mansi Gandhi[1,2,3]

[1] *Department of Chemistry, School of Advanced Sciences, Vellore Institute of Technology (VIT), Vellore, Tamil Nadu, 632014, India*
[2] *Institute of Chemistry, Hebrew University of Jerusalem, Jerusalem 9190401, Israel*
[3] *School of Innovation in Biodesign, Translational Health Science and Technology Institute*

17.1 Introduction

Maternal nutrient intake in the prenatal period is the most important determinant for fetal growth and development. In addition to the fetus, maternal health is also important and should be ensured. The Millennium Development Goals have targeted significant progress towards the quality and care for both pre-delivery and post-delivery for mother and child [1]. The quality of health of pregnant women is quite an essential timeframe involving transitions with important physical and emotional changes in the body [2]. Maternal nutrition is a crucial component for sustained health, having long-term effects on the mother and the infant. Health professionals (such as obstetrician, midwife, and general practitioner) provide guidance with the requisite information for proper food and supplement intake. However, more emphasis should be given on health and wellbeing, as any negligence can invite a number of diseases. Countless health issues such as periodontal and cardiovascular diseases, diabetes, etc., are a few of the severe complications that arise during this period. Quite a good number of articles are published in "Web of Science" and "Google Scholar" when one searches with keywords such as "polyphenol diet for diabetes mellitus" and "metabolic mechanism", "gestational diabetes mellitus and phytonutrients, vegetable foods, PUFA's, inflammation, hyperglycemia" and their combinations.

Diabetes is a high blood glucose condition caused due to failure of insulin secretion or due to abnormalities of biological functions within the body's systems. It is the world's third silent killer after cancer. The International Diabetes Federation (IDF) estimates over 425 million individuals worldwide were diagnosed with diabetes in 2017. And nearly 537 million people are living with diabetes in 2021,

Natural Products and their Bioactives in Antidiabetic Drug Discovery, First Edition.
Edited by Kanti Bhooshan Pandey, Maitree Suttajit and Pinar Atukeren.

which is predicted to increase to 637 million by 2030 [3]. The prevalence of type 2 diabetes mellitus (T2DM) is significantly increased in adults during their reproductive years. Gestational diabetes mellitus (GDM) is the most common metabolic disorder during pregnancy, and its prevalence is increasing worldwide [4]. Women with GDM are at a high risk of developing T2DM later in life; in addition, the higher baseline body mass index (BMI) and weight gain often found after GDM occurrence, increases the risk of progression from GDM to T2DM [5]. Moreover, uncontrolled GDM is associated with a detrimental intra-uterine environment, which leads to fatal complications and an increased risk for the child of developing obesity and metabolic disorders. Thus, paramount importance is given to identify the appropriate treatments to prevent maternal and fatal complications associated with this disease.

A major point to be highlighted in cases of GDM is the sudden and significant weight gain along with physiological changes in the body. This leads to a huge alteration in BMI of the females, leading to obesity as one of the prominent reasons for GDM [6, 7]. Figure 17.1 shows a schematic diagram portraying the pathways

Figure 17.1 Common and simple pathways to avoid the instances of gestational diabetes mellitus (GDM). It includes proper exercise, food nutrition, medical and regular testing, proper care, supplements (if required), and plenty of fluid intake.

to avoid the chances of GDM using simplistic parameters. GDM has become a metabolic and reproductive disorder as reported by Abbas et al., wherein nearly 15 to 20% pregnancies worldwide are affected by this [8].

The natural ways adopted by society to curb this is via "proper nutrition". The concept of "proper nutrition" consists of an appropriate plant-based food rich diet (polyphenols, carbs, proteins, minerals and vitamins, and fiber). These polyphenols have a principal role in defining the fate of GDM. Plant-derived polyphenols comprise polyphenolic compounds available in our daily dietary constituents. Around 8000 phenolic compounds have been identified based on their biological or chemical arrangement [9]. They have excellent anti-inflammatory and antioxidant features that are directly proportional to the number of hydroxyl groups in their arrangement. The antioxidative characteristic enables them to interfere in the biomolecule's oxidation via reaction with free radicals or by neutralizing them. Among the polyphenolic family, many different varieties exist. This chapter discusses current knowledge about the issue and impact of polyphenols on/and the mechanisms and genes relating to the glucose homeostasis and metabolic alterations. Along with the role of polyphenols, in this context, drug discovery is highlighted based on the different biomarkers of GDM.

17.2 GDM Biomarkers and Metagenomics

There is quite a close connection between insulin sensitivity and resistance with GDM. In pregnant women with GDM, insulin resistance increases the risks for developing manifest T2DM and is associated with complications in both the fetus and mother. A pie chart showing these relationships is represented in Figure 17.2. Their association is studied via three different indices (updated homeostasis model assessment, QUICKI, and McAuley's Indices), the diabetes risk factor, and fetal growth indices, which are important parameters for deriving their relationship.

17.3 GDM Management

17.3.1 Role of Natural Homoeopathy/Nutraceuticals

Polyphenols are the biggest group of plant-derived bioactive compounds [10]. They are found in quite variable quantities in chocolates, fruits, olive oils, vegetables, wine, cereals, nuts, and tea. Polyphenols are micro-nutrients that are very important to protect the body's tissues against oxidative stress and associated pathologies. The beneficial effect of polyphenols or a diet rich in polyphenols such as the Mediterranean diet, on various diseases, are the hot topics of research.

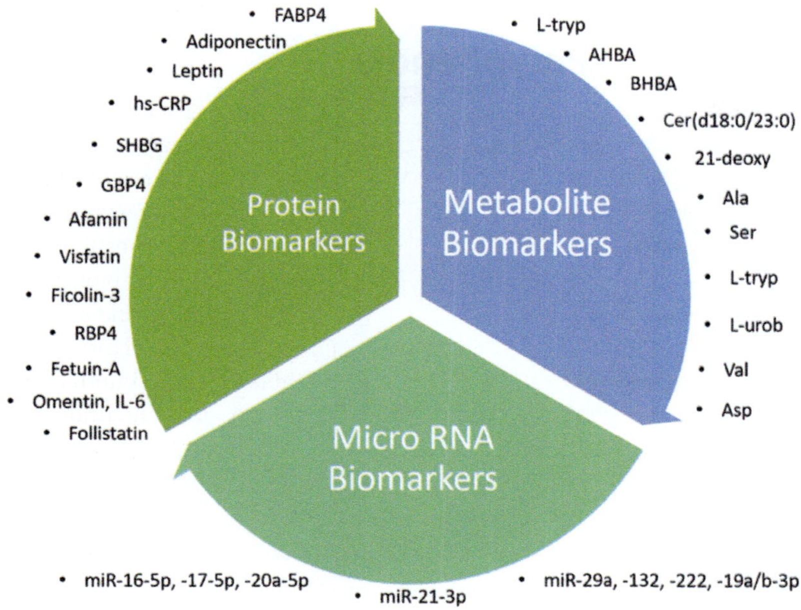

Figure 17.2 Predictive and diagnostic biomarkers for gestational diabetes mellitus (GDM), which are released in the body fluids (blood, urine) from the early stages of pregnancy.

17.3.2 Polyphenols and Oxidative Stress

There is evidence that links polyphenol intake and health promotion as well as prevention of diseases. The structure of polyphenols consists of a polyphenolic ring and they are further classified based on the number of rings and the structural elements that bind to the ring. The majority of polyphenols consists of glycosides, ester, polymers in a hydroxylated form. A description of the types of polyphenols is shown in Figure 17.3.

Catechins belonging to the flavanol family are obtained and extracted from tea, red wine, and cocoa. Different isoforms of catechins, majorly epicatechin, epicatechingallate, epigallocatechin, and epigallocatechin-3-gallate, (EGCG) exist in nature. A recent report shows evidence wherein they provide protection from diabetes mellitus and its other metabolic complications [11]. EGCG inhibits hepato-renal complications by regulating the phosphoenolpyruvate enzymes that help them in gluconeogenesis. The role of catechins is to diminish the levels of fatty acids and cholesterol through inhibition of gene expression of GLUT1 and GLUT4 transporter and later normalize the blood glucose content [12]. Furthermore, a report shows that catechins may help in activation of the

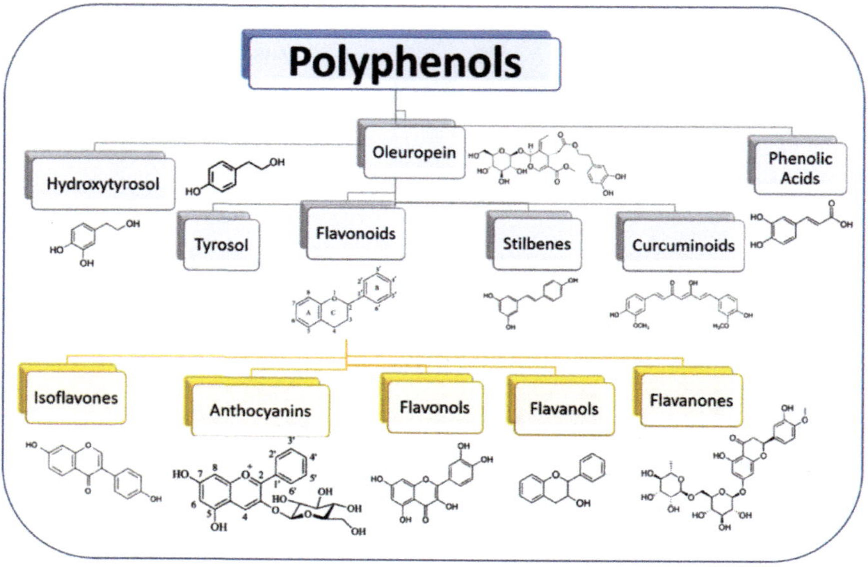

Figure 17.3 Descriptive illustration of the various polyphenol types.

erythroid-2 transcription factor and proteins against oxidative stress, as they activate sirtuin1 (SIRT1) and alleviate cardiovascular problems [13]. Hyperglycemia reduces the lactate accumulation in the brain cortex by elevating its metabolism. Catechins can be a solution to this via increasing the antioxidant potential, reducing peroxidative lipid damage. Additionally, they up-regulate the expression of catalase enzymes [14].

Luteolin is a flavone mostly present in spices and herbs like peppers, onion leaves, basil, broccoli, etc. It possesses various biological functions such as inhibiting protein kinase C (PKC) activity, and it has antioxidant and free radical scavenging activities [15]. Often, luteolin at certain doses (10 and 20 mg/kg) can restore diabetic induced brain impairments such as memory impairments and cognitive functions via antioxidant activity functions. A report involves luteolin neutralization by reactive oxidative species (ROS) by inhibiting the ROS-mediated mitogen-activated protein kinases (MAPK) mechanism and suppressing NADPH oxidase4 and intercellular adhesion expressions molecule1 [16]. Another protein, i.e. p22phox, is an essential component of the NADPH-oxidase enzyme that helps to produce superoxide ion. Presenilin 1 (PS1) protein and amyloid precursor protein (APP) production deteriorates, and is thus potentially anti-amyloidogenic [17].

Quercetin is found in citrus fruits, apples, cherries, blueberries, etc. Quercetin has the ability to cross the blood–brain barrier and protects the brain from injury.

It increases the reduced glutathione (GSH) content in astrocytes and neurons, thus preventing oxidative stress. Its intake (20 mg/kg) increases the level of SIRT1 in young and elderly people. SIRT1 contributes to the regulation of memory via promoting the production of acetylcholine (ACh).

Geinstein, an isoflavone present in soya products, leads to a decline in β-amyloid plaque induced toxicity, preventing neuronal degradation. Its consumption (30 and 60 mg/kg) can decrease the number of apoptotic cells due to reduction in caspases-3 and cytochrome c activity, leading to a decrease in the expression of nitric oxide synthase by inhibiting the inducible gene. Geinstein is reported to improve antioxidative enzyme expression, and reduces ROS generation and thus helps the fight against GDM. Its administration leads to elevation in SIRT1 levels and regulation of decrement in expression of necrosis factor, i.e. NF-ḳB, interleukin 1.

Resveratrol is majorly present in peanuts, grapes, cocoa, and dark chocolate. It is a non-flavonoid polyphenol belonging to a class of stilbenes and an important plant antibiotic. It is synthesized when a fungus infects the plants or under environmental stress conditions. There are two isomeric forms such as "*cis*" and "*trans*". Its "*trans*" form is non-toxic with higher therapeutic efficacy. It shows anti-inflammatory properties (for Alzheimer's disease) and helps in the protection of cellular damage caused by oxidative stress (antioxidant power). Studies show that resveratrol decreases the accumulation of β-amyloid by activating AβPP/PS1 pathways and reducing intracellular proteasomal activity. A dose of 5 mg/mL decreases neurogeneration and reduces the acetylation of PGC-1α and p53, and SIRT1.

17.3.3 Polyphenol Absorption, Metabolism and Elimination

Polyphenols have different chemical structures, synergistic effects, and functionality, which play an important role in defining the path of action. They are further subdivided into different subclasses of flavanols: flavones, flavanones, anthocyanidins and isoflavones. Evaluation of the actual contribution of dietary polyphenolic compounds for absorption, metabolism, and elimination is quite challenging, as many different factors come into play, especially based on the amount of consumption, nutrient content, and bioavailability, while the host-related factors account for genetic profile, obesity, pregnancy, and gut microbiota. As they pass through the body's organs, they are extensively modified through the stomach, small intestine, colon, and liver. Following which, they interact differently with cellular components like membrane, proteins, enzyme, receptors, and transcription factors. A detailed explanantion of polyphenols with their respective modes of action can be found in Table 17.1.

Table 17.1 List of the various polyphenols, food origin, and their mechanisms of action.

S. No.	Polyphenols	Food Domain	Mechanisms of action	References
1	Anthocyanin	Berry drinks like blue and blackberry wine	Decreases plasma glucose and lipid levels in diabetic streptozotocin rats; inhibits dipeptidyl-peptidase IV activity	[22, 23]
2	Procyanidins	Cocoa powder and Theobroma coca	Protects liver cells against an oxidative insult and up-regulates antioxidant enzymes activity via ERK 1/2	[24, 25]
3	Ferulic acid	Cereal	Neutralizes free radicals, stops the production of oxidative radical compounds and increases β-cell mass	[26, 27]
4	Epigallocatech-ingallate (EGCG)	*Camellia sinensis* (Green tea)	Enhances glucose uptake in muscle and fat cells (via GLUT-4 expression); increases insulin secretion; slows down carbohydrate digestion; inhibits carbohydrate absorption by competitively binding with the sodium-glucose transporter-1 (SGLT-1), and protects β-cells from free-radical damage	[28]
5	Flavan-3-ols and tannins-catechins	*Camellia sinensis* (Green tea)	Ameliorates insulin resistance by acting as peroxisome proliferator-activated receptor ligands with a dual alpha/ gamma agonistic effect	[29–32]

17.3.4 Polyphenolic Effects on Glycemic Regulation

There is growing evidence from various animal model research that supports anti-diabetic features, suggesting dietary therapy as a preventive measure and management for glucose homeostasis. Various mechanisms of action in regulation to glucose homeostasis and insulin sensitivity based on *in vitro* and *in vivo* studies are being carried out. The hypoglycemic effects of dietary polyphenols are related to the inhibition of carbohydrate digestion via inhibiting salivary and pancreatic α-amylase and α-glucosidase in the small intestine brush border, inhibiting glucose absorption, and simulating insulin secretion and protection of pancreatic β-cells against glucotoxicity. Polyphenols suppress glucose release from the liver and improve its uptake in peripheral tissues by modulating intracellular signaling. They even inhibit advanced glycation end product formation. The importance of the insulin-sensitizing effect on adiponectin, enhancing AdinoR functions, is quite an interesting counter for insulin resistance as reported by Caselli [18]. Dietary polyphenols present in seed extracts, wine, broccoli, fish oil, linoleic acid, etc., elevate the adiponectin levels in the body resulting in reduced mRNA concentration in leptin 3T3-L1 adipocytes. Their stimulatory effect is observed to be mediated via a histone deacetylase sirtuin (Sirt-1) independent mechanism. This is observed mainly due to phosphoinositide dependent protein-kinase-1/Akt signaling pathway suppression, and furthermore, due to the activation of transcription factor fork-head box O-1 (FOXO1) and AMPK (5′ adenosine monophosphate-activated protein kinase) signaling [19]. Resveratrol reduced the expression and release of IL-6 and TNFα in RAW 264 macrophages and 3T3-L1 cells, and thus reduced extracellular regulated protein kinase ½ (ERK½) and NF-ķB activation [20]. Furthermore, the high glucose induced nitric oxide (NO) reduction is reverted by resveratrol incubation (in bovine artery endothelium cells) because of enhanced endothelium NO synthase mRNA and activity. Hence, this strongly indicates the therapeutic supplement of polyphenols to counteract the diabetic condition [21].

Polyphenols have an important role in defining the uptake of bio-compounds in the body such as vitamins like thiamine (H_2O soluble vitamin B_1), which is crucial during the period of pregnancy for the development of normal fetal growth. Hence, an in-depth understanding of the role of polyphenols and the transportation pathway is also required for regulation across the placenta. Many researchers have carried out studies to understand the short- and long-term effects of different phenolics and their uptake in various cell types (BeWo cells, Caco-2 cells and various cell lines like syncytiotrophoblast). A detailed version is accounted for in the following subtopics.

17.3.5 Dietary Intake of Polyphenols during Pregnancy

The consumption of polyphenols varies greatly between individuals and cultural ethnicities; especially during pregnancy, which is subject to interpretation errors caused by individual differences in absorption and metabolism, genetics, and metabolic changes. A study to evaluate the dietary intake patterns at a low cost with ease of administration to investigate reproducibility and validity of a method is based on a food frequency questionnaire [33]. Studies provide more precise and individual data concerning the intake of subsequent classes of polyphenols required during pregnancy and their potential impact on reproductive health. A publicly accessible database known as "Phenol-Explorer", provides data on the polyphenol content in foods and therefore assists with identifying potential hazards of consuming excess polyphenol-rich food.

The major cause accelerating the issue is heterogeneity, i.e. ethnic as well as inter and intra country differences along with incomplete knowledge of its pathophysiology. Furthermore, standardized guidelines for the metabolic disorder are difficult to arrive at, as are the intervention strategies aimed at preventing and/or reducing the burden of the disorder. Another important ground risk outcome is obesity and maternal overweight.

The search for a cure for GDM has provided increasing evidence that both prevention and management of diabetes are characterized by consumption of plant-based food, especially whole grain, green leafy vegetables, and fruits, and avoidance of animal-based high-fat processed food. The Mediterranean diet follows a plant-based dietary pattern that is beneficial against cardiovascular diseases, neoplastic disease, and reduced mortality. Eating a diet with plant-based foods helps in maintaining a controlled glycemic index, reducing cardiovascular and diabetic risks. The Dietary Approaches to Stop Hypertension (DASH) diet is quite rich in fruits, vegetables, whole grains, and lower amounts of saturated fatty acids (FA), and therefore is beneficial in preventing health problems. The increased level of polyunsaturated fatty acids (PUFA) and low content of trans-fatty acids and phytochemical compounds have established the DASH diet as a better option than conventional medicines in preventing GDM complications.

Polyphenols have shown anti-hyperglycemic, antioxidant, and anti-inflam matory effects. In addition, *n-3* PUFA and essential fatty acids have a source other than non-fish sources majorly in polyphenolic-rich plant foods that is worth considering for the control of GDM. An essential part is to understand the mechanism and attain knowledge of how and where (i.e. mechanisms and respective targets) a given bioactive polyphenol acts. Some biomarkers with their respective occurrence time, sensitivity, and specificity are shown in Table 17.2.

Table 17.2 The potent biomarkers shown with details about their presence (weeks of pregnancy), sensitivity percentage, specificity, and literature.

S. No.	Biomarker	Week of Pregnancy	Sensitivity Percentage	Specificity Percentage	Literature Noted
1	hs-CRP (high sensitivity C-reactive protein)	4th to 6th and 11th to 14th	89	55.3	[34, 35]
2	FABP4 (Fatty Acid Binding Protein 4)	4th to 6th	81.8	71.2	[36]
		11th to 14th	86.2	50.8	
3	SHBG (sex hormone binding globulin)	1st to 12th	85	55	[34, 35, 37]

Table 17.2 (Continued)

S. No.	Biomarker	Week of Pregnancy	Sensitivity Percentage	Specificity Percentage	Literature Noted
4	FGF-21 (Fibroblast Growth Factor-21)	24th to 28th	100	75	[38–40]
6	miR-16-5p (microRNA-16-5p)	4th to 6th	41.6	95.8	[41, 42]
7	miR-17-5p (microRNA-17-5p)	4th to 6th	21.4	95.4	[41]
8	miR-20a-5p (microRNA-20a-5p)	4th to 6th	17.8	95.4	[43, 44]
9	Adiponectin	16th to 18th	80.7	69.1	[45, 46]
		24th to 28th	83.6	56.6	
10	RBP4 (Retinol binding protein 4)	16th to 18th	79	79	[47, 48]
11	Leptin	24th to 28th	81.2	64.2	[49, 50]
12	Ficolin-3	16th to 18th	51	97.2	[51]
13	Ficolin-A	11th to 14th	58.6	76.2	[52]
14	miR-21-3p (microRNA-21-3p)	30th to 36th	52.6	89.3	[53]
15	miR-132 (microRNA-132)	16th	66.6	63.3	[53]

17.5 Conclusion and Future Insights

Polyphenols possess mixed effects on GDM, depending upon their types and doses. The apparent difference between the chronic and acute exposure towards polyphenols *in vivo* may not necessarily be reflective of what is seen *in vitro*. Hence, further studies using villous explants and animal models in support would be interesting and evident. Many successful published reports in this context for the development of extensive food composition tablets consisting of polyphenols and standardized methods for executing experiments have been worked upon. This has made it possible for researchers to conduct observational epidemiological studies and grant substantial results in support of the literature. The bioactive molecule is not solely responsible for the biological effect rather the characterization of metabolite activity is responsible and often needs testing in *in vitro* studies towards gestational diabetes. Many biomarkers in this respect are studied in different phases of the pregnancy period to ensure the management of

GDM and the care of the mother as well as the fetus. Furthermore, identification of the normal physiological range of polyphenols and the metabolites are of utmost importance for determination of its effects achieved from a certain dose via the clinical relevance of results achieved using dose based animal trials. The clinical relevance for the consumption of polyphenolic moieties for dose studies might exceed the normal physiological conditions, making these studies difficult to conclude. As their concentration in the fetus is comparatively lower than in the adult we cannot disregard the potential biological activity of polyphenols as an effective concentration value. Collectively, this chapter concludes that various aspects must be considered for designing a limit to polyphenolic consumption in relation to GDM prevention and its toxicity value for the health of mother and child, irrespective of evaluating the beneficial or adverse effect of polyphenols as a whole.

References

1 Tunçalp WWM, Maclennan C, Oladapo OT, Gülmezoglu AM, Bahl R, et al. Quality of care for pregnant women and newborns - The WHO vision. *BJOG.* 2015; 122:1045–1049.

2 Lagadec N, Steinecker M, Kapassi A, Magnier AM, Chastang J, Robert S, Gaouaou N, Ibanez G. Factors influencing the quality of life of pregnant women: A systematic review. *BMC Pregnancy Childbirth.* 2018; 18(1):455.

3 Saeedi P, Petersohn I, Salpea P, Malanda B, Karuranga S, Unwin N, Colagiuri S, Guariguata L, Motala AA, Ogurtsova K, Shaw JE, Bright D, Williams R. IDF diabetes atlas committee. Global and regional diabetes prevalence estimates for 2019 and projections for 2030 and 2045: Results from the International Diabetes Federation Diabetes Atlas, 9th edition. *Diabetes Res Clin Pract.* 2019; 157:107843.

4 Valero P, Salas R, Pardo F, Cornejo M, Fuentes G, Vega S, Grismaldo A, Hillebrands JL, van der Beek EM, van Goor H, Sobrevia L. Glycaemia dynamics in gestational diabetes mellitus. *Biochim Biophys Acta Gen Subj.* 2022; 1866(7):130134.

5 Sweeting A, Wong J, Murphy HR, Ross GP. A clinical update on gestational diabetes mellitus. *Endocr Rev.* 2022 Sep 26; 43(5):763–793.

6 Song Z, Cheng Y, Li T, Fan Y, Zhang Q, Cheng H. Prediction of gestational diabetes mellitus by different obesity indices. *BMC Pregnancy Childbirth.* 2022; 22(1):288.

7 Gao S, Leng J, Liu H, Wang S, Li W, Wang Y, Hu G, Chan JCN, Yu Z, Zhu H, Yang X. Development and validation of an early pregnancy risk score for the prediction of gestational diabetes mellitus in Chinese pregnant women. *BMJ Open Diabetes Res Care.* 2020; 8(1):e000909.

8 Choudhury AA, Devi Rajeswari V. Gestational diabetes mellitus - A metabolic and reproductive disorder. *Biomed Pharmacother.* 2021 Nov; 143:112183.

9 Ignat I, Volf I, Popa VI. A critical review of methods for characterisation of polyphenolic compounds in fruits and vegetables. *Food Chem.* 2011; 126:1821–1835.

10 Gandhi M, Amreen K. Electrochemical profiling of plants. *Electrochem.* 2022; 3:434–450.

11 Cheng X, Zhang L, Lian YJ. Molecular targets in Alzheimer's disease: From pathogenesis to therapeutics. *Biomed Res Int.* 2015; 2015:760758.

12 Wolfram S, Raederstorff D, Preller M, Wang Y, Teixeira SR, Riegger C, Weber P. Epigallocatechin gallate supplementation alleviates diabetes in rodents. *J Nutr.* 2006; 136(10):2512–2518.

13 Zhang Z, Ding Y, Dai X, Wang J, Li Y. Epigallocatechin-3-gallate protects pro-inflammatory cytokine induced injuries in insulin-producing cells through the mitochondrial pathway. *Eur J Pharmacol.* 2011; 670:311–316.

14 Nunes AR, Alves MG, Tomás GD, Conde VR, Cristóvão AC, Moreira PI, et al. Daily consumption of white tea (Camellia sinensis (L.)) improves the cerebral cortex metabolic and oxidative profile in prediabetic Wistar rats. *Br J Nutr.* 2015; 113:832–842.

15 Wang GG, Lu XH, Li W, Zhao X, Zhang C. Protective effects of luteolin on diabetic nephropathy in STZ-induced diabetic rats. *Evid Based Complement Alternat Med.* 2011; 2011:323171.

16 Xie XY, Kong PR, Wu JF, Li Y, Li YX. Curcumin attenuates lipolysis stimulated by tumor necrosis factor-α or isoproterenol in 3T3-L1 adipocytes. *Phytomedicine.* 2012; 20:3–8.

17 Bruce-Keller AJ, Gupta S, Knight AG, Beckett TL, McMullen JM, Davis PR, et al. Cognitive impairment in humanized APP×PS1 mice is linked to Aβ 1-42 and NOX activation. *Neurobiol Dis.* 2011; 44:317–326.

18 Caselli C. Role of adiponectin system in insulin resistance. *Mol Genet Metab.* 2014; 113:155–160.

19 Wang A, Liu M, Liu X, Dong LQ, Glickman RD, Slaga TJ, et al. Up-regulation of adiponectin by resveratrol: The essential roles of the Akt/FOXO1 and amp-activated protein kinase signaling pathways and DsbA-L. *J Biol Chem.* 2011; 286:60–66.

20 Kang L, Heng W, Yuan A, Baolin L, Fang H. Resveratrol modulates adipokine expression and improves insulin sensitivity in adipocytes: Relative to inhibition of inflammatory responses. *Biochimie.* 2010; 92:789–796.

21 Peng XL, Qu W, Wang LZ, Huang BQ, Ying CJ, Sun XF, et al. Resveratrol ameliorates high glucose and high-fat/sucrose diet-induced vascular hyperpermeability involving Cav-1/eNOS regulation. *PLoS One.* 2014; 9.

22 Valcheva-Kuzmanova SV, Belcheva A. Current knowledge of Aronia melanocarpa as a medicinal plant. *Folia Med (Plovdiv).* 2006; 48(2):11–17.

23 Fan J, Johnson MH, Lila MA, Yousef G, de Mejia EG. Berry and Citrus phenolic compounds inhibit dipeptidyl peptidase IV: Implications in diabetes management. *Evid Based Complement Alternat Med.* 2013; 2013:479505.

24 Martin MÁ, Ramos S, Cordero-Herrero I, Bravo L, Goya L. Cocoa phenolic extract protects pancreatic beta cells against oxidative stress. *Nutrients*. 2013; 5:2955–2968.

25 Zhang W, Zhao J, Wang J, Pang X, Zhuang X, Zhu X, et al. Hypoglycemic effect of aqueous extract of seabuckthorn (Hippophae rhamnoides L.) seed residues in streptozotocin-induced diabetic rats. *Phytother Res*. 2010; 24:228–232.

26 Trojan-Rodrigues M, Alves TLS, Soares GLG, Ritter MR. Plants used as antidiabetics in popular medicine in Rio Grande do Sul, southern Brazil. *J Ethnopharmacol*. 2012; 139:155–163.

27 Fardet A. New hypotheses for the health-protective mechanisms of whole-grain cereals: What is beyond fibre? *Nutr Res Rev*. 2010; 23:65–134.

28 van Woudenbergh GJ, Kuijsten A, Drogan D, van der A DL, Romaguera D, Ardanaz E, et al. Tea consumption and incidence of type 2 diabetes in Europe: The EPIC-interact case-cohort study. *PLoS One*. 2012; 7(5):e36910.

29 Visen P, Saraswat B, Visen A, Roller M, Bily A, Mermet C, et al. Acute effects of Fraxinus excelsior L. seed extract on postprandial glycemia and insulin secretion on healthy volunteers. *J Ethnopharmacol*. 2009; 126:226–232.

30 Islam MS. Effects of the aqueous extract of white tea (Camellia sinensis) in a streptozotocin-induced diabetes model of rats. *Phytomedicine*. 2011; 19:25–31.

31 Toolsee NA, Aruoma OI, Gunness TK, Kowlessur S, Dambala V, Murad F, Googoolye K, Daus D, Indelicato J, Rondeau P, Bourdon E, Bahorun T. Effectiveness of green tea in a randomized human cohort: Relevance to diabetes and its complications. *Biomed Res Int*. 2013; 2013:412379.

32 Mustata GT, Rosca M, Biemel KM., Reihl O, Smith MA, Viswanathan A, Strauch C, Du Y, Tang J, Kern TS, Lederer MO, Brownlee M, Weiss MF, Monnier VM. Paradoxical effects of green tea (Camellia sinensis) and antioxidant vitamins in diabetic rats: Improved retinopathy and renal mitochondrial defects but deterioration of collagen matrix glycoxidation and cross-linking. *Diabetes*. 2005; 54(2):517–526.

33 Miranda AR, Cortez MV, Scotta AV, Soria EA. Dietary intake of polyphenols enhances executive/attentional functioning and memory with an improvement of the milk lipid profile of postpartum women from Argentina. *J Intell*. 2022; 10(2):33.

34 Rodrigo N, Glastras SJ. The emerging role of biomarkers in the diagnosis of gestational diabetes mellitus. *J Clin Med*. 2018 May 23; 7(6):120.

35 Tawfeek MA, Alfadhli EM, Alayoubi AM, El-Beshbishy HA, Habib FA. Sex hormone binding globulin as a valuable biochemical marker in predicting gestational diabetes mellitus. *BMC Womens Health*. 2017; 17(1):18.

36 Patro-Małysza J, Trojnar M, Kimber-Trojnar Ż, Mierzyński R, Bartosiewicz J, Oleszczuk J, Leszczyńska-Gorzelak B. FABP4 in gestational diabetes-association between mothers and offspring. *J Clin Med*. 2019; 8(3):285.

37 Farina A, Eklund E, Bernabini D, Paladino M, Righetti F, Monti G, Lambert-Messerlian G. A first-trimester biomarker panel for predicting the development of gestational diabetes. *Reprod Sci*. 2017; 24(6):954–959.

38 Wang Z, Yuan M, Xu C, Zhang Y, Ying C, Xiao X. FGF21 Serum levels in the early second trimester are positively correlated with the risk of subsequent gestational diabetes mellitus: A propensity-matched nested case-control study. *Front Endocrinol (Lausanne).* 2021; 12:630287.

39 Bonakdaran S, Khorasani ZM, Jafarzadeh F. Increased serum level of FGF21 in gestational diabetes mellitus. *Acta Endocrinol (Buchar).* 2017; 13(3):278–281.

40 Jia J, Wei W, Yu F, Liu R, Shen Y, Zhang R, et al. Circulating levels of fibroblast growth factor 21 in gestational diabetes mellitus: A meta-analysis. *Endocr J.* 2021; 68:345–352.

41 Filardi T, Catanzaro G, Mardente S, Zicari A, Santangelo C, Lenzi A, Morano S, Ferretti E. Non-coding RNA: Role in gestational diabetes pathophysiology and complications. *Int J Mol Sci.* 2020; 21(11):4020.

42 Karin-Kujundzic V, Sola IM, Predavec N, Potkonjak A, Somen E, Mioc P, Serman A, Vranic S, Serman L. Novel epigenetic biomarkers in pregnancy-related disorders and cancers. *Cells.* 2019; 8(11):1459.

43 Pheiffer C, Dias S, Rheeder P, Adam S. Decreased expression of circulating miR-20a-5p in South African women with gestational diabetes mellitus. *Mol Diagn Ther.* 2018; 22:345–352.

44 Liu ZN, Jiang Y, Liu XQ, Yang MM, Chen C, Zhao BH, Huang HF, Luo Q. MiRNAs in gestational diabetes mellitus: Potential mechanisms and clinical applications. *J Diabetes Res.* 2021; 2021:4632745.

45 Madhu S, Bhardwaj S, Jhamb R, Srivastava H, Sharma S, Raizada N. Prediction of gestational diabetes from first trimester serum adiponectin levels in Indian women. *Indian J Endocrinol Metab.* 2019; 23:536–539.

46 Bogdanet D, Reddin C, Murphy D, Doheny HC, Halperin JA, Dunne F, et al. Emerging protein biomarkers for the diagnosis or prediction of gestational diabetes—a scoping review. *J Clin Med.* 2021; 10(7):1533.

47 Chan TF, Chen HS, Chen YC, Lee CH, Chou FH, Chen IJ, Chen SY, Jong SB, Tsai EM. Increased serum retinol-binding protein 4 concentrations in women with gestational diabetes mellitus. *Reprod Sci.* 2007; 14(2):169–174.

48 Krzyzanowska K, Zemany L, Krugluger W, Schernthaner GH, Mittermayer F, Schnack C, Rahman R, Brix J, Kahn BB, Schernthaner G. Serum concentrations of retinol-binding protein 4 in women with and without gestational diabetes. *Diabetologia.* 2008; 51(7):1115–1122.

49 Bawah AT, Seini MM, Abaka-Yawason A, Alidu H, Nanga S. Leptin, resistin and visfatin as useful predictors of gestational diabetes mellitus. *Lipids Health Dis.* 2019; 18(1):221.

50 Pérez-Pérez A, Toro A, Vilariño-García T, Maymó J, Guadix P, Dueñas JL, et al. Leptin action in normal and pathological pregnancies. *J Cell Mol Med.* 2018; 22:716–727.

51 Yuan XS, Shi H, Wang HY, Yu B, Jiang J. Ficolin-3/adiponectin ratio for the prediction of gestational diabetes mellitus in pregnant women. *J Diabetes Investig.* 2018; 9:403–410.

52 Lorenzo-Almorós A, Hang T, Peiró C, Soriano-Guillén L, Egido J, Tuñón J, Lorenzo Ó. Predictive and diagnostic biomarkers for gestational diabetes and its associated metabolic and cardiovascular diseases. *Cardiovasc Diabetol.* 2019; 18(1):140.

53 Guan CY, Tian S, Cao JL, Wang XQ, Ma X, Xia HF. Down-regulated MiR-21 in gestational diabetes mellitus placenta induces ppar-α to inhibit cell proliferation and infiltration. *Diabetes Metab Syndr Obes.* 2020; 13:3009–3034.

18

Medicinal and Aromatic Plants in the Prevention of Gestational Diabetes and Associated Consequences

Current Insights

Jigeesha Mishra[1], Kanti Bhooshan Pandey[2,3], and Shailendra Kumar Srivastava[1,]*

[1] *Jacob Institute of Biotechnology & Bio-Engineering, Sam Higginbottom University of Agriculture, Technology & Sciences, Prayagraj-211007, India*
[2] *CSIR-Central Salt & Marine Chemicals Research Institute, Bhavnagar-364002, India*
[3] *Academy of Scientific and Innovative Research (AcSIR), Ghaziabad-201002, India*
* *Corresponding author*

18.1 Introduction

Gestational diabetes mellitus (GDM) is a pregnancy associated asymptomatic metabolic disorder, and may be defined as severe carbohydrate intolerance of variable levels with onset or first recognition during pregnancy. The prevalence of GDM is constantly rising worldwide. As per the International Diabetes Federation (IDF) report in 2019, 20 million or 16% live births were affected by hyperglycemia in pregnancy, documenting that GDM affected one in six births [1]. The effective management of GDM during gestation becomes essential due to its adverse ramifications on pregnancy outcome, maternal and fetal health during gestation, postpartum, and even much later in life [2, 3]. GDM pregnancies are at increased risk of fetal complications such as cesarean section, macrosomia, shoulder dystocia, respiratory distress syndrome, neonatal hyperbilirubinemia, hypoglycemia, and perinatal mortality [4, 5]. Furthermore, the affected mother is at sevenfold higher risk of developing type 2 diabetes in a later period of life [6]. Allopathically GDM is managed with synthetic drugs; however, an herbal preventive option is considered to be safer due to fewer or no side effects. Studies in last few decades have provided experimental evidence that adherence to plant-derived natural medication showed an inverse correlation with the development and progression of many degenerative disorders including diabetes [7–10].

Natural Products and their Bioactives in Antidiabetic Drug Discovery, First Edition.
Edited by Kanti Bhooshan Pandey, Maitree Suttajit and Pinar Atukeren.

Plants have been fulfilling nearly all basic human necessities since ancient times by serving as source of oxygen, food, and materials for construction. Investigation of the innumerable therapeutic properties possessed by plants has established them as a potential source of medicines. Medicinal and aromatic plants (MAPs) are a reservoir of therapeutic phytochemicals that make them felicitous for the treatment or management of health disorders. An array of reports has documented that MAPs exhibit potent anti-diabetic properties [11, 12]; however, a limited number of plants have been studied for their effects and mechanisms of action for anti-hyperglycemic effects in GDM [13]. In this chapter, we discuss important MAPs and the bioactive compounds obtained from them responsible for the protective role against GDM along with their mechanism(s) of action.

18.2 Available Therapies to Manage GDM

The disorders occurring during pregnancy have maternal as well as fetal ramifications that require their urgent management/treatment. GDM is one of such disorders that adversely affects maternal and fetal systems and pregnancy outcomes. Glycemic control during gestational diabetes is attempted by introducing lifestyle changes such as exercise and controlled diets; however, persisting hyperglycemia requires additional treatment. Drugs in the intervention of gestational hyperglycemia include metformin, glibenclamide, and/or insulin therapy [14]. Metformin, a biguanide, is most widely prescribed as an oral anti-diabetic drug. Its mechanism of action may be AMP-activated protein kinase (AMPK) dependent or/and AMPK independent to improve insulin sensitivity. However, it is known that metformin crosses the placenta, raising concerns of its adverse effects on the physiology of the developing fetus [15, 16]. A report also suggests that metformin may increase the rate of preterm births [17]. Glibenclamide (glyburide) is another oral hypoglycemic agent for hyperglycemia management. It is a sulfonylurea drug that acts on both insulin secretion and on insulin action. Glibenclamide does not cross the placenta; however, meta-analyses and clinical trials suggest that glibenclamide treated patients had higher birth weight and higher rates of macrosomia and neonatal hypoglycemia than insulin and metformin treated patients [14]. Insulin is the oldest pharmacotherapy for diabetes that is still used. However, the mode of its administration, multiple doses per day, and high cost makes it a difficult option. Due to such untoward concerns, recently an increased inclination towards natural remedies for GDM management has been observed. The anti-diabetic properties of plants have always been used empirically in folklore medicine, paving the way for studies on their anti-diabetic potential during pregnancy. Interestingly, not all plants having an anti-hyperglycemic effect are safe for diabetic treatment during pregnancy [18]. This chapter incorporates a number of medicinal plants, their pharmacological activities, and their anti-diabetic action, which are beneficial in the prevention and management of GDM.

18.3 Important MAPs in the Prevention of GDM and its Associated Complications

18.3.1 Abelmoschus Esculentus

Abelmoschus esculentus is a flowering plant of the Malvaceae family and is cultivated abundantly in tropical and sub-tropical regions; it is also widely consumed around the world. Due to its widespread cultivation, it has received a variety of vernacular names such as "lady's finger" or "okra" in English; "bhindi", "bhendi", or "bindi" in Hindi; and "gumbo" or "gombo" in Portuguese and Spanish. Interestingly, along with immature pods, other parts such as leaves, seeds, and flowers of *A. esculentus* have also been used for culinary purposes in various cuisines throughout the world. In addition, the stem and root of *A. esculentus* are also considered useful for various health promotive purposes [19].

Substantial evidence suggests the high nutritional and pharmacological values of *A. esculentus*. The immature *A. esculentus* pods are an opulent source of carbohydrates (7g/100g of raw okra), dietary fibers (3.2g/100g of raw okra), protein (1.93g/100g of raw Okra), vitamin C, vitamin K, folate, and minerals such as potassium, calcium, and magnesium [20]. Additionally, the pods are rich in polyphenolic content and have been reported to contain copious amounts of oligomeric catechin, hydroxycinnamic derivatives, flavanol derivatives, and quercetin derivatives such as quercetin-3-*O*-sophoroside, quercetin-3-*O*-glucoside, quercetin-3-*O*-gentiobiopyranose, and quercetin-4′-*O*-methy-3-*O*-β-D-glucopyranoside [21].

The high bioactive components of *A. esculentus* contribute to its pharmacological potential and antioxidant capacity, thereby associating it with an array of health advantages including decreased susceptibility to chronic non-communicable diseases such as hypertension and obesity [20]. *A. esculentus* has been extensively reported for its anti-diabetic and anti-hyperlipidemic properties [22, 23]. Tian et al. have investigated the hypoglycemic, anti-hyperlipidemic, anti-hypercholesterolemic effects of *A. esculentus* extract on GDM rats; the study has concluded that the treatment of *A. esculentus* restored the level of several biochemical parameters such as glycosylated hemoglobin (HbA1c), fasting blood glucose (FBG), fasting insulin (FINS), total cholesterol (TC), triglyceride (TG), low density lipoprotein (LDL), high density lipoprotein (HDL), and free fatty acids (FFA) to the normal pregnancy range. The study proposed that the improved blood glucose level in *A. esculentus* treated GDM models may be due to its ability to improve antioxidant status and suppress oxidative stress and insulin resistance, which prevail in GDM pregnancies [24].

Okada et al. have reported that supplementation of *A. esculentus* ethanolic seed extract resulted in up-regulation of adiponectin production in a culture medium of mouse adipocytes, which is associated with anti-diabetic activity [25]. An *in vitro* study performed by Ahmed and Kumar documented the inhibitory effects of aqueous, methanolic, and ethanolic extracts of *A. esculentus* pods against α-glucosidase and α-amylase, the intestinal enzymes implicated in the breakdown of complex carbohydrates, and their rapid absorption in the gastro-intestinal (GI) tract, consequently leading to a spike in blood sugar levels [26]. The inhibitory effect of *A. esculentus* seed extract against α-glucosidase may be credited to its isoquercetin and quercetin-3-*O*-β-glucopyranosyl-(1′→6′)-glucoside content [27]. Another attribute contributing to the better glycemic control by consumption of *A. esculentus* is its high dietary fiber content since foods rich in dietary fiber decelerate glucose absorption in the GI tract, thereby obstructing the rise in blood glucose levels [28].

18.3.2 Hibiscus Rosa-sinensis

Hibiscus rosa-sinensis is another therapeutically important plant of the family Malvaceae, commonly referred as "shoe flower plant", "rose mallow", or "Chinese hibiscus". It is an evergreen, glabrous, ornamental shrub that grows in tropical conditions and is distributed throughout India. *H. rosa-sinensis* has gained a therapeutic reputation in the traditional medicine practice of numerous countries including India. Ethnomedicinally, *H. rosa-sinensis* is used in the management and treatment of various morbidities. The flowers possess cardio-protective properties and are reported to be used in ischemic diseases, additionally, they are used as an emollient, aphrodisiac, and demulcent, as well as its decoction being used in bronchial catarrh and fever in Indian traditional medicinal systems [29, 30].

The flowers, leaves, stem, and roots of *H. rosa-sinensis* have been documented to elicit multitudinous therapeutic properties such as antioxidant, anti-inflammatory, anti-hypertensive, wound healing, hepatoprotective, hypoglycemic, hypolipidemic, hypocholesterolemic, and immunomodulatory activities [31, 32]. The anti-diabetic activity of *H. rosa-sinensis* has been reported in multiple studies. Afiune et al. studied the effects of the aqueous extract of flowers of *H. rosa-sinensis* on pregnant diabetic rats and reported its beneficial effect. The treatment of *H. rosa-sinensis* flower aqueous extract normalized the pre-implantation loss and reduced the post-implantation loss rates and percentage of small for pregnancy age (SPA) fetuses; furthermore, the treated pregnant diabetic rats showed increased percentage of adequate for pregnancy age (APA) fetuses, and maternal and fetal weights [33]. In addition, the study reported increased superoxide dismutase (SOD) activity, increased HDL, and decreased TG and VLDL in treated pregnant diabetic rats in comparison to the non-treated pregnant diabetic rats.

The study further reported a decrease in coronary artery risk index (CRI) and atherogenic index (AI) in treated diabetic rats. The effects may be attributed to improvement in the lipid profile of the supplemented streptozotocin induced gestational diabetic rats [33].

Another study performed to investigate the effect of aqueous flower extract of *H. rosa-sinensis* reported similar effects on pre- and post-implantation loss rates and increased maternal weight gain and gravid uterus weight in pregnant diabetic Wistar rats in comparison to the non-treated group. Furthermore, the authors claimed that the treatment improved the reproductive performance of diabetic rats and did not exhibit any teratogenic effects [34]. Severe diabetes is associated with decreased anabolic and accelerated catabolic processes with pronounced weight loss [35]. Many studies have repeatedly documented that treatment of aqueous flower extract of *H. rosa-sinenis* significantly ameliorated multiple biochemical parameters that were disturbed in diabetic pregnancy [33, 34, 36].

18.3.3 Cinnamomum Zeylancium

Cinnamomum zeylancium or "true cinnamon", belonging to the family Lauraceae, is a spice with culinary, therapeutic, and cosmetic usage dating back for centuries. This aromatic, tropical, evergreen plant is native to southern parts of India. The use of cinnamon in the Ayurvedic medicinal system for the treatment of digestive, respiratory, and gynecological infirmities has been reported [37]. *C. zeylancium* is widely known for the use of its bark as a spice and as a good source of volatile oils. In addition to the bark, its leaves are also used as a spice, while other parts of the plant are also beneficial and commercially viable in various industries. The major compounds in essential oil obtained from the bark, leaves, and root of *C. zeylancium* are cinnamaldehyde (Ci), eugenol, and camphor, respectively [38]. Paranagama et al. identified 35 constituents along with Ci as a major component in *C. zeylancium* bark oil [39]. Fifty-three bioactive components have been identified in cinnamon leaf oil with eugenol (81–84.5%) as the major component [40], while (E)-cinnamyl acetate is the major component in volatile oils derived from fruits and flowers, thus bestowing individual parts with different properties [41, 42].

C. zeylancium has been reported to exhibit hypoglycemic, anti-microbial, anti-hypertensive, antioxidant, and anti-inflammatory properties, which are utilized in traditional medicine practices to attenuate infirmities associated with diabetes, bacterial infections, and hypertension [37]. Ci – the major component of *C. zeylancium* bark oil – has been repeatedly reported for its therapeutic properties. Hosni and colleagues have reported that Ci obtained from fresh bark of *C. zeylancium* successfully normalized the fasting serum glucose and insulin levels in gestational hyperglycemic rats [43]. Ci treatment also positively affected the reproductive outcome by decreasing the implantation loss and increasing the number of live fetuses;

furthermore, the same normalizing effect of Ci treatment on GDM mothers were reported for fetal parameters such as fetal weight, glucose, and insulin levels, which were significantly altered in non-treated GDM rats. The study also reported the suppressed adipose tissue PPARγ mRNA expression in GDM rats, when compared to the euglycemic rats, the administration of Ci in GDM rats showed a significant rise in the PPARγ expression [43]. PPARγ is a crucial regulator of lipid storage in the adipose tissue, and glucose and lipid metabolism. Moreover, PPARγ activation also enhances insulin sensitivity by increasing the production of adiponectin, which is reported to decline in pregnant diabetic rats [43, 44].

18.3.4 Saussurea Lappa

Saussurea lappa (Asteraceae) is an important medicinal plant, used in the Indian traditional medicinal system. *S. lappa* is widely distributed in the Himalayan region and is indigenous to India, Pakistan, and China. It is a tall perennial herb that grows up to 1–2 m in height and is commonly known as "Kushtam", "Kuth", or "Costus". Therapeutically, the most utilized part of the plant is its long stout root that has a characteristic pungent smell and a bitter taste. The use of *S. lappa* has been reported for the treatment of about 43 diseases in traditional medicinal systems, propelling research towards its phytochemical constitution, pharmacological uses, and mechanism of action [45]. The plant presents a high terpene concentration and other bioactive constituents such as anthraquinones, flavonoids, and alkaloids. Costus oil is an essential oil, which is extracted from *S. lappa* roots, with dehydrocostus lactone being its main constituent that accounts for 46.75%. Costunolide (9.26%), a sesquiterpene lactone, is another important compound found in costus oil [45, 46]. *S. lappa* extracts in various solvents have been reported to exhibit an array of pharmacological activities such as hypoglycemic, hypolipidemic, anti-cancer, anti-inflammatory, anti-hepatotoxic, immunomodulatory, and anti-ulcer properties [47–49].

Chaturvedi and co-workers have reported reduced blood glucose levels in the alcoholic root extract of *S. lappa* treated rats [50]. Another study performed by Raafat et al. to evaluate the effect of *S. lappa* treatment to gestational diabetics, has elucidated that the bioactive lactones extracted from the plant root of *S. lappa* did not cause hypoglycemia in non-diabetic pregnant rats but was effective in normalizing the blood glucose level in pregnant diabetic conditions. Moreover, with an aim to understand and explore the anti-GDM action of the *S. lappa* infrared assisted extracted bioactive lactones, they studied the insulin level and redox status, and reported that the root extract of *S. lappa* demonstrated insulin secretagogue potential and anti-oxidative activity. The *S. lappa*-infrared extract (SL-IR) obtained from the root showed enhancement in insulin secretion and anti-oxidative potential and thus exhibited protection against GDM [51]. *S. lappa* root extract has also been reported to stimulate the β-cells in STZ-induced diabetes to secrete insulin by inhibiting the expression of NO synthase and increase the uptake of glucose that helps in alleviating secretory defects during hyperglycemia [52].

18.3.5 Moringa Oleifera

Moringa oleifera, usually mentioned as "moringa" in the literature, belongs to the Moringaceae family. The plant is indigenous to the Indian subcontinent and grows naturally in the Himalayan foothill region, continues to northern West Bengal and extends to the terai tract of Uttar Pradesh region of India. Various names of *M. oleifera* can be found in multiple languages due to its introduction and naturalization in various countries around the world. It is commonly referred to as "the drumstick tree" in English language due to its characteristic long slender pod shape. In India, it is known from multiple vernacular names such as "shajan" in Hindi, and "munagakaya" in Telugu. Interestingly, the taxon name "moringa" is derived from its Tamil and Malayalam name "murunggi", or "muringa". *M. oleifera* pods are the most consumed edible part of the plant; additionally, seeds, flowers, and leaves are also edible and exhibit high nutritious value. *M oleifera* is referred to as the "miracle tree" because of its multifarious therapeutic properties [53]. It is used extensively and for diverse purposes in many countries including India, ranging from its use for culinary purposes to the treatment of diseases [53]. There is an overwhelming traditional and empirical use of *M. oleifera* in the Asian traditional medicinal system. Among many other disorders, the plant is widely used for the treatment and prevention of diabetes. Increasing recent evidence of its anti-GDM activity can be found in the scientific literature [54]. A study done by Adu et al. demonstrated the effects of aqueous seed extract of *M. oleifera* on the redox environment in the uterus and serum antioxidant status during gestational diabetes in a rat model, and reported increased serum SOD and catalase (CAT) activities when compared to normal rats. The researchers reported that the potent scavenging of peroxyl and superoxyl radicals by *M. oleifera* seed extract may be associated with its inducing effect on SOD and CAT expression. In addition to this, the seed extract of *M. oleifera* has also been reported to ameliorate the oxidative stress burden in the uterus of gestational diabetic rats [54]. Oxidative stress predisposes the mother to hyperglycemia by affecting the integrity of the pancreatic β-cells, creating an unfavorable intrauterine environment for fetal development and increasing the susceptibility for adverse effects on organogenesis. Interestingly, the increased oxidative stress markers and decreased levels of antioxidants in the uterus during diabetic pregnancy were seen to be ameliorated by *M. oleifera* seed extract treatment [54]. Furthermore, treatment also reduced the lipid peroxidation, determined by malondialdehyde content, increased glutathione (GSH) concentration, and glutathione-*S*-transferase (GST) activity in the uterus of pregnant diabetic rats [54].

The state of hyperglycemia during pregnancy adversely affects the fetal pancreatic development, which is essential for proper glucose homeostasis. The transcription factor pancreatic duodenal homebox-1 (PDX-1) is the vital regulator of pancreatic development and maintenance and regeneration of the β-cells. Importantly, PDX-1 also plays a major role in insulin gene transcription. Gestational diabetic rats exhibited down-regulated PDX-1 and insulin gene expression whereas treatment with *M. oleifera* ethanolic leaf extract up-regulated PDX-1 expression in diabetic pregnant rats, suggesting its positive effects in ameliorating maternal oxidative stress and improving fetal pancreatic organogenesis [55].

18.3.6 Morus Nigra

Morus nigra, commonly known as "black mulberry", belongs to the family Moraceae. Native to Southwestern Asia, *M. nigra* is a deciduous flowering plant, bearing edible fruits that are widely used for the preparation of wine, jams, marmalades, and vinegar. Owing to the nutraceutical properties of genus *Morus*, *M. nigra* has been extensively studied for its nutritive and pharmacological properties [56, 57]. Nutritionally, *M. nigra* is rich in essential minerals such as P, K, Ca, Mg, Fe, Mn, Zn, Cu, and also is a rich source of protein and vitamin C [58, 59]. The fruit and other parts of *M. nigra* elicit strong antioxidant and antimicrobial properties [60, 61]. The leaves have been documented to adhere antimicrobial, anti-nociceptive, anti-hyperlipidemic, and anti-diabetic activities; additionally, along with the leaves, the stem and bark extracts have anti-inflammatory potential and root bark extract exhibited anti-nociceptive activity [57].

The antioxidant and antimicrobial activities of *M. nigra* could be attributed to its high polyphenolic content, which is the highest among the three widely studied species of this genus including *M. alba* and *M. rubra*. *M. nigra* is recognized for its hypoglycemic and organ-protective activities, and as a consequence is used for the prevention of hepatic, renal, neurological, and metabolic diseases including diabetes mellitus. A study performed by Volpato et al. has demonstrated the positive effects of aqueous leaf extract of *M. nigra* on maternal-fetal outcome and lipid profile in pregnant diabetic rats [62]. The study reported anti-hypercholesterolemic and anti-hyperlipidemic effects of *M. nigra* leaf extract on GDM rats. With regard to the maternal-fetal outcome, *M. nigra* leaf extract supplementation to the gestational diabetic rats reduced placental index and weight. Furthermore, treatment with *M. nigra* leaf extract lowered the incidence of skeletal anomalies such as cleft palate, supernumerary rib, bipartite ossification of vertebral centrum, sternebra agenesis, incomplete ossification of sternebra, and abnormally shaped sternebra, and visceral anomalies such as microphthalmia, hydroureter, and hydronephrosis in fetuses of GDM rats when compared to the non-treated diabetic group [61, 62]. The diabetic condition is a state of increased free-radical generation, which has been implicated in fetal malformations in gestational diabetic rats. The decrease in the rate of fetal anomalies has been reported by treatment of *M. nigra* leaf extract, which may be associated with the presence of antioxidants in *M. nigra* leaf extract, acting against the increased oxidative stress in hypoglycemic conditions [62, 63].

18.3.7 Orthosiphon Stamineus

Orthosiphon stamineus is a perennial herb belonging to the Lamiaceae family. It grows in tropical and temperate regions particularly in Southeast Asia including India. Its extensive ethnopharmacological use has prompted studies on its phytochemical composition and therapeutic benefits. *O. stamineus* has gained importance in traditional medicinal practice due to its pharmacological potency in treating chronic ailments such as diabetes, hypertension, osteoporosis, microbial infections, and obesity [64–66]. Various extracts of *O. stamineus* contain three main types of phytochemicals: polymethoxylated flavonoids, phenylpropanoids, and terpenoids [67]. Traditionally, the plant is used in diabetic management; however, recently its possible use in treating GDM successfully has attracted researchers in this area. A study performed by Lokman et al. reported the ameliorative effects of *O. stamineus* aqueous leaf extract on gestational diabetic rats [66]. They reported that the treatment of *O. stamineus* leaf extract to pregnant diabetic rats ameliorated the detrimental effects of hyperglycemia on various biochemical parameters such as blood glucose, cholesterol, insulin, ghrelin, and glucagon-like peptide 1 (GLP-1). Further dose dependent studies reported that *O. stamineus* administration is non-toxic, even at quite high doses [66, 68].

Ghrelin and GLP-1 are the peptides involved in glucose regulation in mammals. Ghrelin has been associated with glucose homeostasis and insulin regulation, whereas GLP-1 activity stimulates insulin secretion by acting on the islets of Langerhans, and is speculated to be liable for decreased blood glucose level [69, 70]. Studies have proposed that *O. stamineus* extract affects these peptides and might be a factor in the anti-diabetic activity of this plant along with the rich polyphenolic content [66, 71].

18.3.8 Oxytenanthera Abyssinica

Oxytenanthera abyssinica, or "lowland bamboo", is one of the multitudinous varieties of bamboo, belonging to the family Gramineae and sub-family Bambusoideae. It is a tropical drought-resistant shrub and is capable of growing in any type of soil, except in clayey soil and salty soil, and can grow up to 6–10 m in height and 4–5 cm in diameter. The plant dies after flowering (once in its lifetime) and cannot regrow spontaneously in the same area.

The stem of *O. abyssinica* is used for furniture, paper, and in construction industries, whereas the leaves are medicinally apt and are used in the treatment of diabetes, rheumatism, and colics, edema, and polyuria; and the rhizomes are used to treat dysentery in traditional medicine practice in Africa and India [72, 73]. Additionally, the roots exhibit a potency to treat various skin disorders [72]. The studies performed to elucidate the phytochemical content of the plant have revealed the presence of flavonoids, alkaloids, polyphenols, saponins, and tannins in abundant amounts [72]. It has been assumed that the presence of various phytochemicals can be attributed to its therapeutic potential and promotes its use in folklore medicine practices. A study performed by Ezeja et al. demonstrated the anti-diabetic potential of *O. abyssinica* methanolic leaf extract on alloxan-induced diabetic mice [74] and reported that *O. abyssinica* leaf extract was able to reduce glucose level after an oral glucose tolerance test (OGTT) in treated mice. The anti-diabetic effects of *O. abyssinica* extrapolated the research on its ramifications on gestational diabetes. In 2013, Yessoufou and colleagues documented that the ethanolic leaf extract of *O. abyssinica* exhibited significant anti-hyperglycemic effects in diabetic pregnant rats, when injected intraperitoneally [75]. The study also documented substantial vitamin C content and considerable antioxidant capacity of butanolic, ethyl acetate, and aqueous extracts of *O. abyssinica* leaves; however, the mechanism by which *O. abyssinica* leaf extracts exhibited a hyperglycemic effect was uncertain [75]. A few proposed mechanisms behind the anti-diabetic potential of *O. abyssinica* are its anti-oxidative and immunosuppressive properties. Increased oxidative stress is implicated in the occurrence of degenerative diseases during pregnancy including gestational diabetes. The anti-oxidative action of *O. abyssinica* alleviates this stress, and thus is useful in the management and treatment of gestational diabetes [75].

18.3.9 Petroselinum Crispum

Petroselinum crispum is commonly known as "parsley" in European countries. In India it has acquired different names in various languages such as "ajmood" in Hindi, "achu-mooda" in Tamil, and "seema malli" in Malayalam. It belongs to the Apiaceae family. Parsley is a leafy, aromatic herb that is annual in tropical and subtropical climates, and biennial in a temperate climate. Endogenous to the Mediterranean region, parsley was introduced in India and is grown in Uttarakhand and Leh ladakh in northern parts and in Ooty, Tamil Nadu, in southern India.

Although used mainly for garnishing to add extra flavor in multiple cuisines around the world, *P. crispum* demonstrates high nutritious and multiple therapeutic properties. The ethnopharmacological use of *P. crispum* can be found in multiple cultures around the world. The plant exhibits pharmacological properties such as antioxidant, anti-diabetic, immunomodulatory, genitourinary, and cardiovascular activities [76]. *P. crispum* is one of the few plants that have been studied for their effects on GDM. A study on *P. crispum* has demonstrated the positive effects of aqueous leaf extract of this plant on placental and fetal morphology in terms of placental weight, placental index, and fetal weight on diabetic pregnant rats [77]. These three parameters were reported to be significantly altered in non-treated diabetic pregnant rats at the 7th day of gestation, while treatment with *P. crispum* aqueous extract restored these alterations closer to the normal in contrast to the non-treated diabetic mothers. The study also reported that *P. crispum* aqueous extract induced anti-hyperglycemic effects on diabetic pregnant rats. Hyperglycemia during pregnancy adversely affects fetal organogenesis. A histopathological study of fetal kidney tissue of GDM rats has shown decreased collagen, malformed or atrophied glomeruli, and development of fibroblasts, bleeding around the kidney, and delamination of the capsule, while *P. crispum* treatment partially ameliorated abnormalities such as collagen content, glomerular restoration, and fewer atrophies; no renal hemorrhage and capsular delimitation were reported in the treated group [77, 78].

The anti-diabetic activity of *P. crispum* has been attributed to its anti-oxidative potential. Flavonoids are major phytochemicals of this plant that are responsible for the anti-oxidative potential of this species. The leaves of *P. crispum* contain apigenin, cosmosiin, oxypeucedanin hydrate, apiin, and kaempferol in good amount [79, 80]. *P. crispum* seeds are a good source of essential oils and myristicin and apiol are the major reported constituents and contributors of the antioxidant capacity of the *P. crispum* seed essential oil [81].

18.3.10 Bridelia Ferruginea

The *Bridelia* genus constitutes of over 60 species, many of them native to Africa, Asia, and Australia, and are commonly grown in tropical and sub-tropical regions. *Bridelia* belongs to the Phyllanthaceae family. *Bridelia ferruginea* is one of the most studied species of this genus in reference to its pharmacological potency and ethnomedicinal uses. The plant is a gnarled shrub that can grow up to 6–15 m high under suitable conditions. Various parts of this plant such as fruits, leaves, bark, and roots are used in the preparation of decoctions. The decoction of the stem bark of *B. ferruginea* is empirically used in the treatment of gynecological disorders including sterility, while the leaf decoction is used in diabetes [82]. Fabyi and co-workers studied the phytochemical composition of the leaves of *B. ferruginea* and reported the presence of pentacyclic terpenoids, which can be associated with its high pharmacological potency and its extensive use in traditional medicinal practices [82].

B. ferruginea has been shown to have the potential to ameliorate pregnancy-induced glycemic imbalance. A study performed by Taiwo et al. reported the effects of aqueous leaf extract of *B. ferruginea* on the glucose level of pregnant rats. The study reported a higher glycemic level in pregnant rats when compared

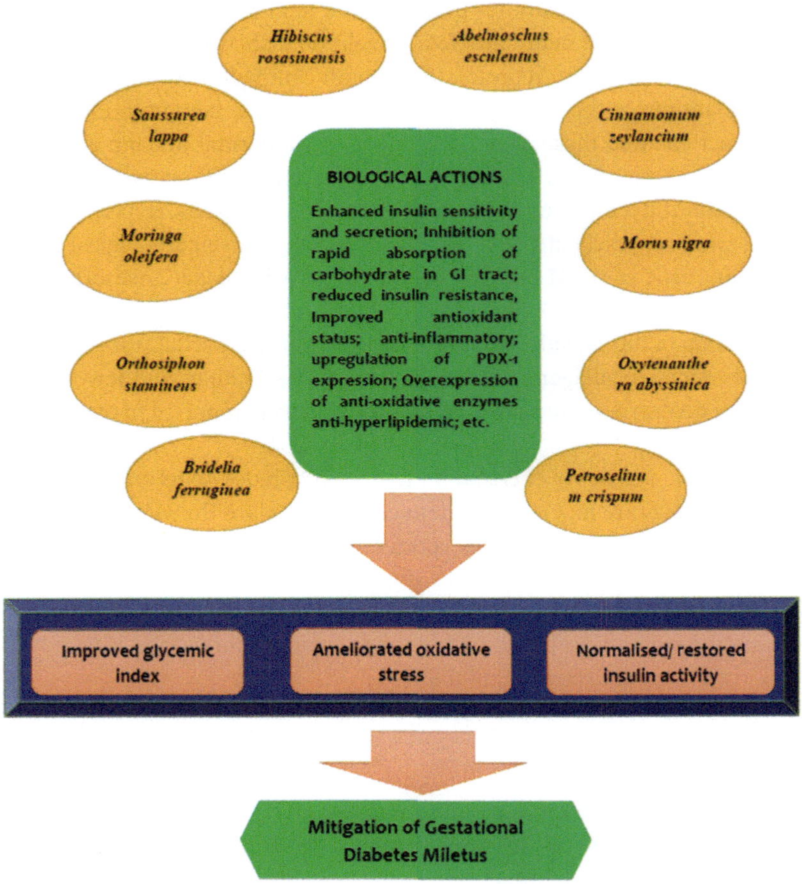

Figure 18.1 Diagrammatic representation of the role of MAPs in mitigating GDM. MAPs are responsible for generating a series of changes and ameliorating the adverse effects of gestational diabetes.

to non-pregnant rats. Pregnant rats fed with *B. ferruginea* aqueous leaf extract exhibited a lower peak in blood sugar level after 90 min of OGTT. Furthermore, the glycemic response index (GRI) of treated mothers was higher than the GRI of non-treated mothers [83]. A recent study conducted by Afolabi in 2018 reported the dose dependent inhibitory effects of *B. ferruginea* leaf extract on the activity of α- amylase and β-glucosidase – the enzymes responsible for carbohydrate metabolism – and provided evidence of anti-hyperglycemic action of *B. ferruginea*, which may be extended to GDM models in the future [84]. Figure 18.1 shows the major biological mechanisms elicited by the discussed MAPs in mitigating GDM consequences.

18.4 Conclusion

The rising rates of unhealthy lifestyles, obesity, and advanced maternal age have contributed to the increasing prevalence of GDM. The management of GDM is crucial to protect the mother and offspring from its deteriorative impact. Available synthetic drugs have direct and indirect side effects. In folklore medicine practices, MAPs have been used empirically for the treatment of chronic diseases such as diabetes since ancient times. Moreover, in low-income countries, gestational diabetic women self-medicate with known anti-diabetic local plants due to unaffordable drug costs. MAPs exhibit strong anti-hyperglycemic properties and many of them/their parts/products have shown potent protection against GDM and its symptoms. However, most of the studies performed are on animal models. More studies on human populations may provide an exact picture of the role of MAPs and GDM management.

Acknowledgments

JM and SKS acknowledge SHUATS, Allahabad, for providing support for the study. KBP is thankful to the director, CSIR-CSMCRI, Bhavnagar for encouragement.

References

1 International Diabetes Federation (2019). *IDF Diabetes Atlas*. 9th edn. Brussels, Belgium: International Diabetes Federation. (Accessed on March 08, 2023) https://diabetesatlas.org/upload/resources/material/20200302_133351_IDFATLAS9e-final-web.pdf

2 Armengaud JB, Ma RCW, Siddeek B, Visser GHA, Simeoni U. Offspring of mothers with hyperglycaemia in pregnancy: The short term and long-term impact. What is new? *Diabetes Res Clin Pract*. 2018; 145:155–166.

3 Nijs H, Benhalima K. Gestational diabetes mellitus and the long-term risk for glucose intolerance and overweight in the offspring: A narrative review. *J Clin Med*. 2020; 9:599.

4 Dornhorst A, Bailey PC, Anyaoku V, Elkeles RS, Johnston DG, Beard RW. Abnormalities of glucose tolerance following gestational diabetes. *Q J Med*. 1990; 77:1219–1228.

5 Coughlan MT, Vervaart PP, Permezel M, Georgiou HM, Rice GE. Altered placental oxidative stress status in gestational diabetes mellitus. *Placenta*. 2004; 25:78–84.

6 Melchior H, Kurch-Bek D, Mund M. The prevalence of gestational diabetes. *Dtsch Arztebl Int*. 2017; 114:412–418.

7 Dey L, Attele AS, Yuan CS. Alternative therapies for type 2 diabetes. *Altern Med Rev.* 2002; 7(1):45–58.

8 Hui H, Tang G, Go VL. Hypoglycemic herbs and their action mechanisms. *Chin Med.* 2009; 4:11.

9 Tiwari BK, Pandey KB, Abidi AB, Rizvi SI. Therapeutic potential of Indian medicinal plants in diabetic condition. *Ann Phytomed.* 2013; 2:37–43.

10 Gupta RC, Chang D, Nammi S, Bensoussan A, Bilinski K, Roufogalis BD. Interactions between antidiabetic drugs and herbs: An overview of mechanisms of action and clinical implications. *Diabetol Metab Syndr.* 2017; 9:59.

11 Grover JK, Yadav S, Vats V. Medicinal plants of India with anti-diabetic potential. *J Ethnopharmacol.* 2002; 81:81–100.

12 Mishra N, Pandey KB, Rizvi SI. Overview of novel herbal supplements in diabetes. In: Bagchi D, Preuss HG, Swaroop A, editors. *Nutraceuticals and Functional Foods in Human Health and Disease Prevention.* USA: CRC Press, Taylor & Francis Group, 2015.

13 Damasceno DC, Leal-Silva T, Soares TS, Moraes-Souza RQ, Volpato GT. Medicinal plants for diabetes treatment during pregnancy. *Curr Med Chem.* 2017; 24:404–410.

14 Balsells M, García-Patterson A, Solà I, Roqué M, Gich I, Corcoy R. Glibenclamide, metformin, and insulin for the treatment of gestational diabetes: A systematic review and meta-analysis. *BMJ.* 2015; 350:h102.

15 Bertoldo MJ, Faure M, Dupont J, Froment P. Impact of metformin on reproductive tissues: An overview from gametogenesis to gestation. *Ann Transl Med.* 2014; 2:55.

16 Rena G, Hardie DG, Pearson ER. The mechanisms of action of metformin. *Diabetologia.* 2017; 60(9):1577–1585.

17 Rowan JA, Hague WM, Gao W, Battin MR, Moore MP. MiG Trial investigators. Metformin versus insulin for the treatment of gestational diabetes. *N Engl J Med.* 2008; 359:106.

18 Dallaqua B, Saito FH, Rodrigues T, Calderon IMP, Rudge MVC, Herrera E, et al. Treatment with Azadirachta indica in diabetic pregnant rats: Negative effects on maternal outcome. *J Ethnopharmacol.* 2012; 143:805–811.

19 Durazzo A, Lucarini M, Novellino E, Souto EB, Daliu P, Santini A. Abelmoschus esculentus (L.): Bioactive components' beneficial properties focused on antidiabetic role for sustainable health applications. *Molecules.* 2018; 24:38.

20 Bawa SH, Badrie N. Nutrient profile, bioactive components, and functional properties of okra (Abelmoschus esculentus (L.) Moench). In: Watson RR, Preedy VR, editors. *Fruits, Vegetables, and Herbs: Bioactive Foods in Health Promotion.* Cambridge, MA, USA: Academic Press, 397; 2016.

21 Arapitsas P. Identification and quantification of polyphenolic compounds from okra seeds and skins. *Food Chem.* 2008; 110:1041–1045.

22 Sabitha V, Ramachandran S, Naveen KR, Panneerselvam K. Antidiabetic and antihyperlipidemic potential of Abelmoschus esculentus (L.) Moench. in streptozotocin-induced diabetic rats. *J Pharm Bioallied Sci.* 2011; 3:397–402.

23 Durazzo A, Lucarini M, Novellino E, Souto EB, Daliu P, Santini A. Abelmoschus esculentus (L.): Bioactive components' beneficial properties—Focused on antidiabetic role—For sustainable health applications. *Molecules*. 2019; 24(1):38.

24 Tian ZH, Miao FT, Zhang X, Wang QH, Lei N, Guo LC. Therapeutic effect of okra extract on gestational diabetes mellitus rats induced by streptozotocin. *Asian Pac J Trop Med*. 2015; 8:1038–1042.

25 Okada Y, Okada M, Sagesaka Y. Screening of dried plant seed extracts for adiponectin production activity and tumor necrosis factor-alpha inhibitory activity on 3T3-L1 adipocytes. *Plant Foods Hum Nutr*. 2010; 65:225–232.

26 Ahmed BT, Kumar SA. Antioxidant and antidiabetic properties of Abelmoschus esculentus extract – an in vitro assay. *Res J Biotechnol*. 2016; 11:34–41.

27 Thanakosai W, Phuwapraisirisan P. First identification of α-glucosidase inhibitors from okra (Abelmoschus esculentus) seeds. *Nat Prod Commun*. 2013; 8:1085–1088.

28 Khatun H, Rahman A, Biswas M, Islam AU. Water-soluble fraction of Abelmoschus esculentus L interacts with glucose and metformin hydrochloride and alters their absorption kinetics after coadministration in rats. *ISRN Pharm*. 2011; 2011:260537.

29 Shanmugasundaram ER, Sundaram P, Srinivas K, Shanmugasundaram KR, Shankaran JR. Double-blind cross-over study of modified Anna Pavala Sindhooram in patients with hyperlipidemia or ischemic heart disease. *J Ethnopharmacol*. 1991; 31(1):85–99.

30 Gauthaman KK, Saleem MT, Thanislas PT, Prabhu VV, Krishnamoorthy KK, Devaraj NS, et al. Cardioprotective effect of the Hibiscus rosa sinensis flowers in an oxidative stress model of myocardial ischemic reperfusion injury in rat. *BMC Complement Altern Med*. 2006; 6:1–8.

31 Pillai SS, Mini S. Hibiscus rosa sinensis Linn. petals modulates glycogen metabolism and glucose homeostasis signalling pathway in streptozotocin-induced experimental diabetes. *Plant Foods Hum Nutr*. 2016; 71:42–48.

32 Al-Snafi AE. Chemical constituents, pharmacological effects and therapeutic importance of Hibiscus rosa-sinensis-A review. *J Pharm*. 2018; 8:101–119.

33 Afiune LAF, Leal-Silva T, Sinzato YK, Moraes-Souza RQ, Soares TS, Campos KE, et al. Beneficial effects of Hibiscus rosa-sinensis L. flower aqueous extract in pregnant rats with diabetes. *PLoS One*. 2017; 12:e0179785.

34 Volpato GT, Afiune LA, Leal-Silva T, Campos KE, Damasceno DC. Reproductive and fetal repercussions of treatment with an aqueous extract of flowers of Hibiscus rosa sinensis in diabetic pregnant rats. *Placenta*. 2015; 36:513.

35 Pupim, L. B., Heimbürger, O., Qureshi, A. R., Ikizler, T. A. and Stenvinkel, P. (2005). Accelerated lean body mass loss in incident chronic dialysis patients with diabetes mellitus. *Kidney Int*. 68(5): 2368–2374. https://doi.org/10.1111/j.1523-1755.2005.00699.x

36 Silva TL, Afiune LAF, de Moraes Souza RQ, de Sousa Soares T, Carneiro TB, Américo MF, de Campos KE, Volpato GT. Effect of Hibiscus rosa sinensis aqueous extract treatment on biochemical parameters in diabetic pregnant rats. *Diabetol Metab Syndr.* 2015; 7(Suppl 1):A77.

37 Ranasinghe P, Pigera S, Premakumara GA, Galappaththy P, Constantine GR, Katulanda P. Medicinal properties of 'true' cinnamon (Cinnamomum zeylanicum): A systematic review. *BMC Complement Altern Med.* 2013; 13:275.

38 Gruenwald J, Freder J, Armbruester N. Cinnamon and health. *Crit Rev Food Sci Nutr.* 2010; 50:822–834.

39 Paranagama PA, Wimalasena S, Jayatilake GS, Jayawardena AL, Senanayake UM, Mubarak AM. A comparison of essential oil constituents of bark, leaf, root and fruit of cinnamon (Cinnamomum zeylanicum Blum) grown in Sri Lanka. *J Natl Sci Foundation Sri Lanka.* 2001; 29:147–153.

40 Mallavarapu GR, Ramesh S, Chandrasekhara RS, Rao BR, Kaul PN, Bhattacharya A. Investigation of the essential oil of cinnamon leaf grown at Bangalore and Hyderabad. *Flavour Fragr J.* 1995; 10:239–242.

41 Jayaprakasha GK, Rao LJ, Sakariah KK. Chemical composition of the volatile oil from the fruits of Cinnamomum zeylanicum Blume. *Flavour Fragr J.* 1997; 12:331–333.

42 Jayaprakasha GK, Jagan Mohan Rao L, Sakariah KK. Chemical composition of the flower oil of Cinnamomum zeylanicum blume. *J Agric Food Chem.* 2000; 48:4294–4295.

43 Hosni AA, Abdel-Moneim AA, Abdel-Reheim ES, Mohamed SM, Helmy H. Cinnamaldehyde potentially attenuates gestational hyperglycemia in rats through modulation of PPARγ, proinflammatory cytokines and oxidative stress. *Biomed Pharmacother.* 2017; 88:52–60.

44 Feige JN, Gelman L, Michalik L, Desvergne B, Wahli W. From molecular action to physiological outputs: Peroxisome proliferator-activated receptors are nuclear receptors at the crossroads of key cellular functions. *Prog Lipid Res.* 2006; 45:120–159.

45 Zahara K, Tabassum S, Sabir S, Arshad M, Qureshi R, Amjad MS, et al. A review of therapeutic potential of Saussurea lappa - An endangered plant from Himalaya. *Asian Pac J Trop Med.* 2014; 7:S60–9.

46 Dhillon RS, Kalsi PS, Singh WP, Gautam VK, Chhabra BR. Guaianolide from Saussurea lappa. *Phytochemistry.* 1987; 26:1209–1210.

47 Lee MG, Lee KT, Chi SG, Park JH. Costunolide induces apoptosis by ROS-mediated mitochondrial permeability transition and cytochrome C release. *Biol Pharm Bull.* 2001; 24(3):303–306.

48 Gokhale AB, Damre AS, Kulkami KR, Saraf MN. Preliminary evaluation of anti-inflammatory and anti-arthritic activity of S. lappa, A. speciosa and A. aspera. *Phytomedicine.* 2002; 9:433–437.

49 Yaeesh S, Jamal Q, Shah AJ, Gilani AH. Antihepatotoxic activity of Saussurea lappa extract on D-galactosamine and lipopolysaccharide-induced hepatitis in mice. *Phytother Res.* 2010; 24:S229–32.

50 Chaturvedi P, Tripathi P, Pandey S, Singh U, Tripathi YB. Effect of Saussurea lappa alcoholic extract on different endocrine glands in relation to glucose metabolism in the rat. *Phyther Res.* 1993; 7:205–207.

51 Raafat K, El-Darra N, Saleh F, Rajha HN, Louka N. Optimization of infrared-assisted extraction of bioactive lactones from Saussurea lappa L. and their effects against gestational diabetes. *Pharmacogn Mag.* 2019; 15:208–218.

52 Li WL, Zheng HC, Bukuru J, De Kimpe N. Natural medicines used in the traditional Chinese medical system for therapy of diabetes mellitus. *J Ethnopharmacol.* 2004; 92(1):1–21.

53 Mahmood KT, Mugal T, Haq IU. Moringa oleifera: A natural gift-a review. *J Pharm Sci Res.* 2010; 2:775–781.

54 Adu F, Yusuf J, Ajao FO, Odedele LO. Effect of aqueous moringa seed extract on oxidative stress in Alloxan- induced gestational diabetic rats. *Med Res Arch.* 2017; 5.

55 Olabanji OJ, Paul CW, Chinnah TI. Effect of Moringa oleifera leaf extract on foetal pancreatic development in streptozotocin-induced diabetic wistar rats. *Int J Innov Sci Res Tech.* 2020; 5(2):886–890.

56 Miljković M, Nikolić S, Nikolić B, Arsić B. Morus species through centuries in pharmacy and as food. *Adv Technol.* 2014; 3:111–115.

57 Lim SH, Choi CI. Pharmacological properties of Morus nigra L. (Black Mulberry) as a promising nutraceutical resource. *Nutrients.* 2019; 11:437.

58 Ercisli S, Orhan E. Chemical composition of white (Morus alba), red (Morus rubra) and black (Morus nigra) mulberry fruits. *Food Chem.* 2007; 103:1380–1384.

59 Sánchez-Salcedo EM, Mena P, García-Viguera C, Martínez JJ, Hernández F. Phytochemical evaluation of white (Morus alba L.) and black (Morus nigra L.) mulberry fruits, a starting point for the assessment of their beneficial properties. *J Funct Foods.* 2015; 12:399–408.

60 Khalid N, Fawad SA, Ahmed I. Antimicrobial activity, phyto chemical profile and trace minerals of black mulberry (morus nigra l) fresh juice. *Pakistan J Bot.* 2011; 43:91–96.

61 Rodrigues EL, Marcelino G, Silva GT, Figueiredo PS, Garcez WS, Corsino J, et al. Nutraceutical and medicinal potential of the Morus species in metabolic dysfunctions. *Int J Mol Sci.* 2019; 20:301.

62 Volpato GT, Calderon IM, Sinzato S, Campos KE, Rudge MV, Damasceno DC. Effect of Morus nigra aqueous extract treatment on the maternal-fetal outcome, oxidative stress status and lipid profile of streptozotocin-induced diabetic rats. *J Ethnopharmacol.* 2011; 138(3):691–696.

63 Reece EA, Eriksson UJ. Congenital malformations: Epidemiology, pathogenesis, and experimental methods of induction and prevention. In: Reece EA, Coustan DR, Gabbe SG, editors. *Diabetes in Women: Adolescent, Pregnancy, and Menopause.* Philadelphia: Lippincott Williams & Wilkins, 2004; pp. 1–8.

64 Seyedan A, Alshawsh MA, Alshagga MA, Mohamed Z. Antiobesity and lipid lowering effects of Orthosiphon stamineus in high-fat diet-induced obese mice. *Planta Med.* 2017; 83(8):684–692.

65 Bokhari RA, Lau SF, Mohamed S. Orthosiphon stamineus (Misai Kucing) ameliorated postmenopausal osteoporosis in rat model. *Menopause.* 2018; 25(2):202–210.

66 Lokman EF, Saparuddin F, Muhammad H, Omar MH, Zulkapli A. Orthosiphon stamineus as a potential antidiabetic drug in maternal hyperglycemia in streptozotocin-induced diabetic rats. *Integr Med Res.* 2019; 8:173–179.

67 Ameer OZ, Salman IM, Asmawi MZ, Ibraheem ZO, Yam MF. Orthosiphon stamineus: Traditional uses, phytochemistry, pharmacology, and toxicology. *J Med Food.* 2012; 15:678–690.

68 Chin JH, Abas HH, Sabariah I. Toxicity study of Orthosiphon stamineus Benth (Misai Kucing) on Sprague Dawley rats. *Trop Biomed.* 2008; 25:9–16.

69 Adeghate E, Ponery AS. Ghrelin stimulates insulin secretion from the pancreas of normal and diabetic rats. *J Neuroendocrinol.* 2002; 14(7):555–560.

70 Alamri BN, Shin K, Chappe V, Anini Y. The role of ghrelin in the regulation of glucose homeostasis. *Horm Mol Biol Clin Investig.* 2016; 26:3–11.

71 Nadkarni P, Chepurny OG, Holz GG. Regulation of glucose homeostasis by GLP-1. *Prog Mol Biol Transl Sci.* 2014; 121:23–65.

72 Ibeh BO, Maxwell E, Bitrus HJ. Phytochemical compositions and in vitro antioxidant capacity of methanolic leaf extract of Axonopus Compressus (P. Beauv.). *European J Med Plants.* 2013; 3:254–265.

73 Kaikini AA, Dhande SR, Kadam VJ. Overview of Indian medicinal tree: Bambusa Bambos (Druce). *Int Res J Pharm.* 2013; 4:52–56.

74 Ezeja MI, Omeh YS, Mbagwu C. Antidiabetic potentials of the methanol leaf extract of Oxytenanthera abyssinica. *Int J Diabetes Dev Ctries.* 2014; 34:116–120.

75 Yessoufou A, Gbenou J, Grissa O, Hichami A, Simonin A-M, Tabka Z. Anti-hyperglycemic effects of three medicinal plants in diabetic pregnancy: Modulation of T cell proliferation. *BMC Complement Altern Med.* 2013; 13:77.

76 Farzaei MH, Abbasabadi Z, Ardekani MR, Rahimi R, Farzaei F. Parsley: A review of ethnopharmacology, phytochemistry and biological activities. *J Tradit Chin Med.* 2013; 33:815–826.

77 Rabou MAA, Ahmed Eid F. Possible protective role of parsley extract on the diabetic pregnant rats and their fetuses. *Pak J Biol Sci.* 2017; 20:552–562.

78 Rabou MAA. Therapeutic effects of parsley extract in streptozotocin-induced gestational diabetic rats. *J Biol Sci.* 2018; 18:364–372.

79 Chaves DS, Frattani FS, Assafim M, de Almeida AP, de Zingali RB, Costa SS. Phenolic chemical composition of Petroselinum crispum extract and its effect on haemostasis. *Nat Prod Commun.* 2011; 6:961–964.

80 Gadi D, Bnouham M, Aziz M, Ziyyat A, Legssyer A, Bruel A, et al. Flavonoids purified from parsley inhibit human blood platelet aggregation and adhesion to collagen under flow. *J Complement Integr Med.* 2012; 9:19.

81 Zhang H, Chen F, Wang X, Yao H-Y. Evaluation of antioxidant activity of parsley (Petroselinum crispum) essential oil and identification of its antioxidant constituents. *Food Res Int.* 2006; 39:833–839.

82 Fabiyi OA, Atolani O, Olatunji GA. Identification of some chemical constituents of the leaves of Alstonia boonei and Bridelia ferruginea. *Selcuk J Agric and Food Sci.* 2016; 30:21–28.

83 Taiwo IA, Adewumi OO, Odeigah PG. Assessment of Bridelia ferruginea benth for its therapeutic potential in pregnancy-induced impaired glucose tolerance in rats. *Int J Med Biomed Res.* 2012; 1:49–55.

84 Afolabi OB, Oloyede OI, Agunbiade SO. Inhibitory potentials of phenolic-rich extracts from Bridelia ferruginea on two key carbohydrate-metabolizing enzymes and Fe2+-induced pancreatic oxidative stress. *J Integr Med.* 2018; 16:192–198.

19

Enhancement of Insulin Sensitivity and Management of Lipid Disorders during Gestational Diabetes Mellitus

Role of Capsaicin

Banu Orta-Yilmaz[1,*], *Ahu Korkut*[2], *and Yasemin Aydin*[1]

[1] *Department of Biology, Faculty of Science, Istanbul University, 34134, Vezneciler, Fatih, Istanbul, Turkey*
[2] *Department of Obstetrics and Gynecology, Faculty of Medicine, Eskisehir Osmangazi University, Turkey*
[*] *Corresponding author*

19.1 Introduction

Gestational diabetes mellitus (GDM) is a metabolic abnormality that occurs during pregnancy in women. It is also defined as carbohydrate intolerance, covering 18% of pregnancies [1, 2]. In the majority of pregnant women with GDM, disorders such as spontaneous abortion, hyperglycemia, urinary tract infection, hypoglycemia, and chronic anemia are observed, as well as effects such as congenital anomalies, developmental disorders, and premature birth in the fetus [3]. It was observed that the development of GDM in pregnant individuals, whose lipid, amino acid, and glucose metabolisms were examined, caused a decrease in maternal insulin sensitivity by creating an insufficient insulin response. It has been observed that decreased maternal insulin sensitivity causes an increase in the nutrient requirement of the fetus, resulting in an increased risk of fetal overgrowth and adiposity [4, 5]. In a Canadian survey, which included gynecologists, obstetricians, and maternal-fetal medicine specialists, the risk of preterm birth caused by changes in insulin requirements of expectant mothers with GDM was evaluated. As a result of the survey conducted with 142 pregnant female participants who did not have any common features other than having GDM, it was stated that decreased insulin sensitivity does not cause preterm birth without definitive evidence [6]. A study of 2639 female participants in China investigated the relationship between diet-related inflammation and the risk of GDM. In the study, the diagnosis of GDM was determined by a glucose tolerance test

Natural Products and their Bioactives in Antidiabetic Drug Discovery, First Edition.
Edited by Kanti Bhooshan Pandey, Maitree Suttajit and Pinar Atukeren.

performed on 24–28-week pregnant women. GDM was observed in 13.1% of 2639 women, and the dietary inflammatory index was observed between 3.15 and 4.45. According to the data obtained, it has been determined that women with pre-pregnancy obesity or obesity predisposition, and a high dietary inflammatory index are associated with GDM [7].

Research on GDM has tended to include natural herbal supplements in this process in order for mothers to have a comfortable pregnancy and to prevent the risk of premature birth for babies. In recent years, capsaicin in hot peppers has been extensively studied in GDM patients, as it has many pharmacological properties such as anticancer, anti-inflammatory, analgesia, and anti-obesity actvities, as well as being effective in improving human glucose and lipid disorders [8]. Studies show that capsaicin reduces postprandial blood sugar and has positive effects on insulin resistance (IR) [2]. In this chapter, we discuss to what extent capsaicin can be effective in the management of insulin sensitivity and lipid disorders in GDM.

19.2 Gestational Diabetes Mellitus

GDM refers to abnormal glucose tolerance detected for the first time in pregnancy. GDM is diabetes that is diagnosed after the 20th week of pregnancy and does not exist in pregestation. GDM is one of the common complications of pregnancy and its prevalence is 7.8% of pregnant women in the United States of America and 2–38% of pregnant women in other countries [9, 10]. Prevalence increases with an increase in mother's age and body mass index (BMI). It is important to scan and detect GDM in order to significantly reduce the negative perinatal results and the risk of death at birth by treatment.

19.2.1 Pathophysiology

GDM is the result of β-cell dysfunction in the background of chronic IR in pregnancy. Pancreatic β-cells secrete and store insulin in response to serum glucose concentration. Disorders occurring at any stage of the β-cell mechanism: proinsulin synthesis, post-translational modifications, insulin storage, sensing of serum glucose concentration, and insulin secretion cause β-cell dysfunction. In addition, GDM-related genes (potassium voltage-gated channel KQT-like 1 and glucokinase) are associated with β-cell function. In animal studies, hyperglycemia was observed in rats undergoing pancreatectomy, despite improvement in β-cell mass within a week, and a relationship between β-cell number and mass and GDM was demonstrated [11]. IR occurs when the biological response of target cells to insulin is reduced. At the molecular level, IR is the failure of insulin signal transduction. Thus, the primary transporter responsible for bringing glucose into the cell for energy production, glucose transporter type 4 (GLUT4), causes insufficient plasma membrane translocation. Binding of insulin to the insulin receptor increases autophosphorylation of tyrosine residues and insulin receptor substrate-1 (IRS-1), which is the first mandatory step that allows activation. Altered expression or phosphorylation of insulin signal transduction regulators such as IRS-1, phosphatidylinositol 3 kinase, and GLUT4 has been observed in GDM [12]. Cell glucose uptake decreased by 54% with insulin stimulation in GDM.

Neurohormonal dysfunction has also played a role in the pathogenesis of IR. The most important neurohormonal metabolic pathway regulators are leptin and adiponectin, which are cell signaling proteins secreted by adipose tissue. During pregnancy, leptin is secreted by the placenta, and in the case of placental IR, leptin secretion from the placenta increases. A state of hyperleptinemia occurs and ultimately contributes to fetal macrosomia by facilitating amino acid transport across the placenta [13]. GDM is associated with a low adiponectin concentration [14]. Adiponectin increases fatty acid oxidation, inhibits gluconeogenesis, and increases insulin secretion by regulating insulin gene expression and insulin exocytosis, regardless of obesity [15]. Adiponectin's secretion is regulated by cytokines such as tumor necrosis factor alpha (TNF-α), interleukin -6 (IL-6), interferon gamma (IFN-γ), and leptin. Adiponectin limits fetal macrosomia by disrupting amino acid transport across the placenta [16]. Adipose tissue secretes cytokines such as TNF-α, IL-6, interleukin 1β (IL-1β) in addition to adipokines (leptin and adiponectin), and plays an important role in the formation of the inflammatory response. These cytokines induce IR by decreasing tyrosine kinase activity on IRS-1 and increasing serine phosphorylation [17, 18]. The concentration of proinflammatory cytokines is increased in GDM [19, 20]. Adipose tissue not only stores excess calories but also releases fatty acids, which contributes to insulin resistance. An obese woman with GDM has metabolic inflammation.

19.2.2 Short- and Long-term Consequences of GDM

GDM has been associated with increased risks of short- and long-term perinatal and postnatal adverse outcomes. In the short term, these outcomes are hypertensive diseases of pregnancy, macrosomic babies, large for gestational age (LGA) newborns, operative delivery, preterm birth, perinatal mortality, fetal/neonatal hypertrophic cardiomyopathy, polyhydramnios, and neonatal respiratory and metabolic problems [21, 22]. A small increase in congenital anomalies has been observed but is mostly associated with T2DM or obesity [23, 24]. In the long term, maternal T2DM and cardiovascular disease development increased. The risk of impaired glucose tolerance, obesity, hypertension, and metabolic syndrome is increased in adolescence and adulthood in children [25, 26].

19.3 Management of GDM

Good glycemic control is the cornerstone of GDM management by reducing the severity and frequency of maternal (preeclampsia) and neonatal (macrosomia, shoulder dystocia) complications [27]. Treatment includes blood glucose monitoring, diet, exercise, and medical therapy; however, in general it is observed that conventional medicines are not very effective in balancing glucose during pregnancy and the related side effects are another matter of concern. Advanced evidence-based research in the field of natural product therapy in the last two decades has increased interest in natural product applications in the management and treatment of health ailments, which is evidenced by enhanced dependency on a Mediterranean type of dietary pattern all over the world. A diet full of leafy vegetables, fruits, and cereals, followed by exercise has shown decreased triglyceride levels, IR, postprandial glucose values, weight gain, and blood pressure values [28–30]. In this chapter, the effect of capsaicin in influencing insulin sensitivity and lipid profile, thereby managing GDM, is discussed.

19.4 Capsaicin and Its Role in the Management of GDM

Capsaicin is the active ingredient obtained from *Capsicum*, which is the most consumed pepper type in the world and provides bitterness to hot peppers. Due to its sharp aroma, it is responsible for protecting the plant against external attacks. Capsaicin ($C_{18}H_{27}NO_3$) is a highly volatile, hydrophobic bioactive compound, which is included in the capsaicinoid group, and also includes dihydrocapsaicin, nordihydrocapsaicin, homocapsaicin, and homodihydrocapsaicin; capsaicin constitutes 70% of capsaicinoids [8, 31]. It is a vanillic acid derivative, interacts with specific receptors, similar to other vanilloids, and causes changes in the metabolic activities of cells. Receptors to which capsaicin specifically binds are called capsaicin receptor or transient receptor potential channel vanilloid type-1 (TRPV1) [8]. In numerous human and animal experiments, it has been revealed that TRPV1 plays a regulatory role in all body systems, especially in weight control, regulation of carbohydrate, protein, and fat metabolism, and the nervous system [32]. The activity of this receptor is directly related to the secretion of neuropeptides in sensory neurons. Accordingly, as a result of capsaicin binding to the TRPV1 receptor, Na^+, Ca^{+2}, and K^+ ion flow in the cell changes and depolarization and neurotranspeptide release are observed in neurons [33, 34]. Although most of the effects of capsaicin on systems are thought to be mediated by the TRPV1 receptor, studies have confirmed that capsaicin has a receptor-independent mechanism of action. In addition to stimulating TRPV1 receptor activation, capsaicin has been linked to nuclear factor-kappa B (NF-κB) inactivation and proliferation-activating receptor-γ (PPAR γ) activation in a mouse obesity study [35].

Many studies have shown that capsaicin administrations reduce tissue damage, decrease blood pressure, increase salivary and gastric secretion, and increase intestinal activity. In addition, in clinical studies of various diseases such as atherosclerosis, diabetic vasculopathy, stroke, angina, chest pain, hypertension, and gastric ulcer, including the treatment of rheumatoid arthritis and vasomotor rhinitis in humans, capsaicin administration has been shown to have beneficial effects [36–38]. A meta-analysis of human studies suggests that capsaicin may be helpful in weight control. According to a study conducted with obese mice, capsaicin administration was shown to increase the expression of adiponectin and adipokine, thereby reducing fat accumulation in adipose tissue [39]. In a study examining the effects of capsaicin in individuals with GDM, findings such as serum insulin, fasting plasma glucose, etc. in pregnant individuals were examined. As a result of the study, it was found that capsaicin has therapeutic effects on lipid metabolism, glucose, and serum insulin levels [2]. In another study, it was shown that glucose intolerance caused by obesity decreases after capsaicin administration both by suppressing inflammatory responses and increasing fatty acid oxidation in adipose tissues and the liver, which are important peripheral tissues that affect insulin resistance [40]. As a result, clinical studies are continuing for different usage areas of capsaicin, a pharmacological agent with very important effects (Table 19.1).

Table 19.1 Clinical effects of capsaicin in metabolic diseases.

Animal model/human	Metabolic disorders	Clinical effects
Rats	Gastric ulceration	Decreased acid secretion, increased alkalinity, and increased gastric blood flow [41, 42].
C57BL/6 J wild-type mice and TRPV1 knockout (TRPV1−/−) mice	Fatty liver	With the stimulation of uncoupling protein 2 (UCP2) in hepatocytes, decreased fatty inflammation and increased glucagon-like peptide-1 (GLP-1) secretion occur [43].
Turkey poults	Damaged lipid metabolism	Decrease in total cholesterol, low-density lipoprotein (LDL-cholesterol), and high-density lipoprotein (HDL-cholesterol) levels [44].
Human	Pain	Reduces neuropathic pain By eliminating the sensitivity of primary sensory neurons, it has a therapeutic effect on postherpetic neuropathy, diabetic, and other neuropathic pain [42, 45–47].

Table 19.1 (Continued)

Animal model/human	Metabolic disorders	Clinical effects
C57BL/6 mice	Visceral obesity	Capsaicin limits the accumulation of fat in adipose tissues and reduces inflammation and improves insulin sensitivity [39].
Diabetic Rats	Diabetes	Reduces high blood glucose [41].
Women with GDM	Gestational diabetes	Regularly improved fasting lipid metabolic disorders with postprandial hyperglycemia and hyperinsulinemia [2, 42].
Ob/ob mice	Diabetic obesity	It contributes to improved glucose homeostasis by increasing short-chain fatty acids, regulating gastrointestinal hormones and inhibiting proinflammatory cytokines. However, it slows down weight gain [48].
Hamster	Gallbladder stone	Significantly reducing the cholesterol/ phospholipid ratio. Prevents gallstone formation [42].

19.4.1 Therapeutic Potential of Capsaicin in Insulin Sensitivity

The most basic factor underlying GDM is the increase in IR caused by the decrease in maternal insulin sensitivity. Decreased insulin sensitivity causes serious irreversible health problems in the immune system, brain functions, and reproductive system. Studies conducted with capsaicin from the past to the present have revealed the curative effect of capsaicin on metabolic disorders such as obesity, IR, impaired glucose tolerance, and hypertension [1]. In particular, it has been observed that capsaicin reduces fasting and postprandial blood sugar and improves insulin and glucose tolerance in diabetes caused by increased blood sugar in obesity. Considering the life histories of women with GDM, it was observed that they had a history of obesity or a family history of obesity, their lifestyles were devoid of physical activity, and healthy nutrition was not included. The development of GDM is inevitable with decreased insulin sensitivity in women with one or more of these features. At this point, capsaicin plays a role in blood sugar regulation by acting on insulin [41].

TRPV1 protein, which is expressed in liver, heart, pancreas and kidneys, is the receptor for capsaicin [49, 50]. The TRPV1 receptor activated by capsaicin is involved in targeting the metabolic syndrome present in the body. TRPV1 is directly activated by high temperatures and capsaicin. Activated and pancreatic TRPV1 play a role in the regulation of serum glucose levels by releasing insulin from β-cells in the islet of Langerhans. As insulin released from β-cells sensitizes

TRPV1 in sensory nerve endings, the release of sensory neuropathies increases. Increased sensory neuropathies' negative feedback is provided by inducing the calcitonin gene related peptide (CGRP) to decrease insulin secretion.

In a study, the possible effects of capsaicin on the blood of streptozotocin (STZ)-induced rats were investigated. As a result of the examinations, it was determined that the rats fed with capsaicin for three weeks decreased their blood sugar level compared to the diabetic control group. However, it was observed that diabetic rats fed with capsaicin for four weeks decreased blood glucose levels in the glucose tolerance test. However, it was determined that there was no significant decrease in the decreased blood glucose level compared to the diabetic rat group. According to the researchers, it is thought that possible increased insulin sensitivity can be improved with capsaicin [51]. In another study, it was determined that capsaicin reduced body weight, fat accumulation, serum leptin levels, and glucose tolerance in rats as a result of experiments performed on diabetic rats [52]. In a study by Yuan et al. [2], the aim was to reveal the effect of capsaicin treatment on blood sugar and possible conditions that may occur during pregnancy in women with GDM. In the study, 44 expectant mothers with gestational diabetes at 22–33 weeks of pregnancy were evaluated. According to the data obtained as a result of the experiment, it was observed that capsaicin has a positive effect on postprandial hyperglycemia and hyperinsulinemia and reduces the incidence of large newborns for gestational age. In various clinical studies, capsaicin has been shown to be used for therapeutic purposes in metabolic abnormalities due to its antioxidant properties. In this context, in a study aiming to determine the inhibition of capsaicin on aldose reductase enzyme activity obtained from different rat tissues, it was determined that the use of low doses of capsaicin would be effective for the prevention of diabetic complications during pregnancy [53].

19.4.2 Advances in the Curative Effects of Capsaicin on Lipid Disorders

Lipids such as cholesterol, cholesterol esters, and triglycerides are carried in the circulation by lipid carriers. Chylomicrons, lipoprotein(a), high-density lipoprotein (HDL), low-density lipoprotein (LDL), very low-density lipoprotein (VLDL), and intermediate-density lipoproteins (IDL) are lipoproteins found in plasma that act as lipid transporters. Studies show that changes in circulating total cholesterol (TC) and triglyceride concentrations are associated with changes in lipoprotein levels [54]. As a result of changes in cholesterol and lipoprotein fractions, various lipid disorders such as hyperlipidemia, hypolipidemia, and hypertriglyceridemia occur [55]. High-fat diets and diabetes are cited as the leading causes of elevated lipid levels in plasma. However, changes in plasma lipoprotein levels are observed in the plasma of pregnant women according to certain periods, and one of the important reasons for these changes is gestational diabetes [56].

In current studies, it has been found that serum LDL-C, VLDL-C, and TC levels increase more, and HDL concentrations decrease in women with GDM compared to women who have had a normal pregnancy [57]. Although drug therapy is used in the solution of the increase in lipid metabolism in GDM, diet is very important during the disease process. Nutritional therapy with a diet containing low-density cholesterol is one of the most commonly applied treatment methods. Many *in vitro, in vivo*, and clinical studies have shown that capsaicin has curative roles in diabetes, cancer, obesity, and lipid peroxidation [58–60]. In addition, feeding patients with food products containing capsaicin contributes to the reduction of gallstone formation. One of the most important features of capsaicin is that it acts as a hypocholesterolemic and reduces TC and low-density lipoprotein cholesterol (LDL-C) levels in the plasma [59]. In addition, capsaicin can reduce myocardial and aortic cholesterol concentrations [61]. Increased cholesterol levels in the blood damage the erythrocyte integrity. Consumption of capsaicin-containing foods reduces lipid fractions and thus protects the membrane integrity of erythrocytes [59]. In addition to many known beneficial properties, it is very important for patients with GDM to be fed with food products rich in capsaicin. Capsaicin can improve lipid disorders that occur as a result of GDM. Capsaicin reduces total cholesterol concentrations in plasma through different mechanisms. One of these mechanisms is the increased excretion of bile acids. Cholesterol is mainly converted to bile acids by cholesterol-7α-hydroxylase (CYP7A1), and capsaicin facilitates the excretion of bile acids. Another mechanism of action of capsaicin is the regulation of the transcription and activity of the CYP7A1 enzyme. Previous studies have shown that capsaicin significantly increases CYP7A1 activity [59, 62]. Capsaicin may play a role in the prevention of obesity, a disease associated with hyperlipidemia, diabetes, and GDM. In a study on mice, it was reported that capsaicin increased the expression of adiponectin and its receptor, thereby regulating metabolism in diabetic obese mice [63]. The mechanism involved in the inhibition of adipogenesis by capsaicin is that capsaicin increases the activation of TRPV1 channels [63, 64].

A study investigating the effect of capsaicin on the prevention of IR and lipid disorders in GDM women found a decrease in total cholesterol and triglyceride concentrations in women with GDM fed with capsaicin. In addition, there was a decrease in postprandial serum insulin, plasma glucose, and IR levels compared to the capsaicin-only group [2]. In a study investigating the possible ameliorative role of capsaicin on lipid disorders and oxidative stress in male Wistar rats, it was determined that the amount of LDL decreased in capsaicin-administered rats. However, the protective role of capsaicin on lipid peroxidation was determined by measuring malondialdehyde (MDA) concentration, and it was determined that the concentration of MDA decreased in rats after treatment with capsaicin in the lungs, kidneys, and liver [60]. In a study with mice, capsaicin was found to prevent obesity and dyslipidemia. In addition, capsaicin caused a decrease in serum TC, TG, and LDL-C concentrations in mice [65]. In a study by Lee et al. [60], the effects of capsaicin on visceral obesity were investigated in male C57BL/6 mice fed a high-fat diet. A decrease in the levels of lipid droplets and adipocytes was observed in the tissues of capsaicin-treated mice. In addition, it was determined that there was a significant decrease in plasma glucose, cholesterol, and triglyceride concentrations in the group treated with capsaicin. When the protective role of capsaicin against oxidative stress in human erythrocytes is examined, studies have shown that capsaicin significantly reduces the amount of lipid peroxidation and carbonyl groups [66]. In another study, the curative effects of capsaicin were examined in rats whose plasma cholesterol levels were increased by a high-fat diet. The results indicated that rats fed capsaicin had significant reductions in hepatic triglyceride and hepatic lipid peroxide levels [67]. Many previous studies examining the effects of capsaicin on lipid disorders have revealed that capsaicin causes reductions in total cholesterol, LDL-cholesterol, and HDL-cholesterol levels [44, 61].

19.5 Conclusion

It has been established that capsaicin can be a significant anti-inflammatory and therapeutic supplement in the case of diabetes mellitus, particularly GDM, both in the treatment of the mother and in the prevention of the teratogenicity of the unborn child due to diabetes mellitus. Capsaicin may prevent hyperglycemia and hyperinsulinemia by increasing insulin sensitivity, and it may also improve lipid profile by modulating total cholesterol, HDL cholesterol, and LDL cholesterol. For this reason, capsaicin obtained from hot peppers, which is a natural substance and antioxidant, if consumed easily during pregnancy, may play an important role in preventing gestational diabetes as well as other issues that could affect the unborn child.

References

1. Zhang S, You Y, Liu J, Qin X, Liu X. Hypoglycaemic effect of capsaicinoids via elevation of insulin level and inhibition of glucose absorption in streptozotocin-induced diabetic rats. *J Funct Foods*. 2018; 51:94–103.
2. Yuan LJ, Qin Y, Wang L, Zeng Y, Chang H, Wang J, Wang B, Wan J, Chen SH, Zhang QY, Zhu JD, Zhou Y, Mi MT. Capsaicin-containing chili improved postprandial hyperglycemia, hyperinsulinemia, and fasting lipid disorders in women with gestational diabetes mellitus and lowered the incidence of large-for-gestational-age newborns. *Clin Nutr*. 2016; 35(2):388–393.

3 Öztürk FY, Altuntas Y. Gestasyonel diabetes mellitus/Gestational diabetes mellitus. *Sisli Etfal Hastan Tip Bul.* 2015; 49(1):1.

4 Catalano PM, Kirwan JP, Haugel-de Mouzon S, King J. Gestational diabetes and insulin resistance: Role in short-and long-term implications for mother and fetus. *J Nutr.* 2003; 133(5):1674–1683.

5 Mecacci F, Carignani L, Cioni R, Bartoli E, Parretti E, La Torre P, Scarselli G, Mello G. Maternal metabolic control and perinatal outcome in women with gestational diabetes treated with regular or lispro insulin: Comparison with non-diabetic pregnant women. *Eur J Obstet Gynecol Reprod Biol.* 2003; 111(1):19–24.

6 Venu VKP, Saifeddine M, Mihara K, Faiza M, Gorobets E, Flewelling AJ, Hollenberg MD. Metformin prevents hyperglycemia-associated, oxidative stress-induced vascular endothelial dysfunction: Essential role for the orphan nuclear receptor human nuclear receptor 4A1 (Nur77). *Mol Pharmacol.* 2021; 100(5):428–455.

7 Zhang Z, Wu Y, Zhong C, Zhou X, Liu C, Li Q, Chen R, Gao Q, Li X, Zhang H, Zhang Y, Cui W, Hao L, Wei S, Yang X, Yang N. Association between dietary inflammatory index and gestational diabetes mellitus risk in a prospective birth cohort study. *Nutrition.* 2021; 87:111193.

8 Sharma SK, Vij AS, Sharma M. Mechanisms and clinical uses of capsaicin. *Eur J Pharmacol.* 2013;720(1–3):55–62.

9 Gregory EC, Ely DM. Trends and characteristics in gestational diabetes: United States, 2016-2020. *Natl Vital Stat Rep.* 2022; 71(3):1–15.

10 Bilous RW, Jacklin PB, Maresh MJ, Sacks DA. Resolving the gestational diabetes diagnosis conundrum: The need for a randomized controlled trial of treatment. *Diabetes Care.* 2021; 44:858.

11 Delghingaro-Augusto V, Nolan CJ, Gupta D, Jetton TL, Latour MG, Peshavaria M, Madiraju SRM, Joly E, Peyot ML, Prentki M, Leahy J. Islet beta cell failure in the 60% pancreatectomised obese hyperlipidaemic Zucker fatty rat: Severe dysfunction with altered glycerolipid metabolism without steatosis or a falling beta cell mass. *Diabetologia.* 2009; 52:1122–1132.

12 Catalano PM. Trying to understand gestational diabetes. *Diabet Med.* 2014; 31:273–281.

13 Masuzaki H, Ogawa Y, Sagawa N, Hosoda K, Matsumoto T, Mise H, Nishimura H, Yoshimasa Y, Tanaka I, Mori T et al. Nonadipose tissue production of leptin: Leptin as a novel placenta-derived hormone in humans. *Nat Med.* 1997; 3:1029–1033.

14 Williams MA, Qiu C, Muy-Rivera M, Vadachkoria S, Song T, Luthy DA. Plasma adiponectin concentrations in early pregnancy and subsequent risk of gestational diabetes mellitus. *J Clin Endocrinol Metab.* 2004; 89:2306–2311.

15 Retnakaran R, Hanley AJG, Raif N, Connelly PW, Sermer M, Zinman B. Reduced adiponectin concentration in women with gestational diabetes: A potential factor in progression to type 2 diabetes. *Diabetes Care.* 2004; 27:799–800.

16 Bouchard L, Hivert MF, Guay SP, St-Pierre J, Perron P, Brisson D. Placental adiponectin gene DNA methylation levels are associated with mothers' blood glucose concentration. *Diabetes*. 2012; 61:1272–1280.
17 Barbour LA, McCurdy CE, Hernandez TL, Kirwan JP, Catalano PM, Friedman JE. Cellular mechanisms for insulin resistance in normal pregnancy and gestational diabetes. *Diabetes Care*. 2007; 30(2):112–119.
18 Kim JH, Bachmann RA, Chen J. Interleukin-6 and insulin resistance. *Vitam Horm*. 2009; 80:613–633.
19 Atègbo JM, Grissa O, Yessoufou A, Hichami A, Dramane KL, Moutairou K, Miled A, Grissa A, Jerbi M, Tabka Z, Khan NA. Modulation of adipokines and cytokines in gestational diabetes and macrosomia. *J Clin Endocrinol Metab*. 2006; 91:4137–4143.
20 Fasshauer M, Blüher M, Stumvoll M. Adipokines in gestational diabetes. *Lancet Diabetes Endocrinol*. 2014; 2:488–499.
21 Venkatesh KK, Lynch CD, Powe CE, Costantine MM, Thung SF, Gabbe SG, Grobman WA, Landon MB. Risk of adverse pregnancy outcomes among pregnant individuals with gestational diabetes by race and ethnicity in the United States, 2014–2020. *JAMA*. 2022; 327(14):1356–1367.
22 Ye W, Luo C, Huang J, et al. Gestational diabetes mellitus and adverse pregnancy outcomes: Systematic review and meta-analysis. *BMJ*. 2022; 377:e067946.
23 Parnell AS, Correa A, Reece EA. Pre-pregnancy obesity as a modifier of gestational diabetes and birth defects associations: A systematic review. *Matern Child Health J*. 2017; 21:1105.
24 Stothard KJ, Tennant PW, Bell R, Rankin J. Maternal overweight and obesity and the risk of congenital anomalies: A systematic review and meta-analysis. *JAMA*. 2009; 301(6):636–650.
25 Maftei O, Whitrow MJ, Davies MJ, Giles LC, Owens JA, Moore VM. Maternal body size prior to pregnancy, gestational diabetes and weight gain: Associations with insulin resistance in children at 9–10 years. *Diabet Med*. 2015; 32(2):174–180.
26 Clausen TD, Mathiesen ER, Hansen T, Pedersen O, Jensen DM, Lauenborg J, Schmidt L, Damm P. Overweight and the metabolic syndrome in adult offspring of women with diet-treated gestational diabetes mellitus or type 1 diabetes. *J Clin Endocrinol Metab*. 2009; 94(7):2464–2470.
27 Hartling L, Dryden DM, Guthrie A, Muise M, Vandermeer B, Donovan L. Benefits and harms of treating gestational diabetes mellitus: A systematic review and meta-analysis for the U.S. Preventive Services Task Force and the National Institutes of Health Office of Medical Applications of Research. *Ann Intern Med*. 2013; 159(2):123–129.
28 van Poppel MN, Oostdam N, Eekhoff ME, Wouters MG, van Mechelen W, Catalano PM. Longitudinal relationship of physical activity with insulin sensitivity in overweight and obese pregnant women. *J Clin Endocrinol Metab*. 2013; 98(7):2929–2935.

29 Bo S, Rosato R, Ciccone G, Canil S, Gambino R, Poala CB, Leone F, Valla A, Grassi G, Ghigo E, Cassader M, Menato G. Simple lifestyle recommendations and the outcomes of gestational diabetes. A 2 × 2 factorial randomized trial. *Diabetes Obes Metab.* 2014; 16(10):1032–1035.

30 Muktabhant B, Lawrie TA, Lumbiganon P, Laopaiboon M. Diet or exercise, or both, for preventing excessive weight gain in pregnancy. *Cochrane Database of Syst Rev.* 2015; 6.

31 Rollyson WD, Stover CA, Brown KC, Perry HE, Stevenson CD, McNees CA, Ball JG, Valentovic MA, Dasgupta P. Bioavailability of capsaicin and its implications for drug delivery. *J Control Release.* 2014; 196:96–105.

32 Sun F, Xiong S, Zhu Z. Dietary capsaicin protects cardiometabolic organs from dysfunction. *Nutrients.* 2016; 8:1–13.

33 Vyklicky L, Lyfenko A, Kuffler DP, Vlachova VV. Vanilloid receptor TRPV1 is not activated by vanilloids applied intracellularly. *Neuroreport.* 2003; 14:1061–1065.

34 Di Marzo V. Cannabinoid-vanilloid receptor interactions in pain signaling. *J Neurochem.* 2003; 85:2–4.

35 Kang JH, Kim CS, Han IS, Kawada T, Yu R. Capsaicin, a spicy component of hot peppers, modulates adipokine gene expression and protein release from obese-mouse adipose tissues and isolated adipocytes, and suppresses the inflammatory responses of adipose tissue macrophages. *FEBS Lett.* 2007; 581:4389–4396.

36 McCarty MF, DiNicolantonio JJ, O'keefe JH. Capsaicin may have important potential for promoting vascular and metabolic health. *Open Heart.* 2015; 2(1):e000262.

37 Richards BL, Whittle SL, Buchbinder R. Neuromodulators for pain management in rheumatoid arthritis. *Cochrane Database Syst Rev.* 2012; 1:Cd008921.

38 Singh U, Bernstein JA. Intranasal capsaicin in management of nonallergic (vasomotor) rhinitis. *Prog Drug Res.* 2014; 68:147–170.

39 Lee GR, Shin MK, Yoon DJ, Kim AR, Yu R, Park NH, Han IS. Topical application of capsaicin reduces visceral adipose fat by affecting adipokine levels in high-fat diet-induced obese mice. *Obesity.* 2013; 21(1):115–122.

40 Kang JH, Goto T, Han IS, Kawada T, Kim YM, Yu R. Dietary capsaicin reduces obesity-induced insulin resistance and hepatic steatosis in obese mice fed a high-fat diet. *Obesity.* 2010; 18:780–787.

41 Satyanarayana MN. Capsaicin and gastric ulcers. *Crit Rev Food Sci Nutr.* 2006; 46(4):275–328.

42 Erdost H. Capsaicin. *Uludag Univ Vet Fak Derg.* 2004; 23(1–2–3):149–155.

43 Li Q, Li L, Wang F, Chen J, Zhao Y, Wang P, Zhu Z et al. Dietary capsaicin prevents nonalcoholic fatty liver disease through transient receptor potential vanilloid 1-mediated peroxisome proliferator-activated receptor δ activation. *Pflugers Arch.* 2013; 465(9):1303–1316.

44 Negulesco JA, Noel SA, Newman HAI, Naber EC, Bhat HB, Witiak DT. Effects of pure capsaicinoids (capsaicin and dihydrocapsaicin) on plasma lipid and lipoprotein concentrations of turkey poults. *Atherosclerosis.* 1987; 64(2–3):85–90.

45 Maihöfner C, Heskamp ML. Treatment of peripheral neuropathic pain by topical capsaicin: Impact of pre-existing pain in the QUEPP-study. *Eur J Pain*. 2014; 18(5):671–679.
46 Winter J, Bevan S, Campbell EA. Capsaicin and pain mechanisms. *Br J Anaesth*. 1995; 75(2):157–168.
47 Fattori V, Hohmann MS, Rossaneis AC, Pinho-Ribeiro FA, Verri WA, Jr. Capsaicin: Current understanding of its mechanisms and therapy of pain and other pre-clinical and clinical uses. *Molecules*. 2016; 21(7):844.
48 Song JX, Ren H, Gao YF, Lee CY, Li SF, Zhang F, Chen H, et al. Dietary capsaicin improves glucose homeostasis and alters the gut microbiota in obese diabetic ob/ob mice. *Front Physiol*. 2007; 8:602.
49 Panchal SK, Bliss E, Brown L. Capsaicin in metabolic syndrome. *Nutrients*. 2018; 10(5):630.
50 Suri A, Szallasi A. The emerging role of TRPV1 in diabetes and obesity. *Trends Pharmacol Sci*. 2008; 29(1):29–36.
51 Park BJ, Park JH, Kageyama H, Osaka T, Inoue S, Lee TH. Effect of capsaicin on glucose metabolism in diabetic rats. *Diabetes Res Clin Pract*. 2000; 50(Suppl_1):395.
52 Kwon DY, Kim YS, Ryu SY, Cha MR, Yon GH, Yang HJ, Kim MJ, Kang S, Park S. Capsiate improves glucose metabolism by improving insulin sensitivity better than capsaicin in diabetic rats. *J Nutr Biochem*. 2013; 24(6):1078–1085.
53 Alim Z, Kilinc N, Sengul B, Beydemir S. Mechanism of capsaicin inhibition of aldose reductase activity. *J Biochem Mol Toxicol*. 2017; 31(7):e21898.
54 Feingold KR. Lipid and lipoprotein metabolism. *Endocrinol Metab Clin*. 2022; 51(3):437–458.
55 Adeyanju MM, Saheed IA, Oyelekan OI, Dele-Osibanjo TA, Adelegan AA, Raimi AJ, Olalekan SO, Alabi OS, Alli KM. Sesamum indicum diet prevents hyperlipidemia in experimental rats. *Food Chem*. 2022; 4:100092.
56 Butte NF. Carbohydrate and lipid metabolism in pregnancy: Normal compared with gestational diabetes mellitus. *Am J Clin Nutr*. 2000; 71(5):1256–1261.
57 Ryckman KK, Spracklen CN, Smith CJ, Robinson JG, Saftlas AF. Maternal lipid levels during pregnancy and gestational diabetes: A systematic review and meta-analysis. *BJOG*. 2015;122(5): 643–651.
58 Bley K, Boorman G, Mohammad B, McKenzie D, Babbar S. A comprehensive review of the carcinogenic and anticarcinogenic potential of capsaicin. *Toxicol Pathol*. 2012; 40(6):847–873.
59 Srinivasan K. Biological activities of red pepper (Capsicum annuum) and its pungent principle capsaicin: A review. *Crit Rev Food Sci Nutr*. 2016; 56(9):1488–1500.
60 Lee CYJ, Kim M, Yoon SW, Lee CH. Short-term control of capsaicin on blood and oxidative stress of rats in vivo. *Phytother Res*. 2003; 17(5):454–458.
61 Kawada T, Hagihara KI, Iwai K. Effects of capsaicin on lipid metabolism in rats fed a high fat diet. *J Nutr*. 1986; 116(7):1272–1278.

62 Liang YT, Tian XY, Chen JN, Peng C, Ma KY, Zuo Y, Jiao R, Lu Y, Huang Y, Chen ZY. Capsaicinoids lower plasma cholesterol and improve endothelial function in hamsters. *Eur J Nutr*. 2013; 52(1):379–388.

63 Munoz-Islas E, Elizondo-Martinez CE, Gutierrez-Lopez M, Acosta-Gonzalez RI, Zaga-Clavellina V, Helguera-Repetto AC, Ramirez-Rosas MB, Romero-Sandoval EA, Jimenez-Andrade JM. Effect of experimental gestational diabetes mellitus on mechanical sensitivity, capsaicin-induced pain behaviors and hind paw glabrous skin innervation of male and female mouse offspring. *J Pain Res*. 2021; 14:1573.

64 Zheng J, Zheng S, Feng Q, Zhang Q, Xiao X. Dietary capsaicin and its anti-obesity potency: From mechanism to clinical implications. *Biosci Rep*. 2017; 37(3).

65 Li W, Yang H, Lu Y. Capsaicin alleviates lipid metabolism disorder in high beef fat-fed mice. *J Funct Foods*. 2019; 60:103444.

66 Luqman S, Rizvi SI. Protection of lipid peroxidation and carbonyl formation in proteins by capsaicin in human erythrocytes subjected to oxidative stress. *Phytother Res*. 2006; 20(4):303–306.

67 Hanumanthappa M, Krishnapura S. Hypolipidemic and antioxidant potency of heat processed turmeric and red pepper in experimental rats. *Afr J Food Sci*. 2008; 2(1):1–6.

20

Effects of Natural Products on the Genetics of Gestational Diabetes Mellitus

Onur Baykara

Department of Medical Biology, Cerrahpaşa Medical Faculty, Istanbul University-Cerrahpaşa, Turkey

20.1 Introduction

Diabetes mellitus, more commonly known as diabetes, is a chronic health condition that is characterized by the presence of too much sugar (glucose) in the blood. Insulin is a hormone that is necessary to convert the sugar we obtain from food into energy during the day. If insulin cannot be used at the required levels, deterioration in the biological response, especially in structures such as liver, muscle, and adipose tissue occurs and insulin resistance develops. As a result, the blood sugar level remains high and a higher level of insulin is released from the pancreas in response to this situation. Sugar in the blood cannot be converted into energy and remains at high levels. High sugar rate can cause metabolic syndrome and type 2 diabetes mellitus (T2DM) after a while. In diabetics, insulin is either not produced adequately or it cannot be used by the body effectively, leading to persistent high levels of glucose. In the case of constant high levels of glucose, a number of complications such as damage to eyes leading to blindness, nerves, internal organs including kidney failure, cardiovascular heart diseases (CVD), and excess deposit of fat in liver (steatohepatitis) may occur.

There are several types of diabetes. In type 1 diabetes mellitus (T1DM), the immune cells perceive the pancreatic cells as enemies and attack them, eventually leading to pancreatic insufficiency and stopping the production of insulin. In T2DM, the most common form of diabetes, the body either does not produce or use insulin at all. This type is seen in approximately 95% of all diabetics and can be detected even in early infancy. The third type is gestational diabetes mellitus (GDM), and is only seen in pregnant women, causing high levels of sugar during pregnancy. GDM is generally detected during routine tests of pregnancy and women with GDM

Natural Products and their Bioactives in Antidiabetic Drug Discovery, First Edition.
Edited by Kanti Bhooshan Pandey, Maitree Suttajit and Pinar Atukeren.

usually experience similar symptoms to other types of diabetes. GDM may cause some specific symptoms, such as polyhydramnios (excess amount of amnion fluid), increased growth of the baby, leading to cesarean section or induced labor, pre-eclampsia (a condition characterized with hypertension and proteinuria and is dangerous if left untreated), jaundice in the baby, and loss of the baby. Women with GDM are more likely to develop T2DM and GDM in future pregnancies as genetic risk factors contribute greatly to the pathogenesis [1, 2]. GDM is not fully curable, but it can be taken under control by changing diet and exercise habits.

Developing diabetes is interconnected with lifestyle, eating habits, environmental conditions, as well as genetics. Unlike cancer or some metabolic disorders, the connection between diabetes and genetic background has not been fully clarified. However, it is known that food and nutrients have some certain effects on cell metabolism, including genes and proteins. To change (possibly increase or decrease) the levels of biomarkers like blood/sugar, cholesterol, and a number of proteins, chemicals in food interact with biochemical pathways at the molecular level. Diet is a key factor in controlling personal genetic susceptibility to a number of diseases. Insulin is a major factor in triggering the immune system, and a healthy diet consisting of healthy foods can minimize the need for insulin release because insulin acts as a defense mechanism against too much sugar. Nevertheless, other nutrients such as amino acids and free fatty acids can increase glucose-induced insulin secretion, and various hormones, such as growth hormone, glucagon like peptide-1, estrogen, melatonin, and leptin can function in the regulation of insulin secretion. In this chapter, the genetics of GDM and the possible role of natural products is discussed.

20.2 Genetic Background of Insulin and Gluconeogenesis

Insulin synthesis and secretion is a multi-step process that is regulated at both the transcriptional and translational level, and genes play a major role in these complex steps. Insulin is released from pancreatic β-cells and this release is controlled by electrical activity of β-cell K_{ATP} channels, which are structured as two subunits, namely the sulfonylurea receptor 1 (ABCC8), and potassium inwardly rectifying channel, subfamily J, member 11 (KCNJ11). The mutations in genes of these proteins are associated with insufficient insulin release, eventually leading to diabetes. Normally, insulin binds to the insulin receptor and triggers the phosphorylation of the β-subunit receptor and insulin receptor substrate-1 (IRS1) on six or more tyrosine residues. IRS1 protein is a receptor tyrosine kinase that transduces signals in insulin signaling. When IRS1 is phosphorylated, it provokes the binding and phosphorylation of the regulatory subunit p85α, a positive enhancer of the insulin pathway, of phosphatidylinositol 3-kinase (PI3K) to IRS1. P85α is also interconnected to activating glucose transporters (GLUT) in lipid cells [3]. A polymorphism of the gene (*IRS1*) encoding IRS1 can reduce tyrosine phosphorylation and causes insulin resistance by inhibiting insulin receptor kinase. In addition, an increase in IRS1 phosphorylation impairs binding of insulin to its receptor and as the IRS1 is not phosphorylated, intake of glucose into the cell is disrupted [4]. A certain polymorphism of *IRS1* gene (rs2943641 TT genotype) was associated with higher fasting glucose levels in women with GDM [5]. In GDM, the placenta produces insulin blocking hormones (progesterone and estrogen), and the expression of the *IRS1* gene is suppressed by inhibiting the PI3K pathway. In addition, translocation of insulin induced GLUT-4 is blocked, thus the level of glucose remains high [6].

The *cis*-acting sequences within the 5′ flanking region and *trans*-activators including β-2/neurogenic differentiation 1 (*NeuroD1*), pancreatic and duodenal homeobox- 1(*PDX-1*), paired box gene 6 (*PAX6*) and *MafA*, regulate insulin transcription, while the protein translation is controlled by the stability of preproinsulin mRNA and its untranslated regions [7]. Uncoupling protein 2 (UCP2) is a member of the mitochondrial anion carrier protein family and inhibits insulin secretion from β-cells [8]. Mitochondrial encoded NADH dehydrogenase 1 (*ND1*) is another mitochondrial originated gene and is a part of electron transport in energy metabolism. A mutation in this gene disrupts the insulin secretion by decreasing ATP production.

Lipid and glucose metabolism genes also contribute to the development of diabetes. Peroxisome proliferator activated receptor, gamma (PPARG) is a well-known transcription factor that acts in lipid and glucose metabolism and adipocyte differentiation. This transcription factor controls the expression of target genes by binding to their promoter regions and co-activates with peroxisome proliferator activated receptor-gamma coactivator 1-alpha (PPARGC1A) to regulate genes that are active in energy metabolism. One other glucose metabolism gene is β-3 adrenergic receptor (*ADRB3*) that produces a protein which is expressed in fat tissue and regulates lipolysis and energy expenditure. Glucose transporter 1 (*GLUT1*), as the name implies, is responsible for transporting glucose. There are other genes such as adiponectin (*ADIPOQ*), forkhead box C2 (*FOXC2*), maturity onset diabetes of the young (*MODY*) genes that play a number of different roles in glucose and insulin metabolism. The polymorphisms of these genes and proteins change the expression patterns of target genes and/or proteins and deteriorate insulin synthesis and release [9, 10].

Considering that GDM has a similar genetic pattern to T2DM, scientists have investigated the effect of cordon-bleu protein-like 1 (*COBLL1*) gene, which encodes a protein functional in development and metabolic disorders; and lower serum levels and lower insulin resistance have been associated with a polymorphism of this gene (*COBLL1* rs7607980 C allele) [11].

Another study has performed a genome wide association study (GWAS) to see whether T2DM and GDM were associated and have found that Cdk5 (cyclin dependent kinase) regulatory associated protein 1-like 1(*CDKAL1*), a gene that is involved in T2DM and encodes a protein found in pancreatic β-cells and neuronal tissues, and cyclin-dependent kinase inhibitor 2A (*CDKN2A/2B*), a gene that encodes a protein which is abundantly found in many tissues, were highly associated with the risk of GDM. Furthermore, variations of *IGF2BP2, SLC30A8, HHEX, KCNQ1, MTNR1B, MC4R, FTO, TCF7L2* and many other genes have been found to be associated with the development of GDM, possibly due to their role in the regulation of pancreatic cells [12–15]. Another study contributed to these results and improved the literature by suggesting that *TCF7L2* gene polymorphisms (CTTC: rs4506565, rs7903146, rs7901695, rs12243326); and a *KCNQ1* haplotype (TTT: rs163184, rs2237897, rs2237892) shared common genetic patterns with GDM and T2DM [16].

20.3 Nutrigenetics/Nutrigenomics and the Effects of Nutrients on Genes

The relationship between the effects of nutrients on genes has been extensively studied under the area of nutrigenetics/nutrigenomics. Through scientific and clinical research, nutrigenetics/nutrigenomics play a significant role not only in treating diseases and illnesses but also in promoting health and wellness. It is crucial for the future of both precision healthcare and personalized nutrition [17]. The food and nutrients we eat have the ability to alter the expression of our genes. Early studies have identified folic acid as a crucial regulator of gene expression as folate can directly regulate gene expression by providing the precursors for the synthesis of purines and thymidylate as well as the methylation of DNA. Humans fed with low levels of folate have shown evidence of undermethylation of DNA, a crucial step for the growth of cancer cells. Research using animal or human cells in culture has provided evidence regarding the influence of nutrition on gene expression. Interactions between nutrients, genes, and diseases have been studied for a number of popular food ingredients. In cultured human cells, it has been shown that the vitamins tocopherol and biotin, the mineral zinc, and the phytochemicals flavones and catechins can affect the expression of genes.

A study conducted in Canada has investigated the association between gene expression and the Prudent and Western type dietary patterns. A Prudent dietary pattern consists of fresh fruit and vegetables, healthy food, salad, grilled fish, etc., while Western diet is made of deep fried food, red meat, processed meat like sausages, saturated fat, soft drinks, food rich in carbohydrates, etc. People fed with Western dietary patterns have increased risk of developing cancer, cardiovascular and metabolic diseases, while those fed with a Prudent diet have a lower risk of CVD and obesity. The study has proven that those fed with a Prudent dietary pattern have different gene expressions than those fed with a Western dietary pattern, and that the pathways platelet derived growth factor (PDGF), extracellular signal-regulated kinase 1/2 (ERK) mitogen-activated protein kinase (ERK/MAPK), B-cell receptor, PI3K signaling in B lymphocytes, Rho family GTPases, cyclins which direct cell growth, migration, proliferation, and many cell metabolic events regulation have been up- or down-regulated in men and women. The results have been associated with levels of insulin, where those fed with a healthier diet had lower levels of insulin and those fed with an unhealthy diet had higher levels. The genetic expression differences between men and women have been attributed to sex hormones such as estrogen [18]. Similar studies show that vegetarian or plant-based diets help decrease the risk of developing GDM, and non-vegetarian women with GDM had increased risk for glucose intolerance, while a low calorie vegetarian diet improved insulin sensitivity [19]. These results clearly show that we can control diabetes and alleviate symptoms with proper diet and physical activity.

The ability to study cells in culture provides a way to understand how food additives impact gene expression. There are a number of ways that nutrients might disrupt gene expression. These include alterations to mRNA stability (via vitamin D), DNA methylation (through folic acid), and direct participation in DNA synthesis (by biotin metabolites) (vitamin D).

20.4 The Close Relationship between Nutrients and Gene Expression

How can nutrients affect gene expression? Certain metabolites, amino acids, vitamins, or minerals can act as transcription factors and after entering into the nucleus, they bind to the promoter region of the DNA and can either increase or decrease the transcription. Bioactive components increase or decrease gene expression through either genetic (changes in DNA sequence, chromatin structure) or epigenetic (DNA sequence is not changed but the interacting molecules change the function and/or expression of genes through a number of mechanisms such as histone modification, DNA methylation, acetylation, interaction of miRNAs, etc.) mechanisms [20, 21].

Methionine amino acid and methionine derived-sulfur metabolites found in the transamination pathway can activate the general control non-repressed protein 5 (GCN5), an acetyltransferase that promotes acetylation of the transcriptional co-activator PGC-1α transcriptional co-activator to control the gluconeogenesis in liver. PGC-1α also interacts with the forkhead box protein O1 (FOXO1) and hepatocyte nuclear factor 4α (HNF4α). These two proteins increase the transcription of glucose-6-phosphatase (*G6PC*) and phosphoenolpyruvate carboxykinase (*PCK1*) genes, which encode two proteins that modulate gluconeogenesis [22]. In addition to these proteins, peroxisome proliferator-activated receptor-γ coactivator 1α (*PPAR1α*) can control the transcriptional activity of gluconeogenic genes [23]. Considering that T2DM is a result of uncontrolled gluconeogenesis, it is of major importance to determine the biomarkers of this process to cure and prevent diabetes. Recent research has shown that the small ubiquitin-related modifier (SUMO)-specific protease 2 (SENP2), a catalyzer of desumoylation, reverse process of SUMOylation, a post-translational modification that is essential in cell growth and migration, can control gluconeogenesis by regulating the SUMOylation of the fuel sensor kinase, 5′-AMP-activated protein kinase alpha (AMPKα) [24].

A good example of how nutrients affect genes is vitamin A derivatives found in carrots. Fatty acids like Omega-3 and 6, such as those found in fish, can bind to peroxisome proliferation–activated receptors (PPARs), a class of nuclear transcription factors, which in turn bind to the retinoid X receptor (RXR), a derivative of vitamin A [25].

As just briefly shown here, glucose synthesis, an important factor of development of diabetes, is strictly under the control of genes. A deterioration in the genes, proteins, and/or signaling pathways may interrupt this delicate process, resulting in diabetes. To date, the majority of studies have focused on the genetics and the effects of nutrients on T2DM [26]; however, studies investigating the effect of nutrients on GDM genetics are limited. This subject is discussed in the next section of this chapter.

20.5 Effects of Nutrients on Genes and GDM

Pregnancy is a miracle of life and the mother-to-be requires special care. Her diet must be planned carefully to fulfill the demands of her body. Folate, also known as B9 and folacin, is a vitamin B and is taken into the body in the form of folic acid. Folate is abundantly found in many foods such as grains, legumes, vegetables, and fruits. It is essential for proper DNA and RNA synthesis, production of amino acids, and cell division. Its deficiency, also known as vitamin B9 deficiency, can cause loss of appetite, weakness, and behavioral disorders. In addition to these symptoms seen in healthy people, its deficiency may cause neural tube defect (NTD) and spina bifida in pregnant women. It may develop because of the increasing number of cells of the growing fetus, which shares the mother's nutrients, maternal hormones, and decreased intake and absorption of folate. Different developmental problems may arise from folate insufficiency during pregnancy or infancy due to the fetus' or infant's development of autoantibodies against the folate receptor. Luckily, it can easily be diagnosed, treated and prevented, and in the modern world, almost all pregnant women are prescribed vitamin B9 to prevent complications.

In the liver, folate is converted into tetrahydrofolate (THF) via dihydrofolate reductase. THF is responsible for all biologic functions of folic acid. During reduction from folic acid to dihydrofolate reductase and tetrahydrofolate reductase, one molecule of nicotinamide adenine dinucleotide phosphate (NADPH) is consumed and one molecule of NADP is produced. Tetrahydrofolate's main function is to transport single carbon groups (methyl, methylene or formyl groups) to other molecules to continue synthesis and modification of DNA and RNA. Lack of folate may disrupt this precisely controlled mechanism.

Another important vitamin B is vitamin B12 (also known as cobalamin). It is found in many animal derived foods, such as milk, eggs, yoghurt, animal meat, as well as plants and algae, fortified food and supplements. It is a cofactor in DNA synthesis, amino acid and fatty acid metabolism. Without vitamin B12, we would not be able to synthesize myelin, and the nervous system would develop abnormally. Similarly, without vitamin B12, red blood cells would have problems during the maturation process in bone marrow. Deficiency of vitamin B12 may cause tiredness, headaches, oral ulcers, neurological problems, and memory deficits, anemia, gastrointestinal symptoms, confusion, psychosis, and mania. Vitamin B12 has some certain functions in the cell as a coenzyme. It acts as an isomerase to rearrange atoms, a methyltransferase to transfer methyl groups between two molecules, especially during DNA synthesis, and a dehalogenase that removes a halogen atom from a substrate. Based on the methyltransferase function of vitamin B12, a mutation in the *MTR* gene disrupts production of methionine synthase enzyme, a methyltransferase that is used to produce THF and methionine via using vitamin B12. This loss causes vitamin B12 deficiency, increased levels of homocysteine, so that THF cannot be used in DNA synthesis. As it is understood, vitamin B12 has a crucial role in metabolism, along with folate metabolism, since both vitamins are involved in one-carbon (1C) metabolism cycle.

A recent study has associated low levels of serum folate (<5th percentile) with lower risk of GDM and higher risk of intrahepatic cholestasis of pregnancy (ICP) and pre-eclampsia. On the other hand, high levels of serum vitamin B12 (>95th percentile) have been associated with higher risk of GDM, PE, and ICP [27]. Another study has indicated that taking high amounts of folate supplements before pregnancy significantly reduced the risk of decreased GDM, and use of folate in the pre-pregnancy term could be useful to prevent GDM [28].

Of the fat soluble vitamins, vitamin D has been associated with fetal and maternal immune systems, and bone health and deficiency of vitamin D has been associated with severe side effects such as preterm births, GDM, and pre-eclampsia [29]. Since vitamin D plays roles in glucose homeostasis through a number of mechanisms, it is likely to have an impact on GDM. It has a function in regulating calcium levels, insulin production, and secretion of insulin from the pancreas. Vitamin D, through its metabolites, helps increase insulin and decrease insulin resistance in GDM and there is an inverse relationship between the risk of GDM and vitamin D status in early pregnancy, as vitamin D deficiency increases the risk of developing GDM [30–32]. However, there are controversial results in the literature. It is noteworthy that an important gene and protein that is essential in vitamin D metabolism has no effect on GDM. A case-control study from China has shown that single nucleotide polymorphisms (SNPs) of the vitamin D receptor (*VDR*) gene (rs154410, rs7975232, rs731236, rs2228570, and rs739837) did not have any significant effect on the ethiopathogenesis of GDM [33].

Protein and starch can change the expression of genes. A Chinese study has demonstrated that a diet rich in protein but poor in starch decreases the risk of developing GDM in pregnant women [34]. Betaine, a methyl donor, a methylating agent and a substrate for methionine, is found in grains (especially wheat), broccoli, spinach, shellfish, and beets. A diet poor in betaine may stop insulin synthesis, causing diabetes and metabolic syndrome [35].

Resveratrol, a polyphenic compound found mainly in red grapes, peanuts, mulberries, and some other plants, is an antioxidant and can maintain blood viscosity by inhibiting vasodilation and platelet coagulation. In addition, it can be used to control the body weight of obese rats. An animal study with a GDM rat model has shown that resveratrol significantly increased the blood insulin levels, and blood glucose levels and body weight have decreased. What is more, the levels of IL-6, TNF-α, leptin, and resistin have also regressed compared to the non-GDM rats, which indicate that resveratrol can directly interact with IL-6, TNF-α, leptin, and resistin encoding genes to control insulin levels [36].

Naringenin is a common dietary flavanone that has antioxidant, anti-inflammatory, antihypertensive features and is found in citrus fruits, such as lemons, oranges, grapefruit. It has been shown to reduce the plasma glucose levels in diabetes mellitus. Another animal model study has investigated the effect of naringenin in rats with GDM and shown that naringenin can effectively inhibit apoptosis and decrease oxidative stress in renal cells, increase the expression of inflammatory cytokines such as ICAM-1, TGF-β, TNF-α, and IL-6, and increase the expression of *PPARα*, a fatty acid metabolism related gene [37].

Quercetin is a flavonol found in citrus fruits, apples, green tea, onions, olive oil, broccoli, red wine, and many other foods, and has many anti-inflammatory, antioxidative, and anti-tumoral features. A rodent model of GDM has investigated the effect of quercetin on cardio-metabolic diseases like GDM and shown that quercetin can reduce GDM through reducing nitric oxide synthase-2 (*NOS-2*) expression by modulating the NF-κB by reducing p65 levels, and increasing superoxide dismutase 1 (SOD1) levels [38].

Apart from these, it is noteworthy to mention that many nutrients including dietary flavonoids (e.g. flavonol, quercetin, rutin, kaempferol, isorhamnetin, fisetin, and morin), flavonones (e.g. hesperidin, naringenin, and eriodictyol), flavones (e.g. apigenin, luteolin, tangeritin, chrysin, wogonin, diosmin, and baicalein), isoflavones (e.g. genistein and daidzein), and anthocyanins (e.g. cyanidin, delphinidin, and pelargonidin) have major roles in insulin synthesis, gluconeogenesis, and diabetes but a vast majority of these compounds have not been investigated in women with GDM, but have been investigated in diabetes, mainly in T2DM. These compounds do their jobs by down-regulating the production of proinflammatory cytokines and free radicals, reducing or suppressing oxidative stress, reducing blood and plasma glucose levels, increasing plasma insulin levels, inhibiting glucose uptake, reducing insulin resistance, inducing hypolipidemic activity, reducing lipid peroxidation, protecting against β-cell destruction or apoptosis, suppressing fatty acid synthesis, and many other mechanisms. These compounds take place in diabetes etiopathogenesis by modulating pathways such as glucose transporting, hepatic enzymes, PPARs, tyrosine kinase inhibitors, AMP-activated protein kinases, NF-κB, and β-cell apoptosis. However, these genetic mechanisms and their background need to be investigated for GDM, too [39].

As mentioned above, the etiopathogenesis of GDM is not only influenced by genetic changes, but also epigenetic changes. Briefly, epigenetic changes are changes that do not physically alter the DNA or chromatin structure, but the function of DNA by targeting specific regions on DNA and chromatin by special small molecules. One of these molecules is miRNAs, small, 18–24 nucleotide non-coding RNA sequences that change the expression of a gene by binding to 3′ untranslated regions (UTRs). Even though it is not the subject of this chapter, a growing amount of evidence has shown that the expression of miRNAs can be affected by bioactive dietary components, such as fatty acids, polyphenols, vitamins, and minerals [40]. Quercetin, apigenin, luteolin, kaempferol, chrysin, and many other compounds have been shown to increase or decrease the expression of a plethora of miRNAs in many diseases including cancer and metabolic disorders [20]. However, there is only a limited number of studies focusing on the role of nutrients in GDM genetics and these studies usually have controversial results. An extensive review has summarized the results of 53 studies on miRNAs and GDM and found that the expression of a number of miRNAs including but not limited to miR-9, miR-16, miR-17, miR19a, miR-19b, miR-20, miR-96, miR-145, and others change in GDM. For example, it has been emphasized that the expression of miR-9 has increased in two studies, while it has decreased in two. Likewise, expression of miR-16 has decreased in Mexican women with GDM, whereas three studies from Turkey and South Africa have not found any differences. The other studies share a similar pattern [41]. The differences may be due to differences in race, diet, and environmental conditions. The same review has shown miRNAs that are up-regulated in two or more studies. According to their statements, miRs-195, 330, 342, 520, 657, and 1323 show significant changes in GDM. Their common point is that they all affect fatty acid biosynthesis, glycogen synthesis, insulin secretion, and β-cell development [41]. However, these studies have mostly focused just on mechanisms but not the effect of nutrients on miRNA expression in GDM. Investigating these miRNAs may pave the way to use them as biomarkers of GDM.

On the other hand, some studies have gone deeper to analyze the possible underlying mechanisms. Li et al. have shown that miR-22 and miR-372 were both directly targeting *SLC2A4*, the gene that encodes GLUT4, a glucose transporter, and were down-regulated in GDM as they are involved in the regulation of glucose metabolism via the PI3K/AKT/GLUT4 pathway [42]. Another study has provided evidence that expression of miR-199 is increased in women with GDM by inhibiting the expression of methyl CpG-binding protein 2 (MeCP2) and down-regulating canonical transient receptor potential 3 (Trpc3), partners of the MeCP2-Trpc3 pathway [43]. Again, this study does not investigate the effects of nutrients, but only the mechanism.

Last but not least, molecules such as circular RNAs (circRNAs) have also been investigated in GDM. Like miRNAs, circRNAs are small endogenous molecules that function as sponges for miRNAs to regulate the expression of genes. A number of studies have shown the active roles of circRNA in GDM. circRNAs such as Hsa_circ_0054633 (regulates β-cell proliferation), circ_0008285 (involved in total cholesterol metabolism), circ_0001173 (interacts with glycated hemoglobin), Hsa_circ_102893 (attracts miR-33a-5p and miR-197-3p), and others in GDM are either up- or down-regulated to regulate the signaling pathways and glucose metabolism [44].

20.6 Conclusion

As stated above, this uncharted area needs to be conquered using the newly developed technologies. GDM is an important health problem and investigating its relationship with foods, nutrients, and genetic control will help us to develop new diagnosis and treatment methods.

References

1 Pagel KA, Chu H, Ramola R, Guerrero RF, Chung JH, Parry S, et al. Association of genetic predisposition and physical activity with risk of gestational diabetes in nulliparous women. *JAMA Netw Open*. 2022; 5(8): e2229158.

2 Li M, Rahman ML, Wu J, Ding M, Chavarro JE, Lin Y, et al. Genetic factors and risk of type 2 diabetes among women with a history of gestational diabetes: Findings from two independent populations. *BMJ Open Diabetes Res Care*. 2020; 8(1).

3 Herbst JJ, Andrews GC, Contillo LG, Singleton DH, Genereux PE, Gibbs EM, et al. Effect of the activation of phosphatidylinositol 3-kinase by a thiophosphotyrosine peptide on glucose transport in 3T3-L1 adipocytes. *J Biol Chem*. 1995; 270(43): 26000–26005.

4 Bandyopadhyay GK, Yu JG, Ofrecio J, Olefsky JM. Increased p85/55/50 expression and decreased phosphotidylinositol 3-kinase activity in insulin-resistant human skeletal muscle. *Diabetes*. 2005; 54(8): 2351–2359.

5 Ustianowski P, Malinowski D, Czerewaty M, Safranow K, Tarnowski M, Dziedziejko V, et al. COBLL1 and IRS1 gene polymorphisms and placental expression in women with gestational diabetes. *Biomedicines*. 2022; 10: 8.

6 Wada T, Hori S, Sugiyama M, Fujisawa E, Nakano T, Tsuneki H, et al. Progesterone inhibits glucose uptake by affecting diverse steps of insulin signaling in 3T3-L1 adipocytes. *Am J Physiol Endocrinol Metab*. 2010; 298(4): E881–888.

7 Fu Z, Gilbert ER, Liu D. Regulation of insulin synthesis and secretion and pancreatic beta-cell dysfunction in diabetes. *Curr Diabetes Rev*. 2013; 9(1): 25–53.

8 Chan CB, Harper ME. Uncoupling proteins: Role in insulin resistance and insulin insufficiency. *Curr Diabetes Rev*. 2006; 2(3): 271–283.

9 Robitaille J, Grant AM. The genetics of gestational diabetes mellitus: Evidence for relationship with type 2 diabetes mellitus. *Genet Med*. 2008; 10(4): 240–250.

10 Madhu SV, Raizada N. Genetics of gestational diabetes mellitus - The Indian perspective. *Indian J Endocrinol Metab*. 2022; 26(2): 95–97.

11 Mancina RM, Burza MA, Maglio C, Pirazzi C, Sentinelli F, Incani M, et al. The COBLL1 C allele is associated with lower serum insulin levels and lower insulin resistance in overweight and obese children. *Diabetes Metab Res Rev*. 2013; 29(5): 413–416.

12 Cho YM, Kim TH, Lim S, Choi SH, Shin HD, Lee HK, et al. Type 2 diabetes-associated genetic variants discovered in the recent genome-wide association studies are related to gestational diabetes mellitus in the Korean population. *Diabetologia*. 2009; 52(2):253–261.

13 Benny P, Ahn HJ, Burlingame J, Lee MJ, Miller C, Chen J, et al. Genetic risk factors associated with gestational diabetes in a multi-ethnic population. *PLoS One*. 2021; 16(12): e0261137.

14 Cao M, Zhang L, Chen T, Shi A, Xie K, Li Z, et al. Genetic susceptibility to gestational diabetes mellitus in a Chinese population. *Front Endocrinol (Lausanne)*. 2020; 11: 247.

15 Yahaya TO, Salisu T, Abdulrahman YB, Umar AK. Update on the genetic and epigenetic etiology of gestational diabetes mellitus: A review. *Egypt J Med Hum Genet*. 2020; 21(1): 13.

16 Huerta-Chagoya A, Vázquez-Cárdenas P, Moreno-Macías H, Tapia-Maruri L, Rodríguez-Guillén R, López-Vite E, et al. Genetic determinants for gestational diabetes mellitus and related metabolic traits in Mexican women. *PLoS One*. 2015; 10(5): e0126408.

17 Marcum JA. Nutrigenetics/nutrigenomics, personalized nutrition, and precision healthcare. *Curr Nutr Rep*. 2020; 9(4): 338–345.

18 Bouchard-Mercier A, Paradis AM, Rudkowska I, Lemieux S, Couture P, Vohl MC. Associations between dietary patterns and gene expression profiles of healthy men and women: A cross-sectional study. *Nutr J*. 2013; 12: 24.

19 Schiattarella A, Lombardo M, Morlando M, Rizzo G. The impact of a plant-based diet on gestational diabetes: A review. *Antioxidants (Basel)*. 2021; 10: 4.

20 Mierziak J, Kostyn K, Boba A, Czemplik M, Kulma A, Wojtasik W. Influence of the bioactive diet components on the gene expression regulation. *Nutrients*. 2021; 13: 11.

21 Dalfrà MG, Burlina S, Del Vescovo GG, Lapolla A. Genetics and epigenetics: New insight on gestational diabetes mellitus. *Front Endocrinol (Lausanne)*. 2020; 11: 602477.

22 Tavares CDJ, Sharabi K, Dominy JE, Lee Y, Isasa M, Orozco JM, et al. The methionine transamination pathway controls hepatic glucose metabolism through regulation of the GCN5 acetyltransferase and the PGC-1α transcriptional coactivator. *J Biol Chem*. 2016; 291(20): 10635–10645.

23 Jitrapakdee S. Transcription factors and coactivators controlling nutrient and hormonal regulation of hepatic gluconeogenesis. *Int J Biochem Cell Biol*. 2012; 44(1): 33–45.

24 Dou X, Zhou W-Y, Ding M, Ma Y-J, Yang -Q-Q, Qian S-W, et al. The protease SENP2 controls hepatic gluconeogenesis by regulating the SUMOylation of the fuel sensor AMPKα. *J Biol Chem*. 2022; 298(2): 101544.

25 Ouamrane L, Larrieu G, Gauthier B, Pineau T. RXR activators molecular signalling: Involvement of a PPAR alpha-dependent pathway in the liver and kidney, evidence for an alternative pathway in the heart. *Br J Pharmacol*. 2003; 138(5): 845–854.

26 Berná G, Oliveras-López MJ, Jurado-Ruíz E, Tejedo J, Bedoya F, Soria B, et al. Nutrigenetics and nutrigenomics insights into diabetes etiopathogenesis. *Nutrients*. 2014; 6(11): 5338–5369.

27 Yuan X, Han X, Zhou W, Long W, Wang H, Yu B et al. Association of folate and vitamin B12 imbalance with adverse pregnancy outcomes among 11,549 pregnant women: An observational cohort study. *Front Nutr*. 2022; 9: 947118.

28 Li M, Li S, Chavarro JE, Gaskins AJ, Ley SH, Hinkle SN, et al. Prepregnancy habitual intakes of total, supplemental, and food folate and risk of gestational diabetes mellitus: A prospective cohort study. *Diabetes Care*. 2019; 42(6): 1034–1041.

29 Youness RA, Dawoud A, ElTahtawy O, Farag MA. Fat-soluble vitamins: Updated review of their role and orchestration in human nutrition throughout life cycle with sex differences. *Nutr Metab (Lond)*. 2022; 19(1): 60.

30 Agarwal S, Kovilam O, Agrawal DK. Vitamin D and its impact on maternal-fetal outcomes in pregnancy: A critical review. *Crit Rev Food Sci Nutr*. 2018; 58(5): 755–769.

31 Arnold DL, Enquobahrie DA, Qiu C, Huang J, Grote N, VanderStoep A, et al. Early pregnancy maternal vitamin D concentrations and risk of gestational diabetes mellitus. *Paediatr Perinat Epidemiol.* 2015; 29(3): 200–210.

32 Amraei M, Mohamadpour S, Sayehmiri K, Mousavi SF, Shirzadpour E, Moayeri A. Effects of vitamin D deficiency on incidence risk of gestational diabetes mellitus: A systematic review and meta-analysis. *Front Endocrinol (Lausanne).* 2018; 9: 7.

33 Liu J, Dai Q, Li W, Guo Y, Dai A, Wang Y, et al. Association of vitamin D receptor gene polymorphisms with gestational diabetes mellitus-a case control study in Wuhan, China. *BMC Pregnancy and Childbirth.* 2021; 21(1): 142.

34 Mak JKL, Pham NM, Lee AH, Tang L, Pan X-F, Binns CW, et al. Dietary patterns during pregnancy and risk of gestational diabetes: A prospective cohort study in Western China. *Nutr J.* 2018; 17(1): 107.

35 Cai D, Liu H, Hu Y, Jiang Y, Zhao R. (2017). Gestational betaine, liver metabolism, and epigenetics. In: *Handbook of Nutrition, Diet, and Epigenetics.* (ed. V. Patel and V. Preedy). 1–14. Cham: Springer International Publishing.

36 Zhang G, Wang X, Ren B, Zhao Q, Zhang F. The effect of resveratrol on blood glucose and blood lipids in rats with gestational diabetes mellitus. *Evid Based Complement Alternat Med.* 2021; 2021: 2956795.

37 Zhao C, Yin X, Zhao C. The renoprotective effects of Naringenin (NGN) in gestational pregnancy. *Diabetes Metab Syndr Obes.* 2020; 13: 53–63.

38 Costa, P.C.T.d., de Souza, E.L., Lacerda, D.C. et al. (2022). Evidence for quercetin as a dietary supplement for the treatment of cardio-metabolic diseases in pregnancy: A review in rodent models. *Foods.* 11(18): 2772. https://doi.org/10.3390/foods11182772.

39 Al-Ishaq RK, Abotaleb M, Kubatka P, Kajo K, Büsselberg D. Flavonoids and their anti-diabetic effects: cellular mechanisms and effects to improve blood sugar levels. *Biomolecules.* 2019; 9(9): 430.

40 Kaur S, Kumar S. Crosstalk between food components and microRNAs: Role in metabolism, nutrition, health and diseases. 2020; 7: 1–11.

41 Masete M, Dias S, Malaza N, Adam S, Pheiffer C. A big role for microRNAs in gestational diabetes mellitus. *Front Endocrinol (Lausanne).* 2022; 13: 892587.

42 Li W, Yuan X, He X, Yang L, Wu Y, Deng X, et al. The downregulation of miR-22 and miR-372 may contribute to gestational diabetes mellitus through regulating glucose metabolism via the PI3K/AKT/GLUT4 pathway. *J Clin Lab Anal.* 2022; 36(7): e24557.

43 Guan CY, Cao JL, Zhang L, Wang XQ, Ma X, Xia HF. miR-199a Is upregulated in GDM targeting the MeCP2-Trpc3 pathway. *Front Endocrinol (Lausanne).* 2022; 13: 917386.

44 Fan W, Pang H, Xie Z, Huang G, Zhou Z. Circular RNAs in diabetes mellitus and its complications. *Front Endocrinol (Lausanne).* 2022; 13: 885650.

Index

Note: page numbers in *italics* refer to figures; those in **bold** to tables. The Greek letters α and β are filed under the words alpha and beta. Abbreviations: GDM is used for gestational diabetes mellitus; T1DM and T2DM are used for type 1 and type 2 diabetes mellitus respectively.

Natural Products and their Bioactives in Antidiabetic Drug Discovery, First Edition.
Edited by Kanti Bhooshan Pandey, Maitree Suttajit and Pinar Atukeren.

d

u

Printed and bound by CPI Group (UK) Ltd, Croydon, CR0 4YY
17/09/2026
14964253-0002